Lecture Notes in Electrical Engineering

Volume 1128

The book series *Lecture Notes in Electrical Engineering* (LNEE) publishes the latest developments in Electrical Engineering—quickly, informally and in high quality. While original research reported in proceedings and monographs has traditionally formed the core of LNEE, we also encourage authors to submit books devoted to supporting student education and professional training in the various fields and applications areas of electrical engineering. The series cover classical and emerging topics concerning:

- Communication Engineering, Information Theory and Networks
- Electronics Engineering and Microelectronics
- Signal, Image and Speech Processing
- Wireless and Mobile Communication
- Circuits and Systems
- Energy Systems, Power Electronics and Electrical Machines
- Electro-optical Engineering
- Instrumentation Engineering
- Avionics Engineering
- Control Systems
- Internet-of-Things and Cybersecurity
- Biomedical Devices, MEMS and NEMS

For general information about this book series, comments or suggestions, please contact leontina.dicecco@springer.com.

To submit a proposal or request further information, please contact the Publishing Editor in your country:

China

Jasmine Dou, Editor (jasmine.dou@springer.com)

India, Japan, Rest of Asia

Swati Meherishi, Editorial Director (Swati.Meherishi@springer.com)

Southeast Asia, Australia, New Zealand

Ramesh Nath Premnath, Editor (ramesh.premnath@springernature.com)

USA, Canada

Michael Luby, Senior Editor (michael.luby@springer.com)

All other Countries

Leontina Di Cecco, Senior Editor (leontina.dicecco@springer.com)

** **This series is indexed by EI Compendex and Scopus databases.** **

Shailendra Kumar · Manoj Tripathy ·
Premalata Jena
Editors

Control Applications in Modern Power Systems

Select Proceedings of EPREC 2023, Volume 3

Springer

Editors
Shailendra Kumar
Department of Electrical Engineering
Indian Institute of Technology Bhilai
Bhilai, Chhattisgarh, India

Manoj Tripathy
Department of Electrical Engineering
Indian Institute of Technology Roorkee
Roorkee, Uttarakhand, India

Premalata Jena
Department of Electrical Engineering
Indian Institute of Technology Roorkee
Roorkee, Uttarakhand, India

ISSN 1876-1100 ISSN 1876-1119 (electronic)
Lecture Notes in Electrical Engineering
ISBN 978-981-99-9053-5 ISBN 978-981-99-9054-2 (eBook)
https://doi.org/10.1007/978-981-99-9054-2

This Springer imprint is published by the registered company Springer Nature Singapore Pte Ltd.
The registered company address is: 152 Beach Road, #21-01/04 Gateway East, Singapore 189721, Singapore

Preface

This book entitled *Control Applications in Modern Power System—Select Proceedings of EPREC 2023* provides rigorous discussions, case studies, and recent developments in the emerging areas of control system, especially, load frequency control, wide-area monitoring, control and instrumentation, optimization, intelligent control, energy management system, SCADA systems, design of control strategies is essential for controlling the reactive power and maintains the voltage profiles, etc. The readers would be benefitted in enhancing their knowledge and skills in the domain areas. Also, this book may help the readers in developing new and innovative ideas. The book can be a valuable reference for beginners, researchers, and professionals interested in developments in control system.

Bhilai, India — Shailendra Kumar
Roorkee, India — Manoj Tripathy
Roorkee, India — Premalata Jena

Contents

About the Editors

Shailendra Kumar is currently working as an Assistant Professor in the Department of Electrical Engineering Indian Institute of Technology (IIT) Bhilai. He received his B.Tech. degree in Electrical and Electronics Engineering from Uttar Pradesh Technical University, Lucknow, India, in 2010, and the M.Tech. Degree in Power Electronics, Electrical Machine, and Drives (PEEMD) from the Indian Institute of Technology, Delhi, India, in 2015. He completed his Ph.D. degree in the Department of Electrical Engineering from the Indian Institute of Technology Delhi, New Delhi, India in May 2019. He has worked as a postdoctoral research fellow at Petroleum Institute of Research Centre Khalifa University from May 2019 to November 2019. He has also worked as an Assistant Professor at Maulana Azad National Institute of Technology, Bhopal from June 2020 to May 2023. His research interests include charging stations for EVs, power quality, grid integration of renewable energy resources, custom power devices, and microgrids. Dr. Kumar was the recipient of the prestigious POSOCO power system award (in Master as well as Doctoral categories) in 2016 and 2019. He is also a recipient of Prof. Som Nath Mahendra's Student Travel Awards for the IEEE PEDES 2018 conference and the IEEE UPCON Best Paper Award in 2016 and 2018. Dr. Kumar is a senior member of the IEEE and a member of INAE Societies.

Dr. Manoj Tripathy received his B.E. degree in Electrical Engineering from Nagpur University, Nagpur, India, in 1999, the M.Tech. degree in Instrumentation and Control from Aligarh Muslim University, Aligarh, India, in 2002, and the Ph.D. degree from the Indian Institute of Technology Roorkee, Roorkee India, in 2008. He is currently working as a Professor in the Department of Electrical Engineering, Indian Institute of Technology Roorkee, Uttarakhand, India. He is an IEEE senior member, a Member of the Institution andTechnology (IET), and a Life member of the Institution of Engineers (India). His fields of interest are Power System Protection, Automation, Cybersecurity, Smart Grid Protection, Microgrid Protection, Digital/ Numerical Relays, Artificial Intelligence Relays, Relay Testing, Protection Scheme for a Line with FACTS devices, Testing of Electrical Equipment, Calibration of

Meters, Signal and Image Processing problems, Speech Signal Enhancement Techniques. Dr. Tripathy is a reviewer for various international journals in the area of power systems and speech.

Dr. Premalata Jena is currently an Associate Professor at the Department of Electrical Engineering, Indian Institute of Technology Roorkee, Uttarakhand, India. She received her B.Tech. degree in Electrical Engineering from the Utkal University, Bhubaneswar, India, in 2001, her M.Tech. degree in Electrical Engineering from the Indian Institute of Technology Kharagpur, Kharagpur India, in 2006, and her Ph.D. degree from the Indian Institute of Technology Kharagpur, Kharagpur India, in 2011. She has seven years of teaching experience. She is an IEEE member and INAE, Young Associate. She received many awards such as the Women Excellence Award and Early Carrier Research Award, SERB, DST, Government of India, New Delhi, INAE Young Engineer Award, and POSOCO Power System Award, in 2013. She has published many papers in different reputed journals and conferences. Her fields of interest are Smart Grid, Smart Grid Technology, and Protection, Microgrid, Microgrid Protection, Signal Processing application to power system relaying, Power system Protection, Protection Issues with FACTS Devices, Protection Scheme Development for a line with FACTS devices, Disturbance localization, Signal processing application for disturbance localization.

Adaptive Discrete Extended Kalman Filter for Parameter Estimation in IRFOC-Based IMD with Optimized Gains

Mahesh Pudari, Sabha Raj Arya, and Papia Ray

Abstract In this work, an Adaptive Discrete-time model Extended Kalman Filter technique for simultaneous multi-parameters estimation of induction motor drive is proposed. An Ant-lion optimization algorithm for fine-tuning the PI error regulator gains in indirect rotor flux-oriented control (IRFOC) with space vector pulse width modulation switching technique is implemented. The measure of parameters improves the operation of the drive speed and torque performance, based on proper tuning of slip angular frequency. The effectiveness of the standard discrete Extended Kalman Filter technique is influenced by the choice of system and measurement error covariance noise matrices. These matrices are evaluated using the trial-and-error process and are intended to be consistent. However, the operating environment has an impact on these covariance matrices. To improve monitoring reliability and avoid the time consumption by trial-and-error approach for selecting optimum matrices, an adaptive estimating strategy with the capacity of online matrix updating is proposed. Further, to enhance the effectiveness by eliminating the time required for manual tuning of gains of PI in IRFOC are obtained by the Ant-Lion optimization (ALO) technique. The avoidance of local optimum is an additional aspect of ALO. The proposed technique's feasibility and performance are evaluated and verified under various operational scenarios. The simulation results demonstrate the adopted ADEKF technique has a faster response and thus can precisely estimate the parameters of stator resistance, rotor resistance, stator inductance, and rotor inductance.

Index Terms ADEKF · SDEKF · ALO · IRFOC · PI gains · Parameter estimation · And Adaptive Slip

M. Pudari · S. R. Arya (✉)
Department of Electrical Engineering, Sardar Vallabhbhai National Institute of Technology, Surat 395007, India
e-mail: sabharaj1@gmail.com

P. Ray
Department of Electrical Engineering, Veer Surendra Sai University of Technology, Burla, Sambalpur, India
e-mail: papiaray_ee@vssut.ac.in

S. Kumar et al. (eds.), *Control Applications in Modern Power Systems*, Lecture Notes in Electrical Engineering 1128, https://doi.org/10.1007/978-981-99-9054-2_1

Nomenclature

v_{ds}, v_{qs}	Stationary reference frame stator voltages in volts (V)
i_{ds}, i_{qs} and i_{dr}, i_{qr}	Stationary reference frame stator and rotor currents in amperes (A)
i_{de}, i_{qe}	Synchronously reference frame stator currents in amperes (A)
ω_r	Rotor speed (rad/s)
ω_{sl}	Angular slip frequency (rad/s)
ω_{slad}	Adaptive/tuned Angular slip frequency (rad/s)
ω_e	Stator angular speed (rad/s)
^	Estimated value
~	Prediction Parameter

1 Introduction

The indirect rotor field oriented control (IRFOC) based Induction motor drives (IMD) are more predominant in industrial and electric vehicle applications [1] due to its reliability, robustness, cost-effectiveness, ease of implementation, low maintenance, and excellent peak torque capability. The prominent features of this controller are capable of decoupling the control of developed torque and flux by the stator current. The effectiveness of IRFOC-based IMD is heavily sensitive to motor parameter variation. The improper estimation of motor parameters can lead to the deterioration of the dynamic performance of the IRFOC strategy [2]. The deterioration occurs due to erroneous reference current evaluation and computation of inappropriate slip angular frequency in the IRFOC strategy. The slip angular frequency component is evaluated based on motor parameters and reference flux. The effect of a small mismatch of slip angular frequency can lead to disorientation of the field. Moreover, the electrical parameters such as R_s, R_r, L_s, L_r, and L_m are considered constant for a machine, but in real-time (during running) applications due to variations of temperature and flux linkages, the parameters are influenced. The inaccurate motor parameters that impact the control performance have been investigated extensively in the literature [3]. Nowadays, the estimation of parameters plays a major significant role in IMD-based applications like servomechanism, electrical vehicles, and robotics [4]. The estimated parameters are highly favorable for monitoring the temperature and fault tolerance of IMD [5, 6]. Moreover to strengthen the dynamic operation, control accuracy, and design of current, speed, and torque controllers, precise information on electrical parameters is essential.

Various parameter identification strategies are presented in the literature [7–16]. Lee et al. developed an integral calculation-based current injection strategy for the estimation of parameters such as magnetizing inductance and rotor time constant [7]. Siddavatam et al. [8] proposed a recursive least square technique to identify the discrete model IM parameters at standstill. But it requires mathematical calculations and analog–digital filters. Yanhui et. al. [9] designed the identification of parameters at standstill based on the process of vector construction, to avoid the analog–digital filters in RLS. And other self-commissioning parameter estimation methods are also discussed in [10]. The estimated parameters by these strategies are not adopted in the controller and during the running conditions, the estimated parameters are influenced. So to adopt the parameters during the running condition in the controller auto-commissioning parameter estimation techniques are developed [11–15]. A few auto-commissioning techniques are the Recursive least square technique [12], signal injection-based strategy [13], MRAS [14], and observer-based strategies [15]. The implementation and development of these methods are easy and simple. Moreover, the development of its adaptive mechanism is difficult to design. The evolutionary parameter estimation techniques are developed such as particle swarm optimization [16] and genetic algorithm [17]. These all are challenging aspects to develop and consume a lot of computational burdens. Even though, if the emergence of auto commission estimation techniques has led to enhanced accuracy and efficiency, convergence difficulties can emerge in certain practical scenarios. But, the effectiveness of deterministic-based approaches degrades in the presence of noise, since they neglect system noise (produced by uncertainties, disturbances, and measurement of voltage) and the measurement noise (produced by measurement of current). However, the stochastic-based techniques consider the system and measurement noises, which may enhance the estimating capabilities in both dynamic and steady states.

The extended Kalman filter (EKF) seems to be the most frequently used technique of stochastic-based method for the estimation of non-linear states and parameters [18]. But, the accuracy of EKF estimations is closely associated with the selection of the appropriate noise covariance matrices of the system (Q) and measurement (R). However, as the stochastic characteristics of associated noises in the EKF technique are typically unknown, it is difficult to establish the mathematical relationship between the EKF performance and the covariance matrices of noise. Since numerous approaches have been proposed to find out the values of these matrices in the literature, it's possible to separate the research into two groups: based on constant and dynamic matrices. In the first group of research, a trial-and-error approach is frequently used to evaluate the matrices [19]. Due to more time consumption the results are not optimal over the whole speed range by this method. Other research employs heuristic techniques to optimize these matrices offline, such as the differential evolution technique [20] and genetic algorithm [21]. By eliminating the necessity of mathematical equations and their derivatives in favor of fitness functions and fitness values, the heuristic techniques perform better than conventional optimization techniques like Levenberg–Marquardt and Newton-Rapson [22]. In my previous research work, the hybrid PSOGWO optimization technique for tuning of optimum values of Q and R matrices in SDEKF-based IMD is implemented. However, the optimization approaches also consume more time same as the trial-and-error approach and it is an offline process only. But these covariance matrices are impacted by various operating scenarios of IM, so depending upon the operating scenario it is necessary to update the matrices. The second set of research [23–27] with the capability to update these matrices online is reported in the literature. Some authors also developed fuzzy logic-based techniques for the online updating of matrices [23]. This approach ensures an expert with experience and knowledge to provide accurate estimates of effectiveness, hence the design process is somewhat sophisticated. So to avoid all the above difficulties, the novel adaptive-based technique is proposed for the estimation and continuous online updating of Q and R matrices. Based on the adaptive fading method, innovation, or residual adaptive estimation methods the adaptive EKF (AEKF) is employed in [24]. As summaries, the suggested methods in [24, 25] are simpler to build and yield greater performance than in [23]. Although AEKF is successfully implemented in global positioning systems/integrated inertial navigation systems (GPS/INS) [25]. The AEKF technique estimates the online adoption of the Q matrix with a full variance/covariance matrix. For state/parameter estimation of IMD, the ADEKF technique has not been implemented in the literature. And adaptive-based online estimation of both system and measurement noise covariance matrices is complicated and problematic [26]. For IMD parameter estimation, the evaluation of Q is reliable since R may be approximated [27].

Another major factor of the control is a selection of PI error regulator gains in IRFO control, where manual tuning (trial-and-error) is the most common method. It tends to increase the duration of process control. Maintaining the d-q currents and estimated parameters levels absolutely, requires controlling the PI controller gains. Various techniques are discussed to tune the k_{p} and k_{i} gains for proper operation and controlling performance of the induction motor. By using the Nascent methods, we can estimate the k_{p} and k_{i} gains, but it takes more time [28]. So to obtain the optimized values for enhancing the best solution and to avoid the drawbacks of nascent methods, an optimization technique has been introduced. A Metaheuristic optimization method has given better solutions in many applications [29]. In this work, the authors have employed PI error regulator gains identification using a nature-inspired optimization algorithm i.e., Ant-Lion Optimization (ALO), is employed because it can eliminate local minima [30, 31].

The author's main contributions in this research work are as follows: (a) The adaptive discrete extended Kalman filter is proposed to estimate the simultaneous multi-parameters of IMD, to avoid the disorientation of the field. (b) for an accurate evaluation of slip angular frequency, the precise estimated parameters are adopted in IRFOC Control. (c) To validate the estimation performance reliability of the ADKEF technique, the ADEKF approach performances are compared with the SDEKF approach under the same operating conditions. (d) Observability and stability analysis of the ADEKF technique for the Discrete IM model is discussed. (e) In the implemented IRFO Control, an ALO technique is employed to optimize the PI error regulator gains, yielding the precise PI regulator gains with less time. Using MATLAB/Simulink and tuned PI regulator gains, an IRFOC-based control approach is investigated for the aforementioned IMD.

The rest of the research work is organized in this paper as follows. The proposed system configuration is presented in Sect. 2. The analysis of the discretized mathematical model and the Observability of IM is discussed in Sect. 3. The proposed ADEKF technique principle operation and process of parameter estimation along with stability analysis explained in Sect. 4. In Sect. 5, a mathematical description of estimated parameters using ADEKF and optimization algorithm procedure for evaluation of gains is discussed. Finally, Sect. 6 presents simulation results and analysis of the proposed technique.

2 System Configuration

A schematic representation of the proposed indirect rotor flux-oriented control of Induction motor drive multi-parameter estimation using the adaptive discrete Kalman filter technique is given in Fig. 1. It consists of the calculation of slip angular frequency, diode bridge rectifier fed from three-phase supply, voltage source converter with SVPWM switching technique, voltage, and current sensing elements, and ADEKF technique block for parameter estimation. The unit vector (angular position of filed) is obtained from the summation of slip angular gain and rotor speed. To obtain the accurate angular position, the proper tuning of slip speed is necessary, for the accurate electrical parameters are estimated by using the ADEKF technique by providing stationary reference frame voltage and current signals as an input to ADEKF. The estimated speed by using the speed encoder and speed command is compared and given to the speed regulator, which provides the torque component of current (i_{qe}^*), and similarly based on machine rating the reference flux is evaluated, from this flux component of current (i_{de}^*) is calculated. The evaluated torque and flux component current are compared with synchronously reference frame sensed currents (i_{qe} and i_{de}) and injected into current regulators, which gives the synchronous reference frame voltages (v_{qe}^* and v_{de}^*). By using inverse parks transformation stationary reference voltages (V_{ds}^* and V_{qs}^*) are evaluated and provided as input to the switching pulse generation of the SVPWM technique along with sensed DC link voltage (V_{dc}). The development of the ADEKF technique and tuning of slip angular frequency are explained in the upcoming section.

3 Mathematical Model and Observability of Induction Motor

The derivation of a discrete-time mathematical model of IM and the observability analysis of the derived IM model are addressed in the following subsections.

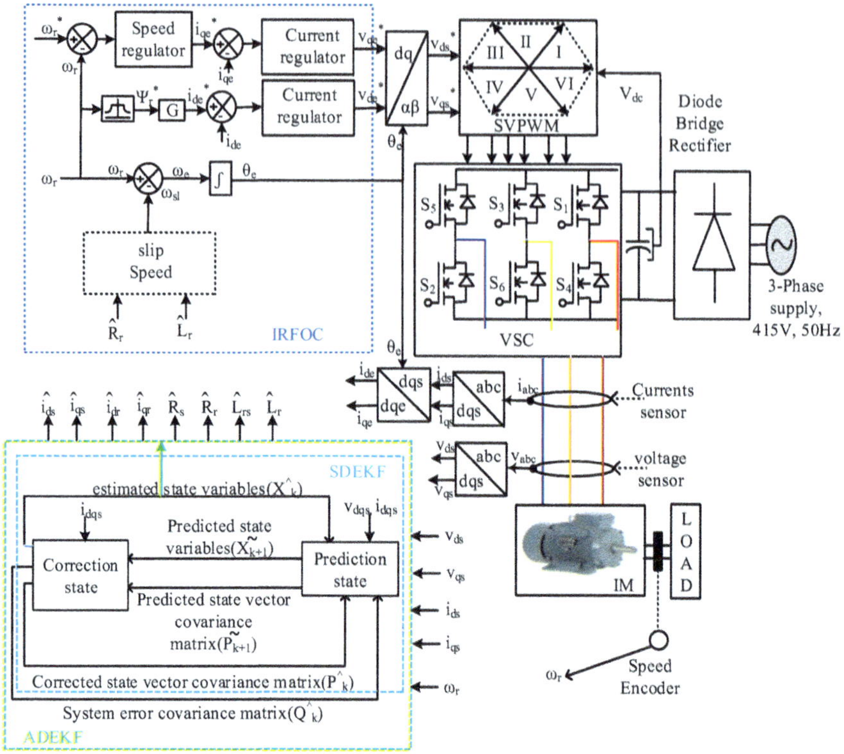

Fig.1 Block Diagram of IRFOC-IMD fed VSC for parameter estimation using ADEKF

3.1 Derivation of Discretized Mathematical Model of IM

The mathematical modeling of induction motor describes the dynamic behavior of electrical and mechanical, which are coupled via the generated torque. It will be utilized to study the steady-state and dynamic behavior of IM. The dynamic discretized model of IM in a stationary reference frame with stator and rotor currents as a state variable in state space form is as follows. The Euler discretization algorithm is used to discretize the continuous model expressions given in [2] with the sampling duration of t_s.

$$\left.\begin{aligned} f_{k+1} &= A_d . f_k + B_d . U_k + v_k \\ z_k &= C_d . f_k + w_k \end{aligned}\right\} \tag{1}$$

where,

$$A_d = \begin{bmatrix} 1 - t_s k_1 & t_s k_3 & t_s k_6 & t_s k_7 \\ -t_s k_3 & 1 - t_s k_1 & -t_s k_7 & t_s k_6 \\ t_s k_4 & -t_s k_5 & 1 - t_s k_2 & -t_s k_8 \\ t_s k_5 & t_s k_4 & t_s k_8 & 1 - t_s k_2 \end{bmatrix}, \quad B_d = \begin{bmatrix} t_s k_9 & 0 \\ 0 & t_s k_9 \\ -t_s k_{10} & 0 \\ 0 & -t_s k_{10} \end{bmatrix},$$

$$C_d = \begin{bmatrix} 1\,0\,0\,0 \\ 0\,1\,0\,0 \end{bmatrix}$$

$$u_k = \begin{bmatrix} v_{ds} \; v_{qs} \end{bmatrix}^T, \; z_k = \begin{bmatrix} i_{ds} \; i_{qs} \end{bmatrix}^T, \; f_k = \begin{bmatrix} i_{ds} \; i_{ds} \; i_{dr} \; i_{qr} \end{bmatrix}^T$$

$$k_1 = \frac{R_s}{\sigma L_s}, \; k_2 = \frac{R_r}{\sigma L_r}, \; k_3 = \frac{L_m^2 \omega_r}{\sigma L_s L_r}, \; k_4 = \frac{L_m R_s}{\sigma L_s L_r}, \; k_5 = \frac{L_m \omega_r}{\sigma L_r}, k_6 = \frac{L_m R_r}{\sigma L_s L_r},$$

$$k_7 = \frac{L_m \omega_r}{\sigma L_s}, \; k_8 = \frac{\omega_r}{\sigma}, \; k_9 = \frac{1}{\sigma L_s}, \; k_{10} = \frac{L_m}{\sigma L_s L_r}, \; \text{and } \sigma = 1 - \frac{L_m^2}{L_s L_r}$$

v_k is the state noise vector and w_k is measurement/output noise vectors with zero mean and white Gaussian vectors.

$$\left. \begin{array}{l} E\{v_k\} = 0 \\ E\{w_k\} = 0 \end{array} \right\} \tag{2}$$

DEKF uses the system and measurement noise covariance matrices (Q and R respectively) to perform recursion instead of the noise vectors v and w. Q and R matrices are evaluated by the noise vector w and v by using Eq. (3).

$$\left. \begin{array}{l} E\{vv^T\} = Q \\ E\{ww^T\} = R \end{array} \right\} \tag{3}$$

3.2 *Observability Analysis of the Discrete-Time IM Model*

To have an estimation of a model's state, the model must have the observability criterion. Numerous researchers have provided the necessary observability requirements for the continuous-time model-based IM system [32]. The observability of the discrete-time model still has not been examined to the best of the researchers' wisdom, because several articles present estimation techniques based on the IM discrete-time model. To be used in the SDEKF and ADEKF techniques of the fifth-order non-linear model of IM represented in (2), it has to fulfill the observability rank criterion that was mentioned earlier in [33, 34]. Taking into consideration the aforementioned,

Theorem: If the rotor current vector revolves in the stator stationary reference frame, the non-linear IM model achieves the observability rank requirement.

Proof: If the rank requirement is achieved, then the non-linear model represented in (2) is observable, i.e., rank $(O) = m = 5$. The matrix 'O' as follows,

$$O = \begin{bmatrix} \frac{\partial h}{\partial f}(f_k) \\ \frac{\partial h}{\partial f}(f_{k+1})\frac{\partial g_d}{\partial f}(f_k) \\ \cdot \\ \cdot \\ \frac{\partial h}{\partial f}(f_{k+4})\frac{\partial g_d}{\partial f}(f_{k+3})\ldots.\frac{\partial g_d}{\partial f}(f_k) \end{bmatrix} \quad (4)$$

where, $g_d = A_d.X_k + B_d.U(k)$ and $h = C_d.f_k$.

The rank of the non-linear observability matrix has 5 at f_k, to fulfill the non-linear observability rank criterion, the non-linear IM model must be $f_k \in R^5$. The dimensions of the observability matrix become 10×5, the rank may be evaluated across any five rows of matrix 'O'. Consequently, the first three factors of O that provide a regular matrix ($O3$) with dimensions 5×5 are utilized.

The corresponding determinant is computed by substituting the $f_k = f$ and $f_{k+1} = f+$, its determinant is evaluated as,

$$\det(O_3) = \gamma((k + \xi f_5 f_5^+)(f_3 f_4^+ - f_3^+ f_4) - \eta(f_5 - f_5^+)(f_3 f_4^+ + f_3^+ f_4)) \quad (5)$$

where,

$$\gamma = \frac{L_m^4 \omega_r^2 t_s^5}{L_r^6 \sigma^4}, \quad k = R_r^2, \quad \xi = L_r^2 \omega_r^2, \quad \eta = L_r \omega_r R_r$$

Now, it is worthwhile noting that f_3 (f_3^+) and f_4 (f_4^+) are the components of the rotor current vectors $\vec{i}_r$ ($\vec{i}_r^+$) along the $\alpha\beta$-axis of the stator reference frame, respectively. Therefore, the $\theta_r(\theta_r^+)$ represents the angle between rotor currents vectors of $\vec{i}_r$ and ($\vec{i}_r^+$) and α-axis. Consequently, the following definitions are feasible:

$$\begin{aligned} f_3 &= i_m \cos(\theta_r), \quad f_3^+ = i_m^+ \cos(\theta_r^+) \\ f_4 &= i_m \sin(\theta_r), \quad f_4^+ = i_m^+ \sin(\theta_r^+) \end{aligned} \quad (6)$$

where, $i_m = \left|\vec{i}_r\right|, \; i_m^+ = \left|\vec{i}_r^+\right|$

Using the above definitions, the final determinant equation can be given as,

$$\det(O_3) = \gamma i_m i_m^+((k + \xi f_5 f_5^+)\sin(\theta_r^+ - \theta_r) - \eta i_m i_m^+(f_5 - f_5^+)\cos(\theta_r^+ - \theta_r)) \quad (7)$$

The above expression demonstrates that for the constant parameter ($f_5 = f_5{}^+$) the rank conditions are satisfied if $\theta_r \neq \theta_r{}^+$, in addition to $\forall i_m \neq 0$, and $\forall i_m^+ = 0$, i.e., if the rotor current vector rotates in the stator stationary reference frame.

4 Adaptive Discrete Extended Kalman Filter Technique

The following subsections discuss the mathematical description of the ADEKF technique derivation and its stability study.

4.1 Derivation of ADEKF Technique

ADEKF technique is implemented to estimate the unknown states (parameters) of the discretized IM model. The procedure to develop the ADEKF algorithm and application of ADEKF for estimation of electrical parameters R_s, R_r, L_s, and L_r are discussed in this section with mathematical analysis.

The Kalman Filter (KF) is derived from the projective theorem, which can make an optimal estimation of the target state in the case of a linear Gaussian model. And it is a recursive method that considers the system's stochastic state space model as well as the measured outputs in multi-input multi-output systems. The system and measurement error noises are perceived to be in the form of white noise. When the error of estimated covariance is minimized, then the state estimation can reach the optimum. This process requires the discrete-time-varying linear model of IM is discussed in Sect. 2 (A).

The ADEKF algorithm mainly includes two stages in the estimation process: one is prediction and another one is correction (filtering). In the prediction process, the next predicted states and error covariance matrices are evaluated, further in the filtering process, considering the predicted states the correction states are evaluated, and by adding both these unknown states/parameters are estimated. The conventional mathematical expressions of the SDEKF technique using the Discrete IM model in (2) can be given as follows.

$$\tilde{f}_{k+1} = A_d \cdot \hat{f}_k + B_d . U_k \tag{8}$$

$$g_{k+1} = \frac{\partial}{\partial f}\left(A_d \cdot \hat{f}_k + B_d . U_k \right)\bigg|_{f=\hat{f}_{k+1}} \tag{9}$$

$$\tilde{P}_{k+1} = g_{k+1} \cdot \hat{P}_k \cdot g'_{k+1} + Q_k \tag{10}$$

$$\lambda_{k+1} = \tilde{P}_{k+1} . h'_{k+1} \cdot [h_{k+1} \cdot \tilde{P}_{k+1} . h'_{k+1} + R]^{-1} \tag{11}$$

$$\hat{f}_{k+1} = \tilde{f}_{k+1} + \lambda_{k+1}[z_{k+1} - \hat{z}_{k+1}] \tag{12}$$

$$\hat{P}_{k+1} = \tilde{P}_{k+1} - \lambda_{k+1}.h_{k+1}.\tilde{P}_{k+1} \tag{13}$$

where 'f' is the state variable vector; 'g' is the gradient matrix to linearize the non-linear system; 'P' state vector covariance matrix (order of 5*5); 'Q' is the system noise covariance matrix (order of 5*5); 'R' output (measurement) noise covariance matrix (order of 5*5); and 'λ' Kalman gain.

A major significant factor while developing the SDEKF technique is a selection of initial constants for covariance matrices Q and R. which affect the system convergence, performance of drive, and stability. The Kalman gain 'λ' in filtering state value will increase as the value of the components of Q increases, which leads to faster filter dynamics but inferior steady-state performance. Similarly, if raising the constant values of R, the current measurements are heavily affected. So, the gain value of the Kalman filter will reduce, owing to poor transient response. And also the initial diagonal system state covariance matrix P_0 signifies the variances or mean square errors with regard to the initial condition. If P_0 owing to changing the magnitudes of the transient are affected. however, transient duration and steady-state constraints will not be affected. In this paper, the system state covariance matrix initial value P_0 is chosen by a trial-and-error approach, since it has nothing to do with a steady-state scenario. The system error covariance matrices Q and R are also chosen by trial-and-error approach. But it is very difficult to calculate both error covariance matrices and if the load/speed variation occurs the Q and R matrices are affected and require updating. So, to avoid this in the proposed ADEKF technique, Q is found automatically from the system using Eq. (16). The ADEKF technique is implemented to estimate the parameter of a dynamic model of the Discrete IMD flow chart as presented in Fig. 2.

To estimate the innovation residual's covariance, it is computed by averaging the preceding residual sequences across a window of length N:

$$\hat{C}_{ez_k} = \frac{1}{N}\sum_{i=i0}^{j} e_{yi}\, e'_{yi} \tag{14}$$

where, the innovation residual in the Kalman filter is given as,

$$e_z = [z_k - \hat{z}_k] = [z_k - H * \tilde{f}_{k+1}] \tag{15}$$

$i_0 = j - N + 1$ is the initial observation in the estimate window. The predicted procedure of noise may be evaluated as:

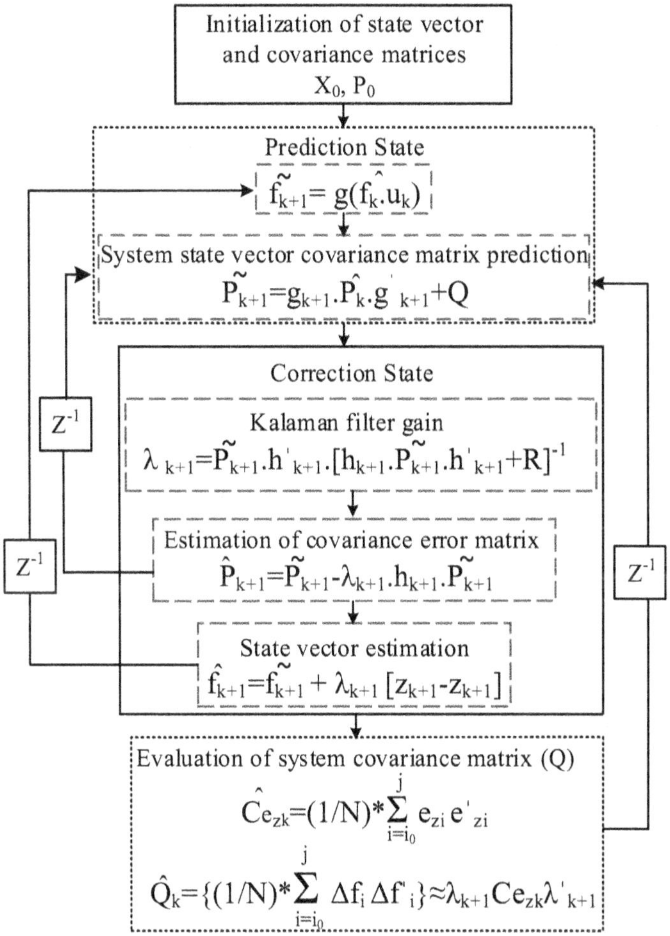

Fig. 2 Flow chart of Proposed ADEKF technique

$$\hat{Q}_k = \left(\frac{1}{N}\sum_{i=i_0}^{j} \Delta f_i \, \Delta f_i'\right) + \hat{P}_{k+1} - g_{k+}1.\, \hat{P}_k \,.g'_{k+1} \tag{16}$$

where, $\Delta f = \hat{f}_k - \hat{f}_{k-1}$.

The expression of (16) may be stated as follows in terms of the innovation sequence under the premise of steady-state:

$$\hat{Q}_k \approx \left(\frac{1}{N}\sum_{i=i0}^{j} \Delta f_i \, \Delta f_i'\right) \approx \lambda_{k+1}\, \hat{C}_{e_{z_k}}\, \lambda'_{k+1} \tag{17}$$

(A) *Stability of ADEKF Technique*

It is necessary to make the following assumptions in order to examine whether the IMD parameter estimation technique ADEKF is stable.

Assumption 1
For all $k \geq 0$,

(a) The $\overline{a}$ and $\overline{h}$ are real numbers with a positive nature in the Discrete IMD model (1), such that $\|A_k\| \leq \overline{a}$ and $\|H_k\| \leq \overline{h}$.
(b) A_k of the Discrete IMD model state vector given in (1) is nonsingular.

Assumption 2

(a) The A_k and H matrices of Discrete IM both fulfill the requirements of uniform observability,
(b) $\overline{r}, r, \overline{q}, q > 0$ are real numbers, such that Q and R matrices are constrained by

$$r\, I_5 \leq R \leq \overline{r}\, I_5$$
$$q\, I_6 \leq Q \leq \overline{q} I_5$$

(c) The initial covariance matrix P_0 is positive definite.

For all $k \geq 0, \overline{p}, p > 0$ are real numbers, so the solution of the Riccati difference expression is constrained by $p\, I_5 \leq P \leq \overline{p}\, I_5$.

The fifth-order Discrete IM mathematical model along with the ADEKF technique and also Assumptions 1 and 2, by considering these the e_k {described as $f_k - \hat{f}_k$}, is exponentially constrained in the mean square sense with probability one when P_0 fulfills the required inequality condition as shown below,

$$\||e_0\|| \leq \varepsilon \ = \min\left(\frac{2S}{\alpha}, \frac{\beta p}{8S\alpha\overline{p}}\right) \tag{18}$$

where $\beta > 0$ is a constant. Detailed information about the proof can be provided in[35].

Furthermore, the stability study was conducted under the hypothesis of Gaussian distributed noises and a positive definite Q and R matrix is acceptable for the ADEKF technique. White noise factors are also included in residual sequences e_k employed in the evolution of $\hat{C}_{e_k}$ and based on its quadratic nature it is positive definite. And the updated Q matrix is also in quadratic nature represented in (17). If P_0 fulfills the given criteria in (18) then it can be finalized that the e_k in the ADEKF technique is exponentially constrained in the mean square sense with probability one.

5 Control Algorithm

For the estimation of electrical parameters, the ADEKF algorithm is applied in this section for a non-linear time-varying system of the extended IM model. The stationary reference frame stator, rotor currents, and estimated electrical parameters of R_s, R_r, L_s, and L_r are regarded as state variables, namely $X = \left[\, i_{ds}\ i_{qs}\ i_{dr}\ i_{qr}\ R_s\ R_r\ L_s\ L_r \,\right]'$. It is an 8th-order non-linear mutually coupled mathematical model, which has the problems of a large number of calculations, poor accuracy, and stability. Therefore, in this work reduced order parallel quadruple extended Kalman method was proposed to estimate the motor parameters. Then it will beneficial to improve the performance and stability of the system of estimation of parameters and reduce the problem of mutual coupling between the parameters. And also the estimation of PI regulator gains using optimization technique is discussed in this subsection.

5.1 Induction Motor Drives Stator Resistance Estimation

For the estimation of the stator resistance parameter, the dynamic model of IM can be extended as follows and the other parameters are considered as known.

$$\hat{f}_{1_{k+1}} = A_{d1}.\hat{f}_{1_k} + B_{d1}.U_k \tag{19}$$

where,

$$A_{d1} = \begin{bmatrix} 1 - t_s l_1 & t_s l_3 & t_s l_6 & t_s l_7 & 0 \\ -t_s l_3 & 1 - t_s l_1 & -t_s l_7 & t_s l_6 & 0 \\ t_s l_4 & -t_s l_5 & 1 - t_s l_2 & -t_s l_8 & 0 \\ t_s l_5 & t_s l_4 & t_s l_8 & 1 - t_s l_2 & 0 \\ 0 & 0 & 0 & 0 & 1 \end{bmatrix}, \quad B_{d1} = \begin{bmatrix} t_s l_9 & 0 \\ 0 & t_s b_9 \\ -t_s b_{10} & 0 \\ 0 & -t_s l_{10} \\ 0 & 0 \end{bmatrix}$$

$$U_k = \left[\, v_{ds}\ v_{qs} \,\right]^T, \ f_{1k} = \left[\, i_{ds}\ i_{qs}\ i_{dr}\ i_{qr}\ R_s \,\right]^T = \left[\, f_1\ f_2\ f_3\ f_4\ f_5 \,\right]^T$$

$$l_1 = \frac{f_5}{\sigma L_s},\ l_2 = \frac{R_r}{\sigma L_r},\ l_3 = \frac{L_m^2 \omega_r}{\sigma L_s L_r},\ l_4 = \frac{L_m f_5}{\sigma L_s L_r},\ l_5 = \frac{L_m \omega_r}{\sigma L_r}, l_6 = \frac{L_m R_r}{\sigma L_s L_r},$$

$$l_7 = \frac{L_m \omega_r}{\sigma L_s},\ l_8 = \frac{\omega_r}{\sigma},\ l_9 = \frac{1}{\sigma L_s},\ l_{10} = \frac{L_m}{\sigma L_s L_r},\ \text{and } \sigma = 1 - \frac{L_m^2}{L_s L_r}$$

5.2 Induction Motor Drives Rotor Resistance Estimation

To estimate the rotor resistance, the dynamic model of IM can be extended as follows and the other parameters are considered as known.

$$\hat{f}_{2_{k+1}} = A_{d2}.\hat{f}_{2_k} + B_{d2}.U_k \tag{20}$$

$$A_{d2} = \begin{bmatrix} 1-t_s m_1 & t_s m_3 & t_s m_6 & t_s m_7 & 0 \\ -t_s m_3 & 1-t_s m_1 & -t_s m_7 & t_s m_6 & 0 \\ t_s m_4 & -t_s m_5 & 1-t_s m_2 & -t_s m_8 & 0 \\ t_s m_5 & t_s m_4 & t_s m_8 & 1-t_s m_2 & 0 \\ 0 & 0 & 0 & 0 & 1 \end{bmatrix}, \quad B_{d2} = \begin{bmatrix} t_s m_9 & 0 \\ 0 & t_s m_9 \\ -t_s m_{10} & 0 \\ 0 & -t_s m_{10} \\ 0 & 0 \end{bmatrix}$$

$$U_k = \left[v_{ds}\ v_{qs} \right]^T, \quad f_{2_k} = \left[i_{ds}\ i_{qs}\ i_{dr}\ i_{qr}\ R_r \right]^T = \left[f_1\ f_2\ f_3\ f_4\ f_6 \right]^T$$

$$m_1 = \frac{R_s}{\sigma L_s}, \quad m_2 = \frac{f_6}{\sigma L_r}, \quad m_3 = \frac{L_m^2 \omega_r}{\sigma L_s L_r}, \quad m_4 = \frac{L_m R_s}{\sigma L_s L_r},$$

$$m_5 = \frac{L_m \omega_r}{\sigma L_r}, \; m_6 = \frac{L_m f_6}{\sigma L_s L_r},$$

$$m_7 = \frac{L_m \omega_r}{\sigma L_s}, \quad m_8 = \frac{\omega_r}{\sigma}, \quad m_9 = \frac{1}{\sigma L_s}, \quad m_{10} = \frac{L_m}{\sigma L_s L_r}, \quad \text{and } \sigma = 1 - \frac{L_m^2}{L_s L_r}$$

5.3 Induction Motor Drives Stator Inductance Estimation

To estimate the stator inductance, the dynamic model of IM can be extended as follows and the other parameters are considered as known.

$$\hat{f}_{3_{k+1}} = A_{d3}.\hat{f}_{3_k} + B_{d3}.U_k \tag{21}$$

$$A_{d3} = \begin{bmatrix} 1-t_s n_1 & t_s n_3 & t_s n_6 & t_s n_7 & 0 \\ -t_s n_3 & 1-t_s n_1 & -t_s n_7 & t_s n_6 & 0 \\ t_s n_4 & -t_s n_5 & 1-t_s n_2 & -t_s n_8 & 0 \\ T_s n_5 & t_s n_4 & t_s n_8 & 1-t_s n_2 & 0 \\ 0 & 0 & 0 & 0 & 1 \end{bmatrix}, \quad B_{d3} = \begin{bmatrix} t_s n_9 & 0 \\ 0 & t_s n_9 \\ -t_s n_{10} & 0 \\ 0 & -t_s n_{10} \\ 0 & 0 \end{bmatrix}$$

$$U_k = \left[v_{ds}\ v_{qs} \right]^T, \quad f_{3_k} = \left[i_{ds}\ i_{qs}\ i_{dr}\ i_{qr}\ L_s \right]^T = \left[f_1\ f_2\ f_3\ f_4\ f_7 \right]^T$$

$$n_1 = \frac{R_s L_r}{f_7 L_r - L_m^2}, \quad n_2 = \frac{R_r f_7}{f_7 L_r - L_m^2}, \quad n_3 = \frac{L_m^2 \omega_r}{f_7 L_r - L_m^2},$$

$$n_4 = \frac{L_m R_s}{f_7 L_r - L_m^2}, \quad n_5 = \frac{L_m \omega_r f_7}{f_7 L_r - L_m^2},$$

$$n_6 = \frac{L_m R_r}{f_7 L_r - L_m^2}, \quad n_7 = \frac{L_m L_r \omega_r}{f_7 L_r - L_m^2}, \quad n_8 = \frac{f_7 L_r \omega_r}{f_7 L_r - L_m^2},$$

$$n_9 = \frac{L_r}{f_7 L_r - L_m^2}, \quad \text{and } n_{10} = \frac{L_m}{f_7 L_r - L_m^2}.$$

5.4 Induction Motor Drives Rotor Inductance Estimation

Similarly, for the estimation of the rotor inductance, the dynamic model of IM can be extended as follows and the other parameters are considered as known.

$$\hat{f}_{4_{k+1}} = A_{d4}.\hat{f}_{4_k} + B_{d4}.U_k \tag{22}$$

$$A_{d4} = \begin{bmatrix} 1-t_s o_1 & t_s o_3 & t_s o_6 & t_s o_7 & 0 \\ -t_s o_3 & 1-t_s o_1 & -t_s o_7 & t_s o_6 & 0 \\ t_s o_4 & -T_s o_5 & 1-T_s o_2 & -t_s o_8 & 0 \\ t_s o_5 & t_s o_4 & t_s o_8 & 1-t_s o_2 & 0 \\ 0 & 0 & 0 & 0 & 1 \end{bmatrix}, \quad B_{d4} = \begin{bmatrix} t_s o_9 & 0 \\ 0 & t_s o_9 \\ -t_s o_{10} & 0 \\ 0 & -t_s o_{10} \\ 0 & 0 \end{bmatrix}$$

$$U_k = \begin{bmatrix} v_{ds} & v_{qs} \end{bmatrix}^T, \quad f_{4_k} = \begin{bmatrix} i_{ds} & i_{qs} & i_{dr} & i_{qr} & L_r \end{bmatrix}^T = \begin{bmatrix} f_1 & f_2 & f_3 & f_4 & f_8 \end{bmatrix}^T$$

$$o_1 = \frac{R_s f_8}{L_s f_8 - L_m^2}, \quad o_2 = \frac{R_r L_s}{L_s f_8 - L_m^2}, \quad o_3 = \frac{L_m^2 \omega_r}{L_s f_8 - L_m^2},$$

$$o_4 = \frac{L_m R_s}{L_s f_8 - L_m^2}, \quad o_5 = \frac{L_m \omega_r L_s}{L_s f_8 - L_m^2},$$

$$o_6 = \frac{L_m R_r}{L_s f_8 - L_m^2}, \quad o_7 = \frac{L_m f_8 \omega_r}{L_s f_8 - L_m^2}, \quad o_8 = \frac{f_8 L_s \omega_r}{L_s f_8 - L_m^2},$$

$$o_9 = \frac{f_8}{L_s f_8 - L_m^2}, \text{ and } o_{10} = \frac{L_m}{L_s f_8 - L_m^2}.$$

5.5 Estimation of IRFO Control-Based IMD Error Regulator PI Gains Using ALO Technique

PI error regulator gains tuning has a significant influence on the effectiveness of any control approach that involves PI controller loops. In other words, PI error regulator gains must be accurately tuned for the better effectiveness of any control technique. Manual trial and error is the most common approach for estimating the approximate gains of PI error regulators. But it consumes more time for tuning, so optimization-based techniques are developed for the optimal tuning of PI error regulator gains [25]. And optimization-based techniques maximize/minimize the system's objective function, based on this concept optimization techniques are implemented to tune the optimal gains of the PI error regulator. In this work, the Ant-Lion optimization (ALO) algorithm is implemented for the evaluation of optimal gains (k_{p}, k_{i}) of the PI error regulator [26]. The objective function is derived from the difference of reference and actual value for optimizing as,

$$\text{Objective function} = \text{ITSE} = \int (t * e^2(t))dt \tag{23}$$

where, $e(t)$ = error = reference value – actual value.

In [27], based on the experience of ants and ant-lions the fundamental expressions of the ALO technique are provided. In general, when ants are searching for food the movement of ants is unpredictable, so a random walk is considered stochastic ants' movement. The mathematical expression is given as,

$$R(i) = [0,\ \text{cumsum}(2r(i_1) - 1), \text{cumsum}(2r(i_2) - 1), \ldots \text{cumsum}(2r(i_n) - 1)] \tag{24}$$

where,

cumsum = cumulative sum, n = maximum no. of iterations, and i = step of random walk (iteration).

The stochastic function $r(i)$ is defined as:

$$r(i) = \begin{Bmatrix} 1 \text{ if rand } > 0.5 \\ 0 \text{ if rand } \leq 0.5 \end{Bmatrix} \tag{25}$$

Rand is the random number provided with uniform distribution of [0, 1]. To limit the random walk within the search space, minimum–maximum-based normalization method is employed. The mathematical representation of the position of ants is as,

$$R_m^i = \frac{(R_m^i - a_m)(d_m - c_m^i)}{(d_m^i - a_m)} \tag{26}$$

where a_m is the minimum random walk of ants (for a maximum random walk of ants is represented as b_m. Other side, ${c_m}^i$ and ${d_m}^i$ represent (trapping of ants) minimum and a maximum of mth variable for ith iteration.

$$c_m^i = al_d^i + c^i$$
$$d_m^i = al_d^i + d^i$$

Here, ${al_d}^i$ represents the ant-lion of ith iteration with dth dimension of a variable.

To model the hunting ability of an ant-lion, a Roulette wheel is developed. As such, ants are intended to be trapped by only one ant-lion. Based on the fitness of the ant-lion, they can construct the trap and if requires ants can move randomly. However when the ant-lion recognizes that the ant was captured and is attempting to flee from the pit, the behavior of the ant-lion is to slide toward the pit's center, so it's preventing the ant from escaping. This is accomplished by optimizing the radios of the pit using the following expression.

$$c^i = \frac{c^i}{M}; \quad d^i = \frac{d^i}{M} \tag{27}$$

Here, $M = 10^w \times \frac{i}{n}$. ('$w$' as constant).

The last stage in hunting is capturing the prey. When ant-lion captures the ant with the best objective function than the chosen ant, the position of the ant-lion is required to update. For this process, the mathematical expression is given as,

$$al_d^i = a_m^i \ \ if \ \ f(a_m^i) > f(al_d^i) \tag{28}$$

The objective function is obtained from one of the performances an index is represented in (23). The objective function is cumulative of integral time squared error1 (ITSE1), ITSE2, and ITSE3. Here, ITSE1 is for the speed PI regulator, ITSE2 is for the d-axis current PI regulator and ITSE3 is for the q-axis current PI regulator. Then, the overall objective function is represented as $F_{\mathrm{obj}} = (0.33 \times \mathrm{ITSE1}) + (0.33 \times \mathrm{ITSE2}) + (0.33 \times \mathrm{ITSE3})$. All PI regulators have been given equal weightage. The maximum number of iterations is considered 30 and the total ants/ant-lions (search agents) is considered 100. The objective function convergence curve is shown in Fig. 3a concerning iterations. The speed, d-axis, and q-axis current PI regulator "k_{p}" and "k_{i}" gain variations with respect to iteration (on the x-axis) is taken on the same plot with left and right y-axis respectively as illustrated in Fig. 3b–d. It is observed that after completion of the optimization process, the objective function reached the steady-state value at 1345.523 after 10 iterations and the stable PI regulators of speed, d-axis, and q-axis current regulator (k_{p}, k_{i}) gain 25.65, 0.413, 13.236, 114.23 and 14.31, 108.02 respectively are settled within the 14 iterations.

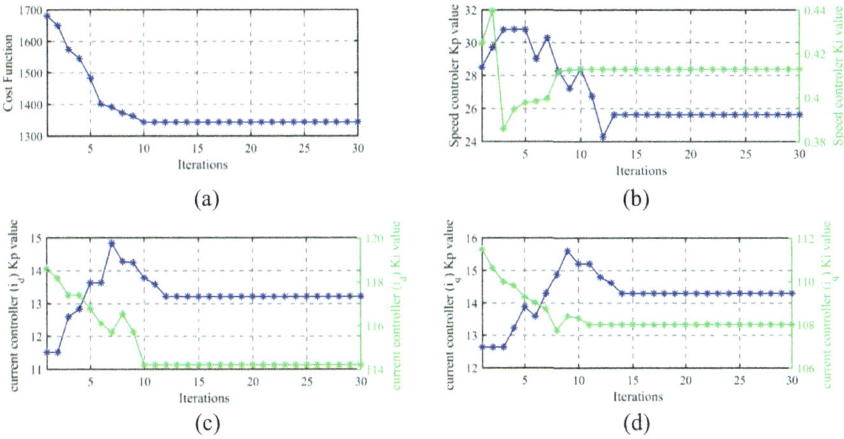

Fig. 3 Performance characteristics of ALO **a** objective function **b** speed PI regulator **c** d-axis current PI regulator **d** q-axis current PI regulator with respect to iterations

6 Results and Discussions

To prove the effectiveness of the proposed technique, the ADEKF-based parameter estimation of the IRFOC-IMD fed VSI system illustrated in Fig. 1 is used and it is developed in the environment of Matlab/Simulink. To develop this model the required parameters and its values are mentioned in Appendex A Table 2. Further, the different operating scenarios are considered for the performance studies such as steady-state and dynamic (speed and load) operations and the resulting estimation performances of both techniques are presented in the following subsections. In these figures, Ref, Act, SDEKF, and ADEKF subscripts represent the reference values, actual values of IMD, the estimated parameters by SDEKF, and estimated parameters by ADEKF techniques respectively. The system noise covariance matrices of 'Q' in SDEKF are evaluated by trial-and-error approach and in ADEKF are evaluated by the controller itself. And the measurement noise covariance matrices of 'R' in SDEKF and ADEKF both techniques are both evaluated by trial-and-error method only. To get satisfactory initial convergence, initial values of the error covariance matrix of P are chosen by trail-and-error approach as $P_0 = \text{diag}[1, 1, 1, 1, 1]$.

6.1 Performance Analysis of ADEKF and SDEKF Under Steady-State Operation

The Estimated electrical parameters of stator resistance (R_s), rotor resistance (R_r), stator inductance (L_s), and rotor inductance (L_r) of IRFOC-based IMD using ADEKF and SDEKF techniques under steady-state operation (with constant load and speed) and by updating the estimated parameters in the calculation of slip angular frequency/ slip speed (ω_{sl}) characteristics results are illustrated in Figs. 4 and 5 respectively. The estimated states of stator and rotor currents (i_{dqs} and i_{dqr}) in stationary (dq) reference frames under steady-state operation using the ADEKF technique are represented in Fig. 6. The performance characteristics of IRFOC-based IMD along with the SVPWM technique are reported in Fig. 7.

Considering the resulting estimation performances presented in Fig. 4, the ADEKF-based version has better performance with lesser error between estimated and actual values. The IMD is operated with 10Nm of load torque and a rated speed of 150 rad/s. In Fig. 4 the estimated parameters using SDEKF have more rise time and error as compared to the ADEKF technique. It observed that more transients are presented in the SDEKF technique as shown in Fig. 4 with zoomed version. From this analysis, the ADEKF technique provides better performance in terms of transients, rise time, and error. While the parallel operation of the ADEKF parameter estimation technique; the other estimated states of currents are the same in all ADEKF blocks, for observation and analysis only one ADEKF stator and rotor current state results are illustrated in Fig. 6. Further, the estimated parameters using SDEKF and ADEKF of rotor resistance and inductance are updated in the IRFO controller for

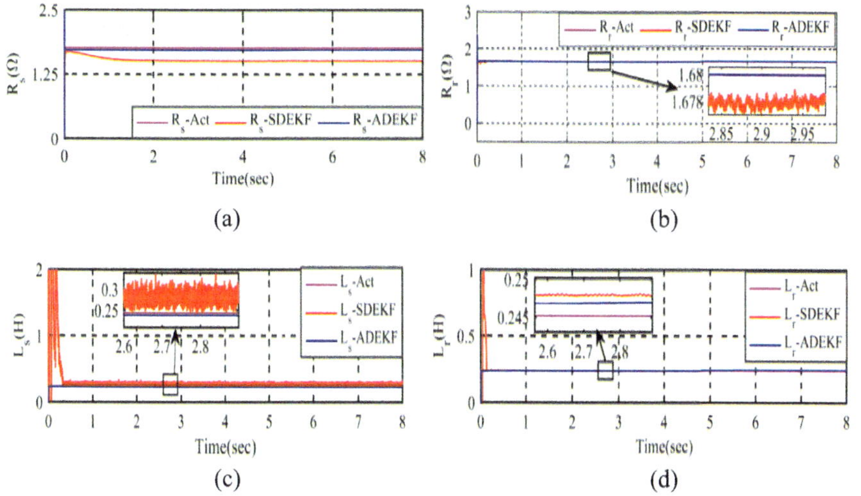

Fig. 4 Estimated electrical parameters of ADEKF and SDEKF techniques under steady-state condition **a** Stator Resistance (R_s) **b** Rotor Resistance (R_r) **c** Stator Inductance (L_s) and **d** Rotor Inductance (L_r)

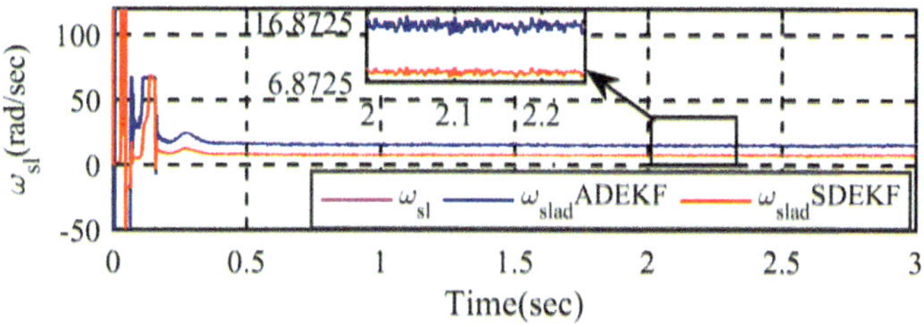

Fig. 5 Performance characteristics of adaptive angular slip speed of IMD of ADEKF and SDEKF techniques under steady-state condition

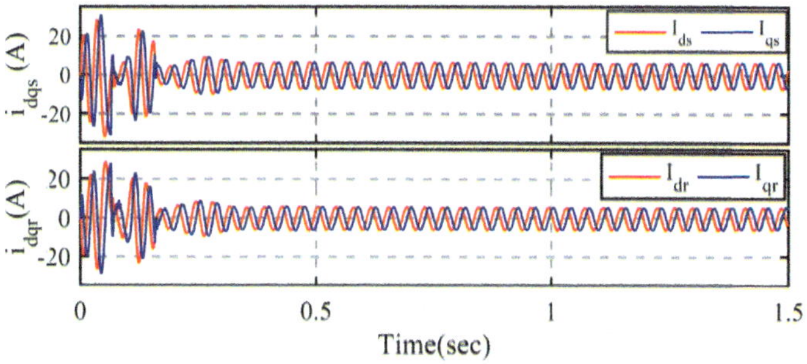

Fig. 6 Estimated states of stator and rotor currents under steady-state condition

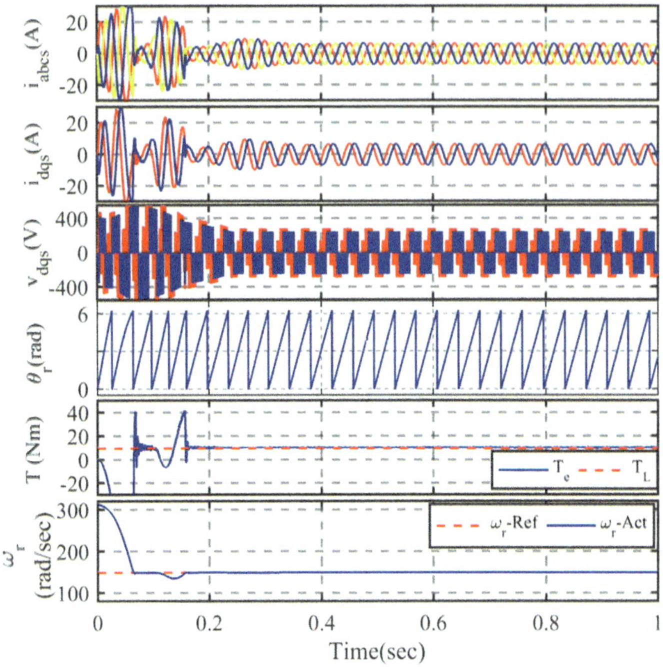

Fig. 7 Performance characteristics of ADEKF-based IFOC-IMD under steady-state condition

tuning of slip angular frequency, the results are observed in Fig. 5. Due to improper (more error) estimation of parameters in SDEKF, the slip speed is affected more. The highly accurate slip speed is evaluated using ADEKF technique-based parameters in IRFOC and it is given the improved performance of IMD characteristics. The performance characteristics of three-phase stator currents (i_{abcs}), stationary reference frame stator currents (i_{dqs}), stationary reference frame stator voltages (v_{dqs}), rotor angular position (θ), electromagnetic and load torque (T_{e} and T_{L}), and rotor reference and actual (estimated using speed encoder) speed (ω_{ref} and ω_{r}), of IMD under constant reference speed and load are shown in Fig. 6 respectively.

The performance comparison percentage error of estimated parameters using SDEKF and ADEKF techniques with actual values of IMD are listed in Table 1. From this data, it is observed that the estimated values of IMD are more precise by using the ADEKF technique as compared to the SDEKF technique. The estimated % errors of stator resistance, Rotor resistance, Stator inductance, and Rotor inductance using the SDEKF technique are 14.5%, 4.16%, 4.36%, and 2.32% respectively more as compared to Similarly, the estimated %errors using ADEKF technique are 2.1%, 0.05%, 2.16%, and 0.69% respectively. The performance of percentage errors using ADEKF is very less as compared SDEKF technique. This proves that the accuracy of the ADEKF technique is best.

Table 1 Performance comparison of estimated parameters using SDEKF and ADEKF with actual values

Parameter	Actual value	SDEKF		ADEKF	
		Estimated value	Percentage error (%)	Estimated value	Percentage error (%)
Stator resistance—R_s (ohms)	1.78	1.521	14.5	1.741	2.1
Stator resistance—R_r (ohms)	1.68	1.61	4.16	1.679	0.05
Stator inductance—L_s (H)	0.2453	0.256	−4.36	0.24	−2.16
Rotor inductance—L_r (H)	0.2453	0.251	−2.32	0.247	−0.69

6.2 *Performance Analysis of ADEKF and SDEKF Under Dynamic Operation of Speed*

To investigate the dynamic performances of the system, initially considered it with dynamic changes in the speed and later in load. The operation of the system started with initially very low speed, after a few fractions of seconds the speed command (reference speed) of IM is 150 rad/s with constant load considered, and during a 2.5–2.7 s time period the dynamic operation is developed. During dynamic operation, the reduced speed of 110 rad/s is applied as a reference speed. Then the dynamic variations in the estimated electrical parameters of Stator Resistance (R_s), Rotor Resistance (R_r), Stator Inductance (L_s), Rotor Inductance (L_r), and other state parameters of stationary reference frame stator and rotor currents (i_{dqs} and i_{dqr}) of using SDEKF and ADEKF techniques are incorporated as seen in Fig. 8. It is observed that the estimated parameters are affected using SDEKF during dynamic operation as compared to the ADEKF technique. More transients are developed and more time takes to reach the steady state after dynamic operation in SDEKF-based technique. If any dynamic changes happen in the ADEKF technique the effect of estimated parameters is very low, because of the adaptive/continuous updating of noise covariance matrices. Transients are very poor during dynamic speed conditions in estimated electrical parameters using ADEKF; at the instant of dynamic operation, only transients are developed and reached steady states very fast. For the evaluation of slip angular frequency, required estimated parameters using SDEKF and ADEKF techniques are adopted and the tuned slip angular frequency (ω_{sl}) results as shown in Fig. 9. During dynamic operation, evaluated ω_{sl} in SDEKF is more as compared to actual and the time taken to reach the steady state is high. But in ADEKF, at the sudden dynamic change of time, only transients are produced and the rest of the time required value slip speed is developed. From this analysis, the performance of IMD using ADEKF is improved and accurate parameters are estimated. Further, the performance characteristics of ADEKF parameter estimation based IRFOC-IMD,

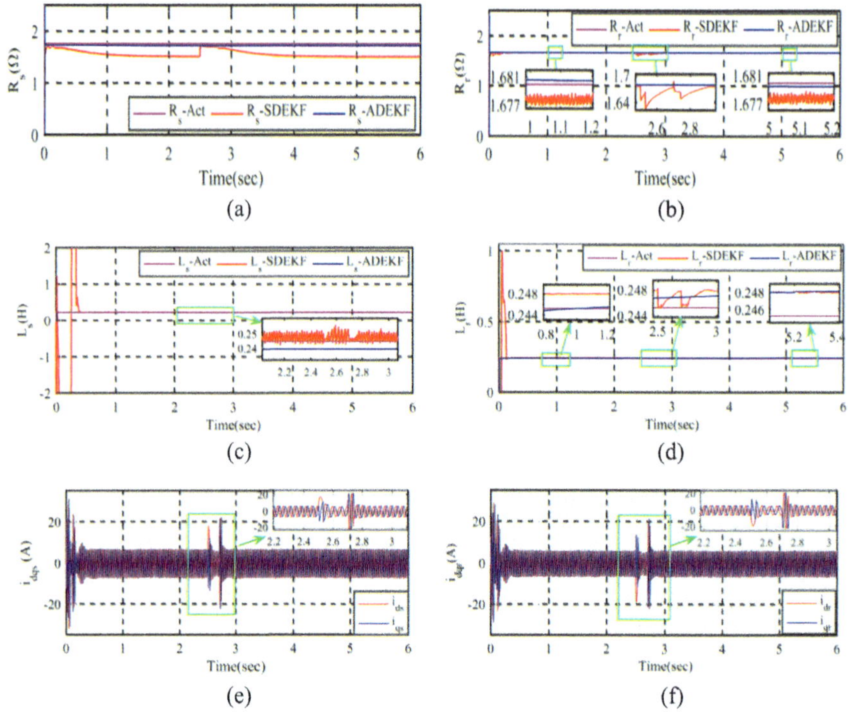

Fig. 8 Estimated parameters of ADEKF and SDEKF techniques during dynamic speed condition **a** Stator Resistance R_s, **b** Rotor Resistance R_r, **c** Stator Inductance L_s, **d** Rotor Inductance L_r, **e** Stator currents i_{dqs}, and *f* Rotor currents i_{dqr}

three-phase stator currents (i_{abcs}), stationary reference frame stator currents (i_{dqs}), stationary reference frame stator voltages (v_{dqs}), rotor angular position (θ), electromagnetic and load torque (T_e and T_L), and rotor reference and actual (estimated using speed encoder) speed (ω_{ref} and ω_r), under the dynamic speed of operation are shown in Fig. 10.

6.3 *Performance Analysis of ADEKF and SDEKF Under the Dynamic Operation of Load*

The dynamic operation is examined with increased and decreased load torque changes during 2.1–2.2 s and 3.9–4 s respectively, initially; the load torque is 10 Nm and the reference speed 150 rad/s is considered. During dynamic operation of 2.1–2.2 s, the increased torque of 15 Nm and 3.9–4 s reduced torque of 10 Nm. The dynamic variations in the estimated electrical parameters of Stator Resistance (R_s), Rotor Resistance (R_r), Stator Inductance (L_s), Rotor Inductance (L_r), and other state

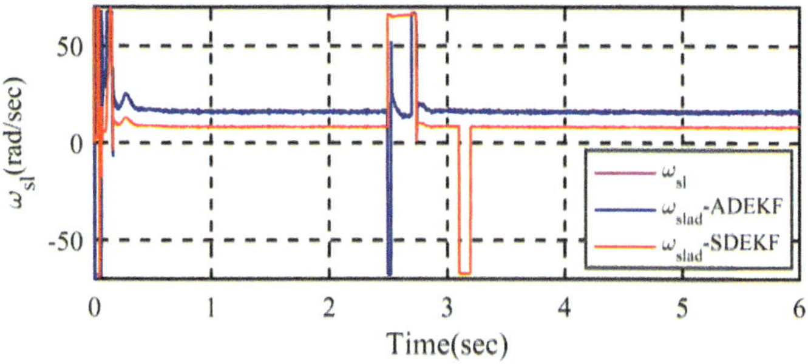

Fig. 9 Performance characteristics of ADEKF and SDEKF techniques during adaptive angular slip speed of IMD during dynamic operation of speed

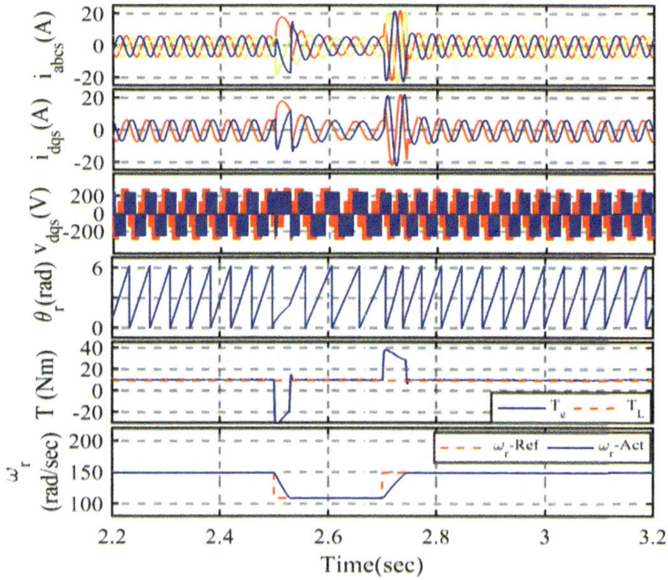

Fig. 10 Performance characteristics of ADEKF-based IRFOC-IMD during dynamic operation of speed

parameters of stationary reference frame stator and rotor currents (i_{dqs} and i_{dqr}) of using SDEKF and ADEKF techniques are shown in Fig. 11.

It is observed that the estimated parameters are highly affected using SDEKF during dynamic operation as compared to ADEKF technique as same as dynamic speed operation. More transients are developed and more time is taken to reach the steady state after dynamic operation in SDEKF-based technique. If any dynamic changes happen in the ADEKF technique the effect of estimated parameters is very

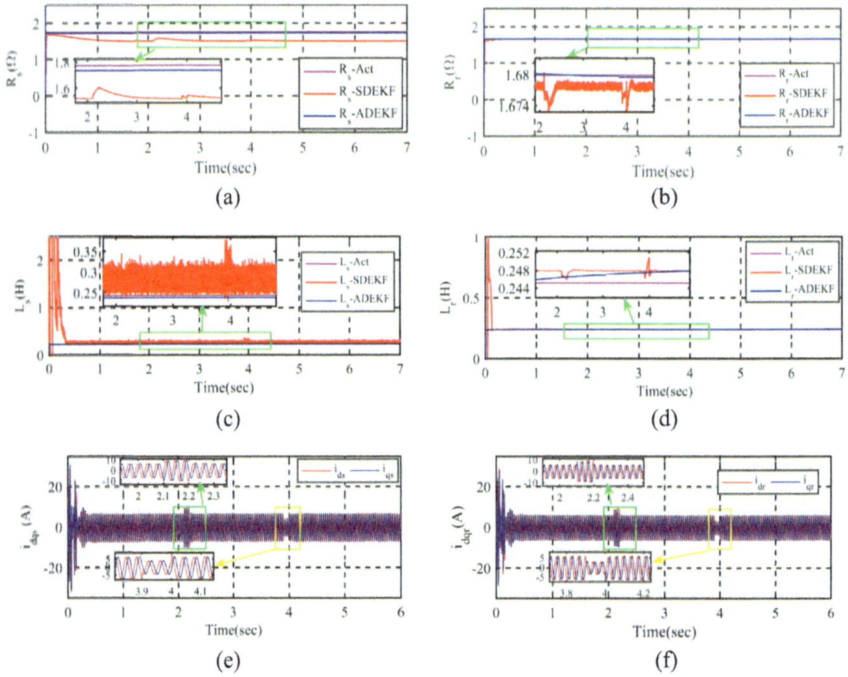

Fig. 11 Estimated parameters of ADEKF and SDEKF techniques during dynamic load condition **a** Stator Resistance R_s, **b** Rotor Resistance R_r, **c** Stator Inductance L_s, **d** Rotor Inductance L_r, *e* Stator currents i_{dqs}, and **f** Rotor currents i_{dqr}

low, because of the adaptive/continuous updating of noise covariance matrices. Transients are very poor during dynamic torque condition in estimated electrical parameters using ADEKF; at the instant of dynamic operation, only transients is developed and reached a steady state very fast. It is observed in Fig. 11 that estimated states of stator and rotor currents during the high amount of load torque high current and low amount of load torque reduced current withdrawn from the source. For the evaluation of slip angular frequency, required estimated parameters using SDEKF and ADEKF techniques are adopted and the tuned slip angular frequency (ω_{sl}) results as shown in Fig. 12. Evaluated ω_{sl} using SDEKF is low as compared to the actual value. But in ADEKF, ω_{sl} is evaluated as the actual value with strong tracking and during dynamic operation due to change of load speed and current affected, so slip angular frequency also changes. Further, the performance characteristics of ADEKF parameter estimation based IRFOC-IMD, three-phase stator currents (i_{abcs}), stationary reference frame stator currents (i_{dqs}), stationary reference frame stator voltages (v_{dqs}), rotor angular position (θ), electromagnetic and load torque (T_e and T_L), and rotor reference and actual (estimated using speed encoder) speed (ω_{ref} and ω_r), under the dynamic load of operation, are illustrated in Fig. 13 respectively. The changes in load torque affected the IM electromagnetic torque, and both are strongly overlapped shown in Fig. 13.

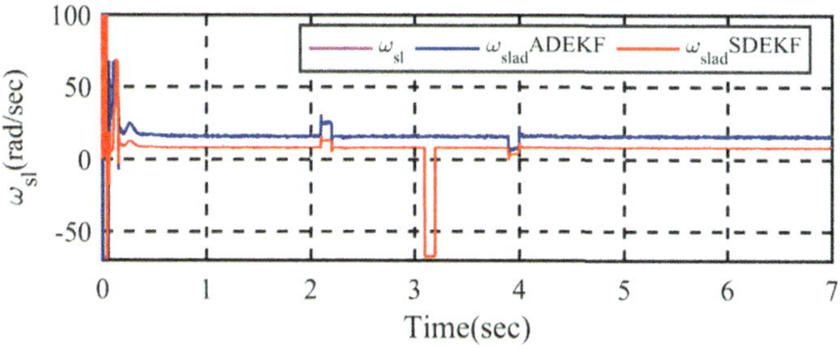

Fig. 12 Performance characteristics of ADEKF and SDEKF techniques adaptive angular slip speed of IMD during dynamic operation of Torque

During dynamic operation of load, the IM rotor speed variations are very less because of an IRFOC; it controls and produces the reference speed.

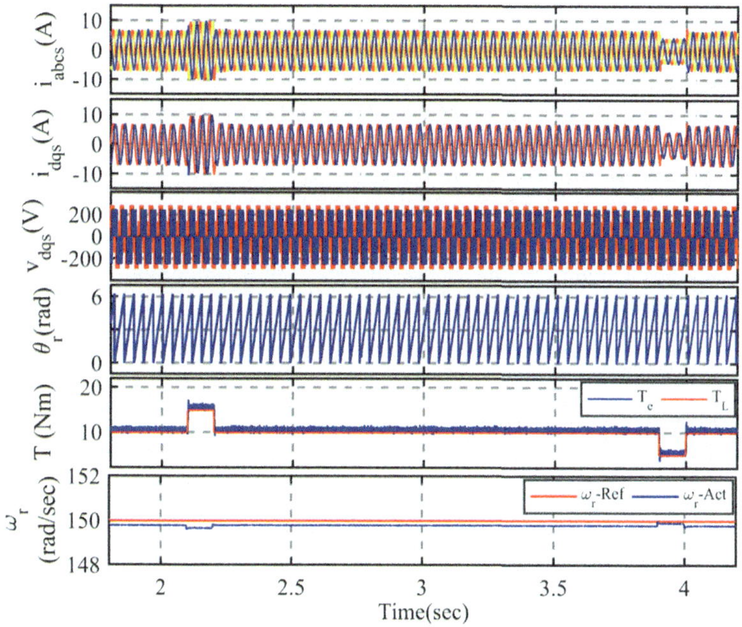

Fig. 13 Performance characteristics of ADEKF-based IFOC-IMD during dynamic operation of a load

7 Conclusion

In this paper, the ADEKF technique is proposed for simultaneously parallel estimation of electrical parameters of i_{ds}, i_{qs}, i_{dr}, i_{qr}, R_s, R_r, L_s, and L_r of IMD and ALO-based optimal PI error regulator gains estimation is implemented in IRFOC. For tuning of slip angular frequency in RFO control IMD, the estimated parameters (R_r and L_r) are adopted. In transient conditions and under parameter perturbations, SDEKF's estimated performance deteriorates because it treats Q and R matrices as constant even though the IM operating circumstances are variable. As a consequence, in order to provide an accurate estimation, these matrices must be regularly updated through actual environments. However, the evaluation of Q and R matrices at an instant of time is complicated. So in this work, the R matrix can be evaluated by the explorative method, and by using the proposed ADEKF technique the Q matrix is updated continuously. To ascertain the superiority or inferiority of the ADEKF technique, the estimation effectiveness of ADEKF is examined by comparing it with the SDEKF technique under a challenging scenario. Moreover, the performance of estimated percentage errors of all four parameters are least in ADEKF as compared to SDEKF. The estimated parameter percentage errors as compared to actual values are R_s, R_r, L_a, and L_r are 2.19%, 0.178%, 2.2, and 0.774% respectively. Further, the optimization-based ALO technique is used for stable PI error regulator gains evaluation. It was noticed that the ALO technique attained all stable PI gains within the 14th iterations, the optimized speed, d-axis, and q-axis current regulator gains 25.65, 0.413, 13.236, 114.23, and 14.31, 108.02 respectively. The proposed ADEKF technique has the potential to produce better estimation accuracy than among SDEKF technique with a slight computational difference by considering estimation results and related errors.

Appendix A

See Table 2.

Table 2 Simulation motor parameter specification

Rated voltage (V)	415 V
Rated power (P)	5hp
Supply frequency (f)	50 Hz
Pole pairs (p)	2
Rated speed (ω)	157 rad/s
Inertia constant (J)	0.011 kg-m^2

References

1. Brooks LA, Castro JL, Castro EL (2014) Speed and position controllers using indirect field-oriented control: a classical control approach. IEEE Trans Industr Electron 61(4):1928–1943
2. Chakraborty C, Maiti S (2021) Induction motor drives principles, control, and implementation. Taylor & Francis, CRC Press
3. Nordin KB, Novotny DW, Zinger DS (1985) The influence of motor parameter deviations in feed forward filed orientation drive system. IEEE Trans Ind Appl IA-21(4):1009–1015
4. Toliyat HA, Levi E, Raina M (2003) A review of RFO induction motor parameter estimation technqiues. IEEE Trans Energy Convert 18(2):271–283
5. Vas P (1993) Parameter estimation, condition monitoring, and diagnosis of electrical machines. Clarendon Press, Oxford
6. Almounajjed A, Sahoo AK, Kumar MK, Bakro MW (2021) Condition monitoring and fault diagnosis of induction motor—an experimental analysis. In: 7th international conference on electrical energy systems (ICEES), pp 433–438
7. Lee SH, Yoo A, Lee HJ, Yoon YD, Han BM (2017) Identification of induction motor parameters at standstill based on integral calculation. IEEE Trans Ind Appl 53(3)
8. Reddy SRP, Lohanathan U (2020) Offline recursive identification of electrical parameters of VSI-Fed induction motor drives. IEEE Trans Power Electron 35(10)
9. He Y, Wang Y, Feng Y, Wang Z (2012) Parameter Identification of an induction machine at standstill using the vector constructing method. IEEE Trans Power Electron 27(2)
10. Odhano SA, Pescetto P, Awan HAA, Hinkkanen M, Pellegrino G, Bojoi R (2019) Parameter identification and self-commissioning in ac motor drives: a technology status review. IEEE Trans Power Electron 34(4):3603–3614
11. Tang J, Yang Y, Blaabjerg F, Chen J, Diao L, Liu Z (2018) Parameter Identification of inverter fed Induction motors: a review. J Energ 11(9):2194
12. Tinazzi F, Carlet PG, Bolognani S, Zigliotto M (2020) Motor parameter-free predictive current control of synchronous motors by recursive least-square self-commissioning model. IEEE Trans Industr Electron 67(11):9093–9100
13. Wei J, Xu H, Zhou B, Gerada C (2019) An Integrated method for 3-phase AC excitation and high-frequency voltage signal injection for sensorless starting of aircraft starter/generator. IEEE Trans Industr Electron 66(7):5611–5622
14. Pei G, Liu J, Li L, Du P, Pei L, Hu Y (2018) MRAS based online parameter identification for PMSM considering VSI nonlinearity. In: IEEE international power electronics and application conference and exposition (PEAC), pp 1–7
15. Liu Y, Fang J, Tan K, Huang B, Wenshuai H (2020) Sliding mode observer with adaptive parameter estimation for sensorless control of IPMSM. J Energ 13(22):5991
16. Bhowmick D, Chowdhury SK (2018) Parameter and loss estimation of three phase induction motor from dynamic model using H-G diagram and particle swarm optimization. In: IEEE 8th power India international conference (PIICON), pp 1–5
17. Bijan MG, Al-Badri M, Pillay P, Angers P (2018) Induction machine parameter range constraints in genetic algorithm based efficiency estimation techniques. IEEE Trans Ind Appl 54(5):4186–4197
18. Carrassi A, Vannitsem S (2011) State and parameter estimation with the extended Kalman filter: an alternative formulation of the model error dynamics. Quaterly J R Meteorol Soc 137:435–451
19. Barut M (2010) Bi-Input-extended Kalman filter-based estimation technique for speed-Sensorless control of IM. J Energy Convers Manag 51(10)
20. Salvatore N, Caponio A, Neri F, Stasi S, Cascella G (2010) Optimization of delayed-state Kalman-filter-based algorithm via differential evolution for sensorless control of induction motors. IEEE Trans Ind Electron 57(1):385–394
21. Jayaramu ML, Suresh HN, Bhaskar MS, Padmanaban S, Subramaniam U (2021) Real-time implementation of Extended Kalman Filter Observer with improved speed estimation for sensorless control. IEEE Access 9:50452–50465

22. Zerdali E, Barut M (2017) The comparisons of optimized Extended Kalman Filters for speed-sensorless control of induction motors. IEEE Trans Ind Electron 64(6):4340–4351
23. Zhang Y, Yin Z, Li G, Liu J, Tong X (2018) A Novel speed estimation method of induction motors using real-time adaptive Extended Kalman Filter. J Electr Eng Technol 13(1):287–297
24. Zerdali E (2019) Adaptive extended Kalman filter for speed-sensorless control of induction motors. IEEE Trans Energy Convers 34(2):789–800
25. Wang W, Liu Z, Xie R (2006) Quadratic extended Kalman filter approach for GPS/INS integration. Aerosp Sci Technol 10(8):709–713
26. Long Z, Zhang X, Peng X, Yang G (2019) An Improved adaptive extended Kalman filter used for target tracking. In: Chinese automation congress (CAC), pp 1017–1022
27. Hashlamon I (2019) A new adaptive extended Kalman filter for a class of nonlinear systems. J Appl Comput Mech 6(1):1–12
28. Nagda NN (2019) Variable fractional order PID Control for precision positioning (A frequency domain approach). Master of Science Thesis, Delft University of Technology
29. Pareek S, Kishnani M, Gupta R (2014) Optimal tuning of PID controller using Meta heuristic algorithms. In: International conference on advances in engineering & technology research (ICAETR-2014), pp 1–5
30. Assiri AS, Hussien AG, Amin M (2020) Ant Lion optimization: variants, hybrids, and applications. IEEE Access 8:77746–77764
31. Mirjalili S (2015) The ant lion optimizer. J Adv Eng Softw 83:80–98
32. Lefebvre G, Gauthier JY, Hijazi A, Lin Shi X, Digarcher VL (2017) Observability-index-based control strategy for induction machine sensorless drive at low speed. IEEE Trans Ind Electron 64(3):1929–1938
33. Alonge F, Cangemi T, Ippolito FD, Fagiolini A, Sferlazza A (2015) Convergence analysis of extended Kalman filter for sensorless control of induction motor. IEEE Trans Industr Electron 62(4):2341–2352
34. Zerdali E, Barut M (2018) Extended Kalman filter-based speed-sensorless load torque and inertia estimations with observability analysis for induction motors. J Power Electron Drives 3(38):115–127
35. Liu X, Li L, Li Z, Fernando T, Iu HHC (2017) Stochastic stability condition for the extended Kalman filter with intermittent observations. IEEE Trans Circuits Syst II Express Briefs 64(3):334–338

Modal Analysis of Series Compensated DFIG Based Windfarm with LQR Controller for Damping of SSR Oscillations

R. Mahalakshmi and K. C. Sindhu Thampatty

Abstract Recently, due to the exhaustion of fossil fuels and its increase in emission of carbon made the world to produce electricity through non-conventional energy sources like solar, wind etc. This paper deals with the problems in connecting the Wind Turbine Generator (WTG) into the grid network using the series compensated long transmission line. Inclusion of series capacitor in the transmission line causes Sub Synchronous Resonance (SSR) which results in electromechanical torque oscillations. The analysis of SSR oscillations by the insertion of different levels of series compensation for the transmission line through Eigen value approach is carried out in this paper. The response of the system is tested and improved using Linear Quadratic Regulator (LQR) controller. For this analysis, the entire wind energy series compensated power system is expressed mathematically using state space equations. The modelled system has 14 state variables. The model with LQR controller is validated using MATLAB/SIMULINK.

Keywords LQR controller · Wind energy · Series compensated transmission line · Sub synchronous resonance · Eigen value analysis

1 Introduction

Generation of electricity from wind energy is increasingly popular nowadays to protect our planet from carbon footprint. Wind farm capacity has increased recently in India and the total capacity has been extended to 41.93 GW [1]. The extraction of energy from wind is not much easier than that from solar energy. The wind power system requires long distance transmission lines for grid integration, as wind farms

R. Mahalakshmi
Department of Electrical Engineering, Amrita School of Engineering, Bangalore, India

K. C. Sindhu Thampatty (✉)
Department of Electrical Engineering, Amrita School of Engineering, Amrita Vishwa Vidyapeetham, Coimbatore, India
e-mail: t_sindh@cb.amrita.edu

S. Kumar et al. (eds.), *Control Applications in Modern Power Systems*, Lecture Notes in Electrical Engineering 1128, https://doi.org/10.1007/978-981-99-9054-2_2

are located far away from the load center or grid. Among the many types of WTG, DFIG based WTG is more advantageous than the other generators [2]. The electric power generated from wind mills must be transferred a long way through the transmission line. Long lines have a high inductive reactance and reduce the power transfer capability because the power is inversely proportional to the inductive reactance. To enhance the power transfer capacity of transmission lines, traditionally the series capacitors are inserted.

The most common issue in inserting series capacitor in the wind turbine-based power system is the series resonance. The capacitor in the line may resonant with the inductance of the line which can cause electrical resonance in the system. If the electrical frequency (f_{elec}) coincides with the mechanical frequency (f_{mech}) of turbine set, then there will be a dangerous effect in the entire system and is called as Sub Synchronous Resonance. Equation (1) describes the condition for which SSR occurs [3, 4].

$$f_{\text{mech}} = f_{\text{sys}} - f_{\text{elec}} \tag{1}$$

where f_{sys} is the system frequency.

The sub synchronous resonance can have a high impact on the wind power integrated power system. This effect causes the electromechanical torque to oscillate exponentially and causes severe damage to the shaft of the turbine and generator set [5]. Hence this has to be analyzed and avoided in the system.

There are various methods to analyze the stability of the system for varying compensation levels like Eigen value approach, frequency scanning method, Electromagnetic Transient Programming approach time domain analysis etc., Refs. [6, 7]. Among all these methods, the very common approach for the SSR analysis is the eigen value approach. As the torque oscillations are produced in the system, it has to be damped out quickly to protect the power system. Detailed research is conducted on the design of new controllers in which the controllers are adjusted to the variations in operating conditions. The damping of SSR using LQR [8], low pass filter with compensation [9], partial feedback linearization [10], particle swarm optimization algorithm [11] and state feedback observed control [12], are discussed. For the Eigen value approach, the wind energy system has to be modelled using state space approach and the entire system is represented using a single state matrix which can explain the complete dynamics of the system. The Eigen values of this single matrix imply whether the system is stable or not for a particular compensation level.

This paper discusses the detailed modal analysis conducted on the study system for the identification of different torsional mode frequencies which can cause the torque oscillations in the system. A detailed design of the Linear Quadratic Regulator (LQR) controller used in the system to stabilize the unstable torsional modes is also explained in the paper.

The organization of the paper in Sect. 2 describes the complete system modelling followed by the LQR controller design in Sect. 3. Section 4 gives the performance analysis of the system with controller under different operating conditions followed by conclusion in Sect. 5.

2 System Modelling for SSR Analysis

The system under consideration is shown in Fig. 1. The scheme comprises of 2 MW DFIG based WTG, 100 km line length and percentage compensation of the line is increased with the help of varying series capacitors. When the compensation level increases, the system oscillates due to SSR. The unstable states can be brought to a stable state by using adaptive controllers. Here, the LQR controller is used for the damping of oscillations produced due to SSR.

The DFIG based WTG is used in the wind systems. The DFIG is more popular as this comes along with the RSC and GSC converters. The entire system consists of many sub systems such as two mass model of wind turbine, DFIG with converters, Grid connected series compensated transmission line (SCTL) etc. Each subsystem is represented in state space form and they are clubbed together to form an integrated system equation. The equation of the system represented in state space form is given in Eq. (2)

$$[\dot{X}_{sys}] = [A_{sys}][X_{sys}] + [B_{sys}][U_{sys}] \tag{2}$$

The output equation is given in Eq. (3)

$$Y_{sys} = C_{sys}X_{sys} + D_{sys}U_{sys} \tag{3}$$

where

$$[X_{sys}] = [X_{wt}X_{DFIG}X_{tl}X_{BBC}]^T$$

State variable matrix is given by

$$[X_{sys}] = [\omega_t\delta_{tg}\omega_g i_{ds}i_{qs}i_{dr}i_{qr}i_{ld}i_{lq}V_{cd}V_{cq}V_{dc}i_{dg}i_{qg}]^T$$

$X_{wt}X_{DFIG}X_{tl}X_{BBC}$ are state variable matrices of wind turbine, DFIG, Transmission line and back to back converters respectively.

The system matrix of the complete system is given in Eq. (4) [13–17].

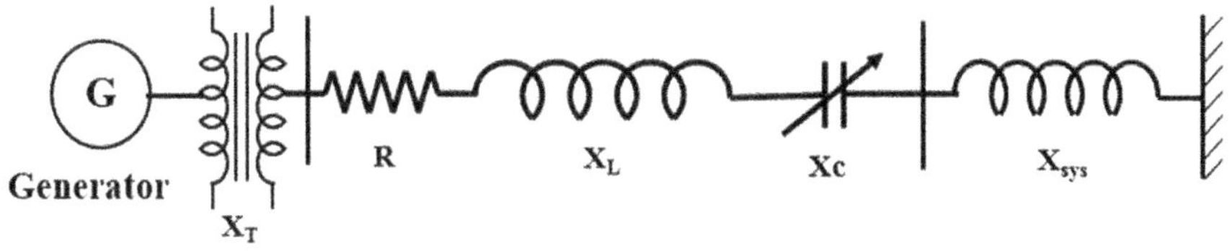

Fig. 1 Single line diagram of the proposed system

$$[A_{\text{sys}}] = \begin{bmatrix} A_{wt} & 0 & 0 & 0 \\ 0 & A_{\text{DFIG}} & 0 & 0 \\ 0 & 0 & A_{tl} & 0 \\ 0 & 0 & 0 & A_{\text{BBC}} \end{bmatrix} \tag{4}$$

The system matrices of the wind turbine, DFIG, SCTL and RSC-GSC are given in the paper [6]. The control variables of the complete system are given in Eq. (5)

$$[U_{\text{sys}}] = [T_\omega V_{ds} V_{qs} V_{dr} V_{qr} V_{bd} V_{bq}]^T \tag{5}$$

In total, there are 14 state variables and 7 control variables. A_{sys} has a size of 14 × 14 and B_{sys} has a size of 14 × 7 size. C will be column matrix and all the elements are zero except one. One element will be unity and this depends on the state variable for which the output response is needed. D_{sys} is zero matrix. The A_{sys} and B_{sys} matrices indicate the behavior of the system. The C_{sys} and D_{sys} are the output matrices that depend on the precise selection of variables.

3 Design of LQR Controller

The block diagram of the plant with the LQR controller is shown in Fig. 2. The LQR controller is an adaptive controller and it tunes the system for dynamically varying operating conditions such as change in wind speed, change in compensation level etc., Ref. [15]. The controller enhances the stability of torsional frequencies by moving the unstable modes to the stable state.

The following steps are involved in developing the system with controller

- Develop a linear model of the system
- Adjust the Q and R matrix for optimal operation
- Find optimal gain set K
- Find closed loop system matrix $A_c = A_{\text{sys}}$-BK
- Find Eigen values with controller and analyze the stability

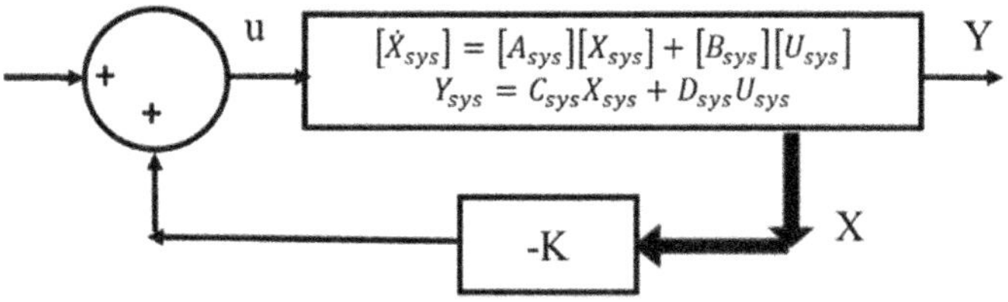

Fig.2 LQR controller

Q and R are the weight matrices on states and control inputs which are tuned for the optimal feedback gain K. The gain K decides the pole location of the particular mode of the plant for stable operation.

The LQR controller feedback gain matrix K can be obtained by feedback law given in Eq. (6)

$$u = -Kx \tag{6}$$

where K is given by,

$K = R^{-1}\left(B^T S + N^T\right)$ and N is null.

The feedback law minimizes the quadratic cost function given in Eq. (7)

$$J(u) = \int_0^\infty \left(x^T Qx + u^T Ru + 2x^T Nu\right)\mathrm{d}t \tag{7}$$

Bound by dynamics of the proposed system

$$\left[\dot{X}_{\mathrm{sys}}\right] = \left[A_{\mathrm{sys}}\right]\left[X_{\mathrm{sys}}\right] + \left[B_{\mathrm{sys}}\right]\left[U_{\mathrm{sys}}\right] \tag{8}$$

The solution S is derived from the following Eq. (9)

$$A^T S + SA - (SB + N)R^{-1}\left(B^T S + N^T\right) + Q = 0 \tag{9}$$

With the controller, the system matrix A is modified to Ac where Ac is given in Eq. (10)

$$Ac = (A - (B * K)) \tag{10}$$

Control matrix with the controller Bc is given as

$$Bc = B$$

As the system matrix A is modified to Ac, the corresponding unstable modes are stabilized as shown in Tables 1 and 2.

Table 1 Eigen values with and without controller for the wind speed of 7 m/s, 80% compensation level (1.0e+02 indicates that the eigen values are multiplied by 100)

Without controller	With controller
1.0e+02 * **2.0155**	1.0e+02 * **−2.0155**
−1.9553	−1.9553
−0.0188 + 0.6087i	−0.0188 + 0.6087i
−0.0188 − 0.6087i	−0.0188 − 0.6087i
−0.0938 + 0.0420i	−0.0941 + 0.0437i
−0.0938 − 0.0420i	−0.0941 − 0.0437i
0.0532 + 0.0296i	**−0.0417 + 0.0289i**
0.0532 − 0.0296i	**−0.0417 − 0.0289i**
−0.0013	−0.0136
0.0223	**−0.0508**
−0.0005 + 0.1450i	−0.0005 + 0.1450i
−0.0005 − 0.1450i	−0.0005 − 0.1450i
−0.0005 + 0.1250i	−0.0005 + 0.1250i
−0.0005 − 0.1250i	−0.0005 − 0.1250i

Table 2 Eigen values with and without controller for the wind speed of 9 m/s at 80% compensation level

Without controller	With controller
1.0e+02 * **2.0172**	1.0e+02 * **−2.0172**
−1.9577	−1.9577
−0.0215 + 0.6294i	−0.0216 + 0.6294i
−0.0215 − 0.6294i	−0.0216 − 0.6294i
−0.0817 + 0.0605i	−0.0870 + 0.0639i
−0.0817 − 0.0605i	−0.0870 − 0.0639i
0.0473 + 0.0344i	**−0.0356 + 0.0546i**
0.0473 − 0.0344i	**−0.0356 − 0.0546i**
−0.0080	−0.0711
0.0227	**−0.0141**
−0.0005 + 0.1450i	−0.0005 + 0.1450i
−0.0005 − 0.1450i	−0.0005 − 0.1450i
−0.0005 + 0.1250i	−0.0005 + 0.1250i
−0.0005 − 0.1250i	−0.0005 − 0.1250i

4 Results and Analysis

4.1 Eigen Value Analysis

The system stability is analyzed with and without LQR controller. The matrices (A_{sys}, B_{sys}, C_{sys} and D_{sys}) are considered for this analysis. The Eigen values for the system with LQR are found out and it is tabulated. As there are 14 state variables (X_{sys}) in the system, there are 14 Eigen values.

Table 1 indicates the set of Eigen values of the whole system with and without LQR controller for the wind velocity of 7 m/s at 80% of the compensation level. From Table 1, it is understood that low frequency oscillations of 2.9 Hz which causes instability in the system have to be moved to the stable state using LQR controller. The positive real part of the Eigen value implies that the system is unstable. Also, the non-oscillatory mode which has no frequency terms creates unstable oscillations. By means of the controller, this mode is also moved to the stable operating points.

Similarly, Table 2 indicates the set of Eigen values of the whole system with and without LQR controller for the wind velocity of 9 m/s at 80% of the compensation level. From Table 2, it is understood that low frequency oscillations of 3.44 Hz which causes instability in the system have been moved to the stable state using LQR controller. Also, the non-oscillatory mode which has no frequency creates unstable oscillations. By means of the controller, this mode is also moved to the stable operating points.

4.2 Zero-Pole Plot Without Controller

Figures 3 and 4 indicate the zero-pole (Z-P) plot or map of the whole system without controller. Figure 3 clearly indicates the poles are placed outside the unit circle.

It is visible that some of the poles lie outside of the unit circle which indicates that the system is unstable. This is caused due to the insertion of the series compensation level of 80% at a wind speed of 7 m/s. The wind speed is increased to 9 m/s and the Z-pole plot is checked with and without LQR controller. Figure 5 indicates the pole zero map of the system with controller. With the use of controller, the poles lying

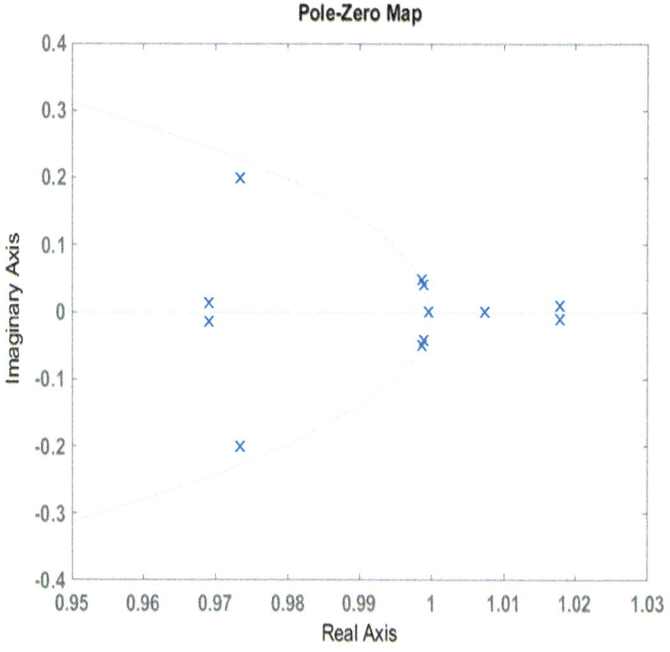

Fig. 3 Z-p plot with no control (Increased)

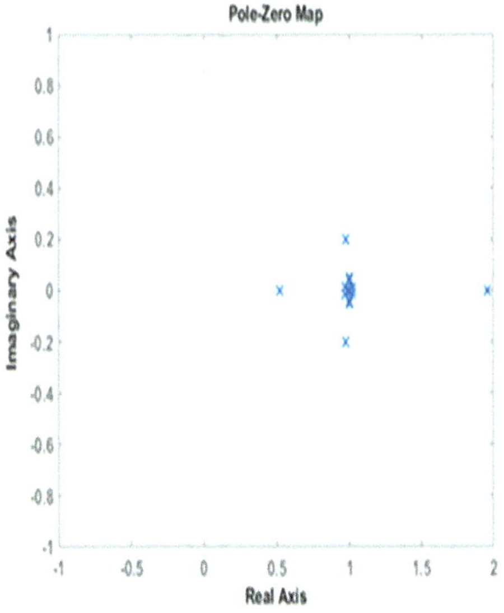

Fig. 4 Z-p plot with no control

outside the unit circle are pushed inside the unit circle. This means that controller action helps in damping low frequency of oscillations (see Figs. 6 and 7).

4.3 *Response of the Torque*

Figures 8 and 9 show the step response of torque (the state variable) without and with the controller. The torque response overshoots and it becomes undamping when the controller is absent. With the presence of controller, the torque becomes stable within less time for the insertion of 80% compensation at the wind speed of 7 m/s.

When the compensation increased to 90%, the step response with the controller was marginally stable as shown in Fig. 10. Under different wind speed conditions, the torque and rotor speed varies as shown in Fig. 11

As the compensation level changes, the oscillations produced in torque also change as shown in Fig. 12. When the controller is connected, the oscillations are damped out easily as shown in Fig. 12

Similarly, Figs. 13 and 14 give the variation in Rotor speed and Electromagnetic torque for the change in compensation level and the wind speed variation respectively with the controller. The response of the controller is very fast under dynamic conditions.

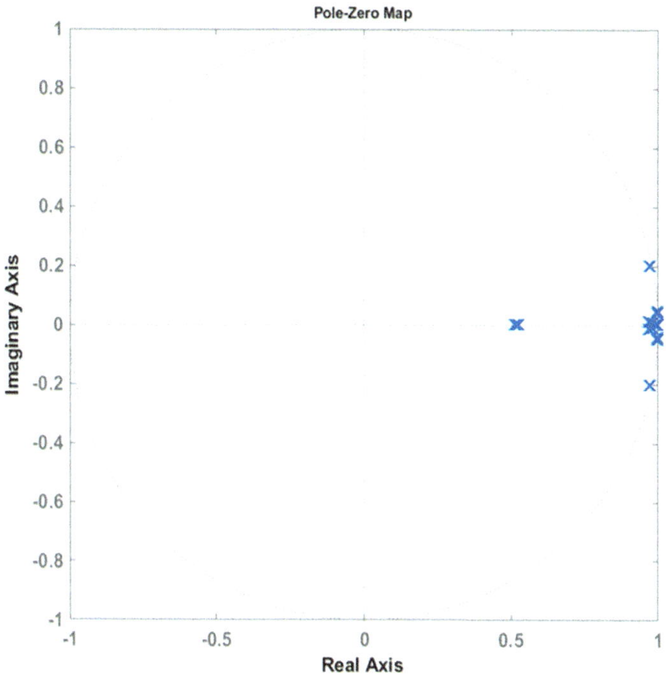

Fig. 5 Z-p plot with control

4.4 *Effect of Change in Stiffness Coefficient*

The response of a system to any disturbance is largely depending on the system inertia and the stiffness coefficient. Hence the damping of SSR oscillations also varies with the stiffness coefficient. The torsional modes of dynamics are greatly depending upon the shaft stiffness coefficient. The standard values of the stiffness coefficient K_{tg} change as the life of the system increases. To analyze the effect of stiffness coefficient on the system stability, the value of K_{tg} is changed from 5 to 50 pu. It has been observed that as the value of K_{tg} changes, there are changes in the torsional modes corresponding to $\lambda_{3,4,5,6}$ and $\lambda_{9,10}$ (which are in bold letters) indicated in Table 3. The system is operated at a wind speed of 9 m/s with 70% series compensation. Table 3. Eigen values for different stiffness coefficients K_{tg} (Wind speed—9 m/s, 70% line compensation).

The frequency of electromechanical modes corresponding to $\lambda_{3,4}$ increases with the increase in compensation level. The torsional modes with respect to $\lambda_{5,6}$ become unstable with the low frequency of oscillations as they have a higher value of positive real value. $\lambda_{11,12}$, $\lambda_{13,14}$ belong to torsional modes and are not changed much for the change in stiffness coefficients. The non-oscillatory modes ($\lambda_{7,8}$) are identified and observed as the least variations in the values.

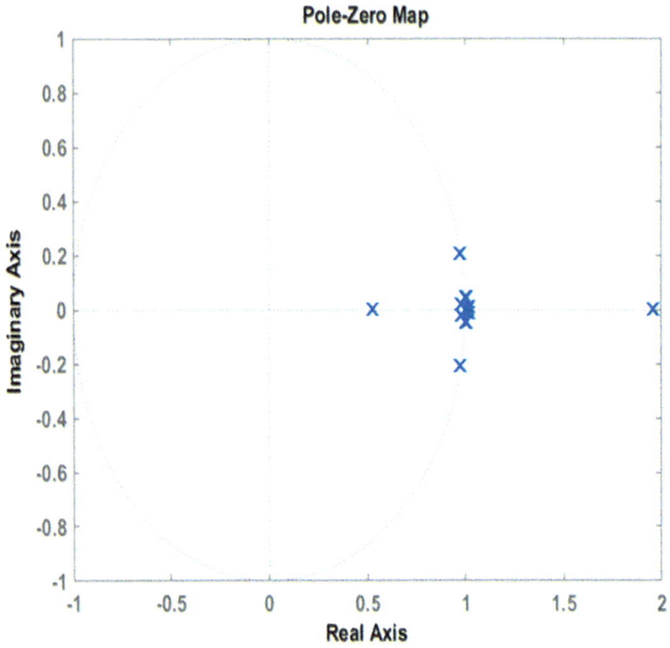

Fig. 6 Z-p plot with no control (9 m/s, 80%)

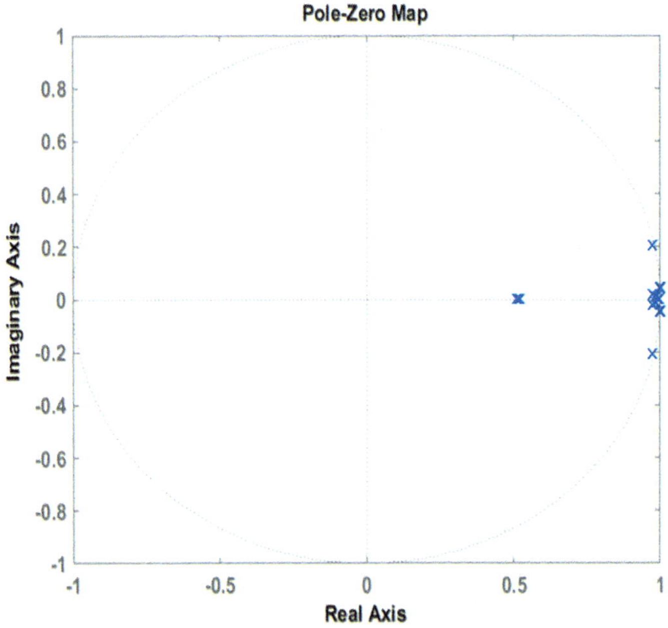

Fig. 7 Z-p plot with control for 9 m/s,80% compensation level

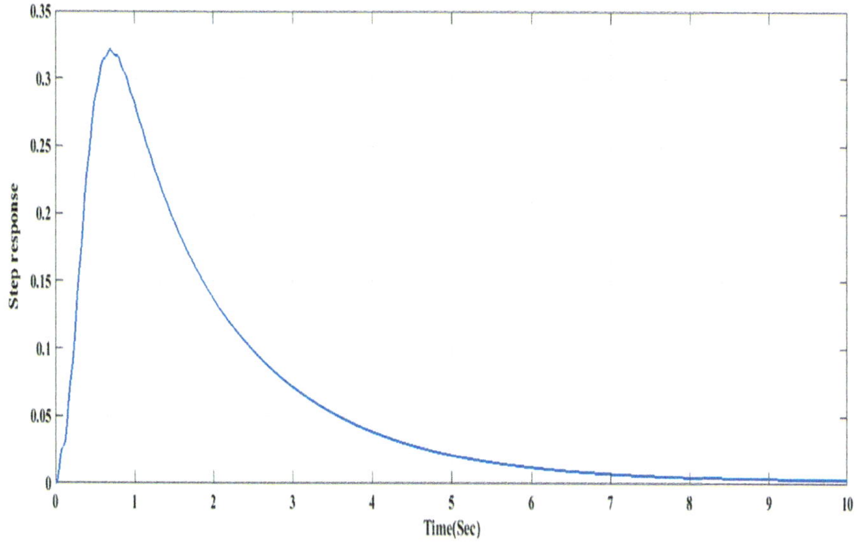

Fig. 8 Step response of the torque without controller (7 m/s, 80% compensation level)

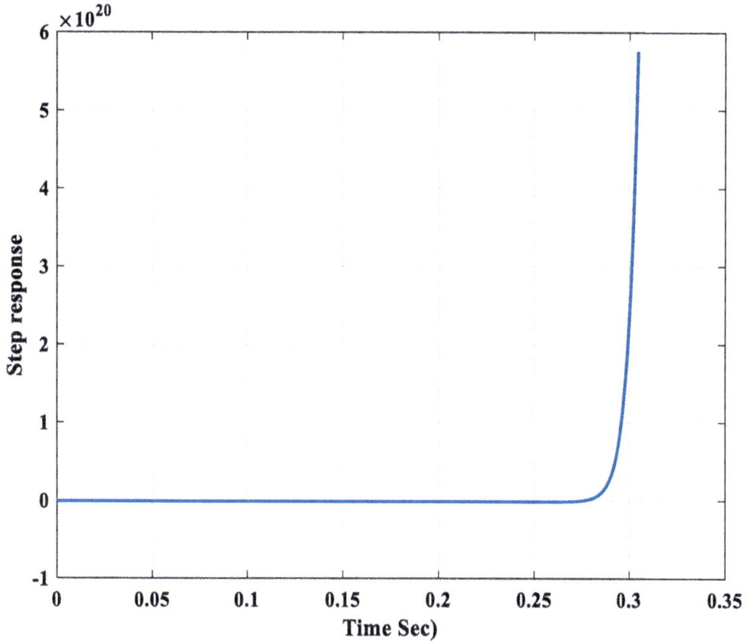

Fig. 9 Step response of the system without controller (7 m/s, 80% compensation level)

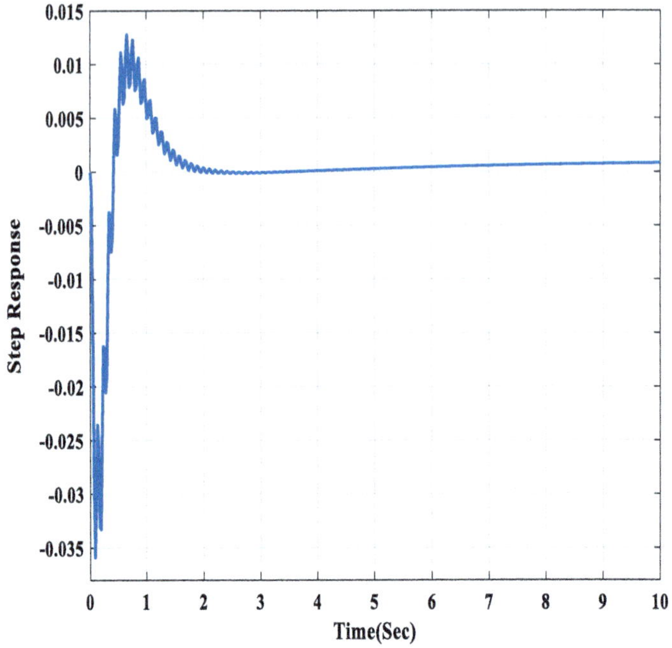

Fig. 10 Step response (7 m/s 90% compensation level)

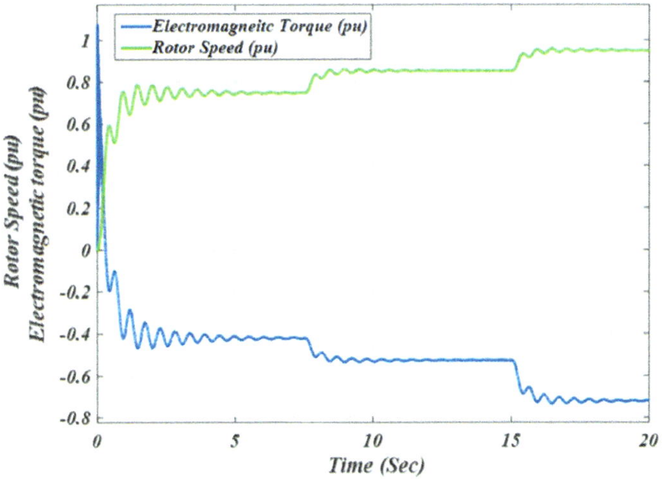

Fig. 11 Response of the system for varying wind speeds

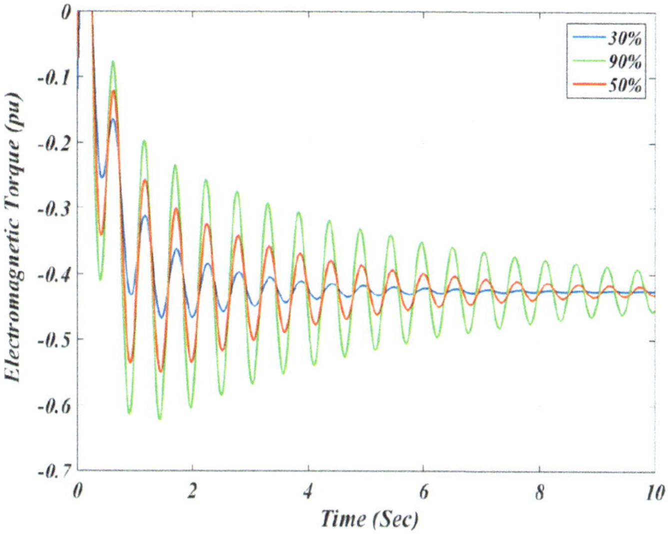

Fig. 12 Torque response (wind speed of 7 m/s)

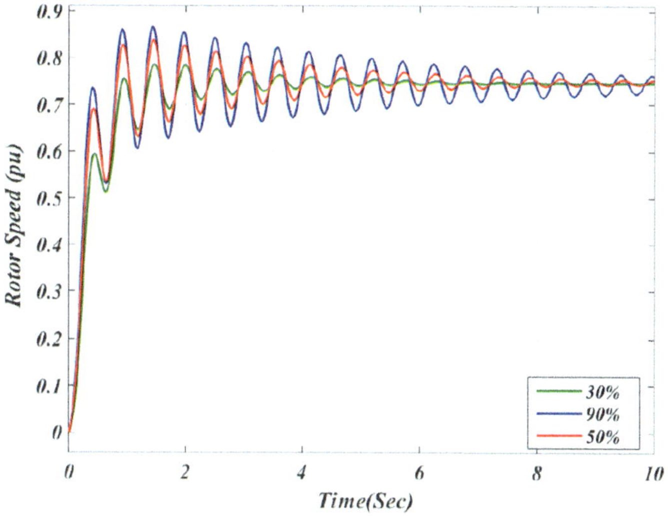

Fig. 13 Rotor speed at various compensation levels (wind speed of 7 m/s)

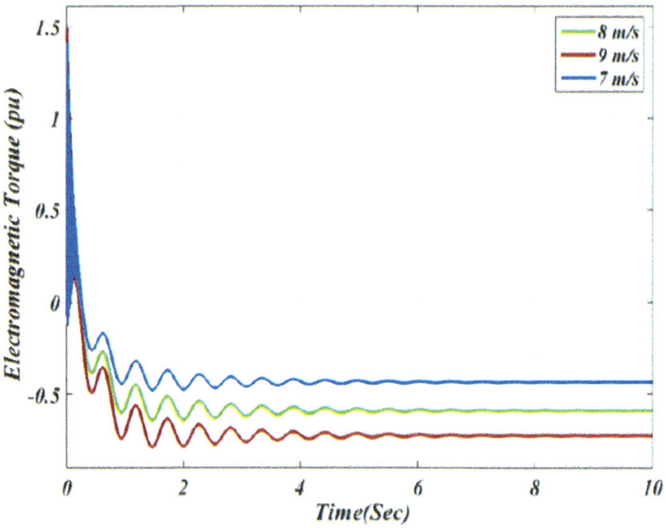

Fig. 14 Electromagnetic torque at various wind speeds (50% compensation level)

Table 3 Eigen values for different stiffness coefficients K_{tg} (Wind speed—9m/s, 70% line compensation)

Eigen values	$K_{tg} = 5$	10	30	50
λ_1, λ_2	$-0.05 \pm 198.85i$	$-0.05 \pm 198.86i$	$-0.06 \pm 198.92i$	$-0.08 \pm 199.03i$
λ_3, λ_4	$-1.33 \pm 44.25i$	$-1.56 \pm 61.51i$	$-1.71 \pm 105.29i$	$-1.72 \pm 135.55i$
λ_5, λ_6	$2.65 \pm 8.84i$	$2.84 \pm 9.01i$	$2.96 \pm 9.12i$	$2.98 \pm 9.14i$
λ_7	−9.67	−9.59	−9.54	−9.53
λ_8	−0.95	−0.95	−0.95	−0.95
λ_9, λ_{10}	$-1.83 \pm 1.22i$	$-1.82 \pm 1.23i$	$-1.83 \pm 1.24i$	$-1.83 \pm 1.24i$
$\lambda_{11}, \lambda_{12}$	$-0.05 \pm 13.66i$	$-0.04 \pm 13.66i$	$-0.04 \pm 13.66i$	$-0.04 \pm 13.66i$
$\lambda_{13}, \lambda_{14}$	$-0.05 \pm 11.66i$	$-0.04 \pm 11.66i$	$-0.04 \pm 11.66i$	$-0.04 \pm 11.66i$

5 Conclusion

An addition of series capacitor is more common to enhance the power transfer capacity of the long transmission line which links the WTG and far away load centers or grid. But this can cause undamped oscillations due to SSR which can cause total system failure. In this paper, the different torsional modes are identified using eigen value approach and an effective method of damping the oscillations produced due to SSR using LQR controller is explained in detail. Modal analysis is used to identify different torsional modes existing in the system and the effect of these torsional modes in the system stability. It has been observed that, under fast dynamic variations of operating conditions, the LQR controller is very effective with the optimal tuning of the gain in real time.

References

1. World Energy Council. http://www.worldenergy.org/
2. Pena R, Clare J, Asher G (1996) IEE Proc.-Electr Power Appl 143(3)
3. Mohan Mathur R, Varma RK (2012) IEEE press. A Wiley Publications
4. IEEE committee Report 1980 Proposed Terms and Definitions for Sub Synchronous Oscillations," *IEEE Trans. on Power Apparatus and Systems*, vol. PAS-99, no. 2, pp. 506–511
5. IEEE Committee Report (1985) Terms definitions and symbols for sub synchronous oscillations. IEEE Trans Power Appar Syst PAS-104:1326–1334
6. Mahalakshmi R, Sindhu Thampatty KC (2020) Int Trans Electr Energy Syst 30(3):1–22
7. Raj KR, Mahalakshmi R (2017)International conference on smart grids, power and advanced control engineering (ICSPACE), pp 93–98
8. Ghafouri M, Karaagac U, Karimi H, Jensen S, Mahseredjian J, Faried SO (2017) 32(6):4934–4942
9. Huang PH, Moursi MSE, Xiao W, Kirtley JL (2015) IEEE Trans Power Syst 30(3):1442–1454
10. Chowdhury MA, Mahmud MA, Shen W, Pota HR (2017) IEEE Trans Energy Convers, 707–719
11. Yao J, Wang X, Li J, Liu R, Zhang H (2019) IEEE Trans Energy Convers 34(2):849–859
12. Mohammadpour HA, Ghaderi A, Mohammadpour H, Santi E (2015) Electr Power Syst Res 123:57–66
13. Fan L, Miao ZL (2015) Modelling and analysis of doubly fed induction generator wind energy systems. Elsevier Ltd.
14. Mohammadpour HA, Santi E (2015) Analysis of sub-synchronous resonance in doubly fed induction generator based wind farms. John Wiley-University of South Carolina
15. Krishnan KCC, Thampatty KCS (2020) IEEE students conference on engineering & systems (SCES), pp 1–6
16. Mahalakshmi R, Thampatty KCS (2016) IEEE region 10 conference (TENCON), pp 930–935
17. Sravya ASL, Yoshita LN, Reddy BDD et al (2020) IEEE international women in engineering (WIE) conference on electrical and computer engineering (WIECON-ECE), pp 200–205
18. Tungal RV, Anand R, Ramprabhakar J (2019) 3rd international conference on trends in electronics and informatics (ICOEI), pp 797–802
19. Krishnan C, Sindhu Thampatty KC (2022) Comparative study of machine learning algorithms applied to the prediction of unstable sub-synchronous resonance (SSR) oscillations in power system. IEEE IAS Global Conference on Emerging Technologies, GlobConET 2022

Optimal Scheduling of Microgrid Using GAMS

Indurekha Shial and Rajat Kanti Samal

Abstract In a power supply system, a wide range of optimization problems exist where different processes are involved. Economic dispatch is an important optimization task in a power system. The information transparency and security of microgrid systems improve by microgrid economic dispatch. It also makes the power grid a very clear, safe, efficient, and reliable development path. Here this paper explains the solution to the economic dispatch problem for the different generating units including both conventional and non-conventional sources in a microgrid using the General Algebraic Modeling System (GAMS). The main purpose of this paper is to optimize the problem and minimize the total power generation cost. The scheduling problem is solved by using GAMS from which the total cost is found to be 119$ at 2.84 MW load.

Keywords GAMS · Microgrid · Renewable sources · Optimal scheduling

1 Introduction

In the power system for power generation, Distributed Generators (DG) are classified into two types dispatchable generators and non-dispatchable generators. Optimal Scheduling of dispatchable generators is a key to the operation of any microgrid economically. Microgrid is suitable and provides an optimal solution toward the electrification of small and isolated village areas in many developed countries. Recently, the concept of microgrid (MG) has been introduced in the distribution network. Microgrid is known as a small network of a synchronized power provider which is customer-controlled from a control center the small fraction of overall required demand and sometimes the overall load demand for a small distribution system during failure of main grid. A microgrid can joined with the main grid, in which they are connected to the main electrical grid or in an island mode, where they are disconnected from the main power grid and the minor scale generators (biomass,

I. Shial (✉) · R. K. Samal
Veer Surendra Sai University of Technology, Burla, India
e-mail: indushial86@gmail.com

S. Kumar et al. (eds.), *Control Applications in Modern Power Systems*, Lecture Notes in Electrical Engineering 1128, https://doi.org/10.1007/978-981-99-9054-2_3

geothermal, wind, and solar) which are the only source of power generation. In this document, there is an idea to realize a simple network consisting of thermal generator, biomass, geo-thermal, wind, solar, and battery system in a microgrid, in spite of the fact that the upcoming focus is to unify the renewable generations and storage. Currently, it is focused in the case at which there is no generation from both renewable sources of generators and storages. There are many optimal scheduling problems which are presented and solved by methods: General Algebraic Modeling System (GAMS) and Genetic Algorithm (GA). The inputs to the microgrid central controller are scheduling and generator set point results. Here, the optimal cost might aid to justify as equipment in the installation of automation in the local load when multiple small size generators are used for supply; then it gives many advantages like increased reliability, improved operation efficiency, high availability, and maintainability [6]. Due to the use of multiple generators in a microgrid, complexity and cost increase, which gives difficulty in maintaining all generators at the same terminal voltage, phase angle, and frequency. In standalone microgrids when both voltage and frequency are regulated, then all DGs are suitably interfaced to the microgrid by means of either a traditional breaker or by use of a power electronics converter. In a microgrid, this does not consider the realization individually when the diesel generators are operated in isolated area. In the field of microgrid optimization, a large number of research topics focus on optimization. To be exact, appropriated power generation scheduling problems are usually non-convex, non-linear multi-objective optimization problems. For traditional optimization algorithms, it is very difficult. Most optimization algorithms include the particle swarm optimization (PSO), convex optimization, Genetic Algorithm (GA), and Ant Colony Optimization algorithm which have different optimization effects in different fields. Highly complex nonlinear problems can be optimized by using GAMS. In microgrids, an optimal scheduling of generation as a mixed-integer second order cone programming problem has been solved by GAMS software [4]. Here electric vehicles and demand response are modeled where by smoothing the load profile and reducing the cost of power purchases from the main grid, the total operating cost reduced. Using differential evolution (DE) algorithm, a renewable energy source of microgrid is optimized and minimizez the cost [5]. Here DE algorithm is based on energy management system and also is compared with particle swarm optimization (PSO) algorithm. A multi-microgrid structure is integrated with power-to-X technologies optimized by robust decentralized optimization method [2]. A microgrid consisting of a fuel cell, photovoltaic(PV) cell, wind turbine, and storage system is used for optimal operation and supervision by using the stochastic optimization method [1]. The main purpose of this paper is to find out problem uncertainty. A novel stochastic framework based on the UT method is also used in some papers to model the uncertainty effects in the optimal solution of power generation in the renewable sources of energy of microgrids. In paper [7] according to power converter rule, the microgrid formed by the control techniques of the power converters is presented. Here we presented different droop control schemes for inductive, resistive, and generic lines. The hierarchical control structure on microgrids also presented for optimizing efficiency and performances, due to which here also the generation cost is reduced. In [8], a comparative analysis study of Quadratic Programming (QP) and GAMS approach is used to solve the ELD problem which is dynamic. In this power system, the transmission losses are considered. In [9], a multi-energy microgrids system, i.e. water energy microgrid is considered a combined scheduling model for the best dispatch of simulating, heating, power gas, and water sources. Here the problematic optimization model is accepted to reduce the total cost including working and emission costs and also reduced the volume of fresh water which is extracted from underground reservoirs. A multi-purpose optimal dispatching model is proposed by the paper [10] where the power generation is distributed in the microgrid. Here the grid-connected mode combines both the environmental benefits and operating cost. The paper [11] presents the solution of the economic load dispatch problems by using general algebraic modeling system (GAMS) and Quadratic Programming (QP). Both methods are proposed for optimizing the problem. In [12], both Genetic Algorithm (GA) and General Algebraic Modeling System (GAMS) are applied to reduce the cost of power generation and emission. It depends on high and volatile diesel prices. Here only diesel generators are used in this microgrid as dispatchable sources. The economics of the system depends on the cost/kwh of electricity supplied on that system. In [14], the microgrid consisting of wind and solar energy sources is considered for analyzing the dispatch rate of power. Due to the use of renewable energy sources, the microgrid reduces total generation cost of the system. To reduce the generation of cost, the most used method is optimal scheduling method. Different algorithms are used to solve the problem to find the best solutions. The operating cost of different endless energy sources like biomass, solar photovoltaic cells, wind power, and fuel cells are very less in comparison to conventional sources. In paper [15] for design and development of the microgrid system, renewable energy sources are used to analyze the cost. In [17], Artificial Fish Swarm Algorithm is used in a microgrid consisting of windmills, solar PV, and CHP. For a DC microgrid formed by wind, PV solar, battery, and load some energy management and control strategies are used to optimize the operation of that microgrid. In the linear programming model, a bi-level mixed-integer is used to present the optimal scheduling of the network where the system is distributed [19]. Microgrids and smart homes are considered in this network; more use of Distributed Energy Sources (DERs) in that network reduced the total operating cost and also increased social welfare and consumer comfort index due to the bi-level method. A centralized energy management system is designed for microgrids in stand-alone mode [21]. In paper [22] like GAMS software, a sufficient stochastic framework is also used to model the correlated uncertainties in the microgrid which includes the renewable energy sources. The main contribution of the paper is to find the result by optimizing the scheduling problem in microgrid using GAMS.

2 Problem Formulation

2.1 *Formulation for Microgrid System*

A microgrid is a very small system where the electric power incorporates generation, transmission, and distribution. It can attain the power equilibrium and allocation of optimal energy over a given field or as an effective source of power or in the network where the load is distributed. A microgrid is a self-contained grid where it uses thermal generators, battery system, and renewable energy sources for power generation. A microgrid can work independently. A microgrid may be grid connected or islanded mode. In the grid connected mode, the microgrid is connected to the distributed system where it exchanges power with a utility grid. In the islanded mode, the microgrid itself is generally small and sufficient to arrange the demand of all loads; it is required to catagorize loads based on their significance and ensure continuous supply to important loads. The main benefit of a microgrid is that it can operate in standalone mode or the main grid disconnection mode. Microgrid has reduced in cost due to the use of renewable energy sources. The structure of the microgrid test system considered is shown in Fig. 1. For a thermal generator, the overall fuel minimization cost function considering valve point loading of the generating unit is given as

$$\texttt{Minimize}\ F_i = \sum_{i=1}^{10} a_i + b_i P_i + c_i P_i^2 + |d_i sin\left(e_i(P_i^{min} - P_i)\right)^2| \quad (1)$$

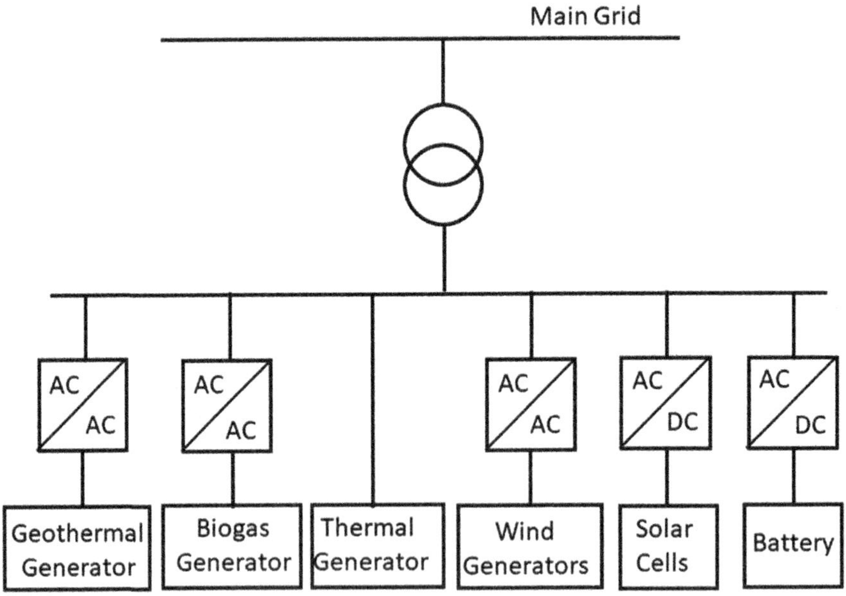

Fig. 1 Microgrid structure

Here, F in (\$/h) is the cost function. The real power output of the ith generator is denoted by P_i while a_i, b_i, and c_i are the cost curve coefficient. Biomass energy is a highly available energy that reduces waste and is of very low cost, but it requires a very large area to operate and store. Approximately 86 percent of modern bio energy is used for heating applications, 9 percent for transport, and rest 5 percent is used for electricity. For biomass generator, the cost function is given as

$$\texttt{Minimize}\ F_i = d_i P_i + e_i P_i \tag{2}$$

Geothermal energy is an exceptionally constant source of energy, with practically no emission, high efficiency, and high investment cost than other renewable sources. It is a natural heat derived from the interior of the earth which is used to generate electric power supply. The study and exploration of GE started in India in 1970. Currently, in 26 countries geothermal electricity generation is used and in 70 countries geothermal heating is used. For geothermal generators, the cost function of the microgrid is given as

$$\texttt{Minimize}\ F_i = \left(\frac{h}{u} + k_i\right) P_i \tag{3}$$

Wind energy is a reliable and infinite renewable energy source. The wind is a consequence of solar energy that is around 2% of the energy of the sun reaching the earth is converted into wind energy. The wind energy installed capacity was 41.93 GW, but recently its generation capacity in India has increased. Like other RE sources, a linear cost function used for the wind power is expressed as

$$\texttt{Minimize}\ F_i = w_i P_i \tag{4}$$

Solar energy can be used for diverse purposes like electricity (photovoltaic cells) or heat (solar thermal). Photovoltaic (PV) cell is one of the most popular renewable energies. Recently, India stands in the fourth position across the globe in solar PV deployment. The installed capacity of solar power has reached around 61.97 GW. A linear cost function that is used for solar PV power is expressed as

$$\texttt{Minimize}\ F_i = v_i P_i \tag{5}$$

For battery systems, the linear cost function is the same as solar PV power. In an electric grid, the battery energy storage system is the fastest responding source. The electric battery is provided to power electrical devices like electrical car and smart phones.

Among all the generators, the thermal generator has a non-linear cost function which makes the whole scheduling problem non-linear in nature. So while choosing the solver for the problem, we have to describe the model as non-linear.

2.2 General Algebraic Modeling System (GAMS) Section to Solution Methodology

GAMS is a high-level powerful mathematical optimization model. It is used for modeling the different types of complex, linear, non-linear, mixed-integer type problems and mixed complementary problems also. Specially GAMs handles large complex problems. It requires more revision to get an accurate model by using other software. Using the same GAMs model file, the models can be developed, documented, and solved simultaneously. GAMs can handle different optimization problems (linear and non-linear) simultaneously. To represent new ideas, GAMs language offers sufficient flexibility. The model statement first collects the equations and makes a group, then levels them to solve. The model is shown in Fig. 2.

Modeling languages of GAMS have a syntax that is generally sets, variables, equations, model statements, and output. The data is very important for modeling. After collecting data as per the procedure of GAMS, it declares the variables and then declares the equations. To code the basic structure of a mathematical model, there is a solution procedure of GAMS. Different components of GAMS are sets, data, variables, equations, model, and output explained below.

GAMS Formulation follows the basic format as given below

- Sets: Declaration, Assignment of members;
- Data (Parameters, tables, Scalars): Declaration, Assignment of values;
- Variables: Declaration, Assignment of types, Assignment of bounds and/or initial values (optional);
- Equations: Declaration, Definition;
- Model and Solve Statements;
- Display Statements (optional).

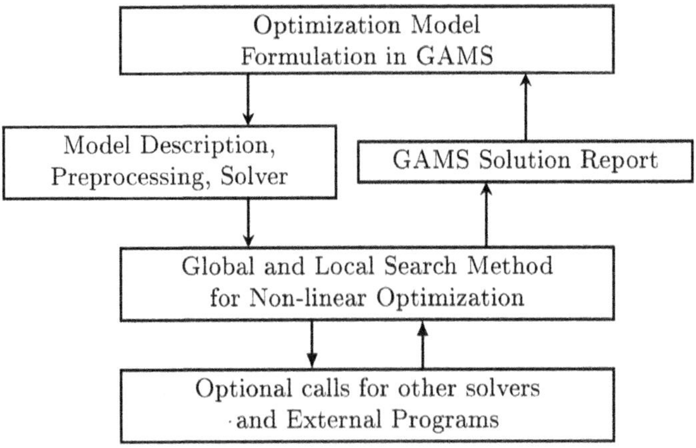

Fig. 2 Solution methodology

3 Results and Discussion

A microgrid test system is considered which contains various generators. Here Table 1 shows the system parameters [23]. After solving the scheduling problem in GAMS, the obtained results are shown in Table 2. The individual generation of each generator for different demands is given in Table 3.

So from the above results by applying GAMS to a six generators microgrid scheduling including thermal generator, biomass generator, geothermal generator, wind generator, solar cell and battery system; the total cost is found to be 119.855

Table 1 System parameters of microgrid

Type of generator	Values of parameter	Type of cost function	Rated capacity in MW
Thermal	a = 10	Non-linear	1.0
	b = 200		
	c = 100		
	t = 33		
	v = 0.0714		
Biogas	d = 180.39	Linear	0.8
	e = 0.56		
Geothermal	u = 58.6	Linear	0.9
	h = 332		
	k = 28.3		
Wind	w = 3.25	Linear	0.8
Solar	v = 3.5	Linear	0.8
Battery	v = 3.45	Linear	0.9

Table 2 Results of microgrid

Load in MW	Cost in $
2.834	119.855
4.834	416.147

Table 3 Individual generation of microgrid

Generator no	Load = 2.834 MW	Load = 4.834 MW
G1	0	0.634
G2	0	0.800
G3	0.334	0.900
G4	0.800	0.800
G5	0.800	0.800
G6	0.900	0.900

$ at 2.834 MW load and 416.147 $ at 4.834 MW load, respectively. Here CPLEX solver is used to solve the scheduling problem. It is observed that for thermal generator and biomass generator at the first case of load demand the microgrid system is not operated, but when the load demand increases then the system starts operating. So it has been found that the microgrid has minimum cost and emission also due to the presence of more number of renewable energy sources. Here, the cost function of the thermal generator is taken as non-linear type at 1.0 MW rated capacity where the cost function of other generation sources that is biogas, geothermal, wind, solar cells, battery system for rated capacity of 0.8 MW, 0.9 MW, 0.8 MW, 0.8 MW, 0.9 MW, respectively. In this microgrid individually, the generations are 0 MW, 0 MW, 0.334 MW, 0.800 MW, 0.800 MW, and 0.900 MW at minimum load 2.834 MW of thermal generator, biomass generator, geothermal generator, wind generator, solar cells, and battery system, respectively. Similarly for those above sources at maximum load 4.834 MW the individual generations are 0.634 MW, 0.800 MW, 0.900 MW, 0.800 MW, 0.800 MW, and 0.900 MW, respectively.

4 Conclusion and Future Scope

In this work, a microgrid system having different generating sources like thermal generator, biomass generator, geothermal generator, wind generator, solar cells, and battery system is optimized for economic scheduling of meeting load demand by implementing the GAMS software. Results of this microgrid system are shown in the above tables. From this above work, it is clear that total cost and emission can be reduced using renewable energy sources in a microgrid system on existing grid. It is also clear that to reduce emission and the increasing environment degradation, the use of clean energy is the only way. Renewable energy source is the clean source which can reduce the overall stress on existing grid. In this six generators microgrid system more reduction in cost and emission can be achieved by using mini hydropower plants in place of thermal turbine generation unit, but it requires more efficient storage capacity battery, also more optimized cost reduction and improvement in energy efficiency can be achieved by including additional electric vehicle in the above microgrid system.

References

1. Taherpoor H, Niknam T, Kavousi-Fard A (2015) A novel stochastic framework for energy management in renewable micro-grids considering uncertainty of measurement and forecasting. J Intell Fuzzy Syst 28:999–1008. https://doi.org/10.3233/IFS-141383
2. Saatloo M, Amin P, Yasin M, Mohammad AM-I, Behnam Z, Kazem M, Mousa A-MA (2021) Robust decentralized optimization of Multi-Microgrids integrated with Power-to-X technologies. Appl Energy 304:1–22. https://doi.org/10.1016/j.apenergy.2021.117635

3. Silva J, López J, Bañol A, Nataly R, Marcos JS, LCP (2021) An optimal stochastic energy management system for resilient microgrids. Appl Energy 300:117435. https://doi.org/10.1016/j.apenergy.2021.117435
4. Shaghaghi-shahr G, Sedighizadeh M, Aghamohammadi M (2020) Optimal generation scheduling in microgrids using mixed-integer second-order cone programming. Eng Optim 52:1–29. https://doi.org/10.1080/0305215X.2019.1695790
5. Tiwari N, Srivastava L (2016) Generation scheduling and micro-grid energy management using differential evolution algorithm. In: 2016 international conference on circuit, power and computing technologies (ICCPCT). Nagercoil, India, pp 1–7. https://doi.org/10.1109/ICCPCT.2016.7530218.
6. Hoda T, Taher N, Abdollah K-F (2012) A novel stochastic framework for energy management in renewable micro-grids considering uncertainty of measurement and forecasting
7. Rocabert J, Luna A, Blaabjerg F, Rodríguez P (2012) Control of power converters in AC microgrids. In: IEEE transactions on power electronics, vol 27, 11th edn, pp 4734–4749. https://doi.org/10.1109/TPEL.2012.2199334.
8. Benhamida F, Ziane I, Souag S, Salhi Y, Dehiba B (2013) A quadratic programming optimization for dynamic economic load dispatch: Comparison with GAMS. In: 3rd international conference on systems and control. Algiers, Algeria, pp 625–630. https://doi.org/10.1109/ICoSC.2013.6750926.
9. Faezeh J, Mohammad AM, Kazem Z, Behnam M-I, Mousa M, Amjad A-M (2022) Multi-energy microgrids: An optimal despatch model for water-energy nexus. In: Sustainable cities and society, vol 77, p 103573, ISSN 2210–6707. https://doi.org/10.1016/j.scs.2021.103573
10. Li BW, Jue XN (2020) Optimal scheduling of microgrid based on improved biogeography-based optimization algorithm, pp 1–5. https://doi.org/10.1145/3424978.3424986
11. Augustine N, Suresh S, Moghe P, Sheikh K (2012) Economic dispatch for a microgrid considering renewable energy cost functions. In: 2012 IEEE PES innovative smart grid technologies (ISGT). Washington, DC, USA, pp 1–7. https://doi.org/10.1109/ISGT.2012.6175747
12. Moshi GG, et al (2012) Optimal generation scheduling of small diesel generators in a microgrid. In: Devendra D, Hari P, Manjaree PB (eds) 2014 IEEE international energy bisen. Solution of large scale economic load dispatch problem using quadratic programming and GAMS: A comparative analysis. J Inf Comput Sci 7:200–211
13. Olivares DE, Cañizares CA, Kazerani M (2011) A centralized optimal energy management system for microgrids. In: 2011 IEEE power and energy society general meeting. Detroit, MI, USA, pp 1–6. https://doi.org/10.1109/PES.2011.6039527
14. Augustine N, Suresh S, Moghe P, Sheikh K (2012) Economic dispatch for a microgrid considering renewable energy cost functions. In: 2012 IEEE PES innovative smart grid technologies (ISGT). Washington, DC, USA, pp 1–7. https://doi.org/10.1109/ISGT.2012.6175747
15. Rehman AU, Zeb S, Khan HU, Shah SSU, Ullah A (2017) Design and operation of microgrid with renewable energy sources and energy storage system: A case study. In: 2017 IEEE 3rd international conference on engineering technologies and social sciences (ICETSS). Bangkok, Thailand, pp 1–6. https://doi.org/10.1109/ICETSS.2017.8324151
16. Sérgio P, Paula F, Vaz AIF (2016) Optimization modeling to support renewables integration in power systems. In: Renewable and sustainable energy reviews, vol 55, pp 316–325, ISSN 1364-0321. https://doi.org/10.1016/j.rser.2015.10.116
17. Prakash Kumar K, Saravanan B, Swarup KS (2016) Optimization of renewable energy sources in a microgrid using artificial fish swarm algorithm. In: Energy procedia, vol 90, pp 107–113, ISSN 1876-6102. https://doi.org/10.1016/j.egypro.2016.11.175
18. Ujwala SR, Shahakar YD, Pote (Patil) PR (2017) Optimization of renewable energy sources connected in microgrid using different control strategies, vol 6, 5th edn
19. Mohsen K, Hossein A, Seyed AN (2022) Microgrids energy management in automated distribution networks by considering consumers' comfort index. Int J Electr Power Energy Syst 139: 108013, ISSN 0142-0615. https://doi.org/10.1016/j.ijepes.2022.108013
20. Kumar A, Hussain D, Khan M (2018) Microgrids technology: a review paper. Gyancity J Electr Comput Sci 3:11–20. https://doi.org/10.21058/gjecs.2018.31002

21. Olivares DE, Cañizares CA, Kazerani M (2011) A centralized optimal energy management system for microgrids. In: 2011 IEEE power and energy society general meeting. Detroit, MI, USA, pp 1–6. https://doi.org/10.1109/PES.2011.6039527
22. Sajad T, Seyed Saeedallah M, Taher N (2016) Stochastic energy management of renewable micro-grids in the correlated environment using unscented transformation. Energy 109:365–377, ISSN 0360-5442. https://doi.org/10.1016/j.energy.2016.04.067
23. Lu X, Cheng L (2021) Day-ahead scheduling for renewable energy generation systems considering concentrating solar power plants. Math Probl Eng 2021:1–14

Design of an On-Grid Floating Solar Photovoltaic System: A Case of Vaigai Dam in Tamil Nadu

Mohamed Salman

Abstract A floating solar photovoltaic system is an emerging energy technology that overcomes land constraints in large-scale photovoltaic installations. This study designed an on-grid floating solar photovoltaic system for the Vaigai dam in Tamil Nadu's Theni district. A 22 MW floating solar photovoltaic plant was designed with 90,045 300Wp polycrystalline solar panels and nine 2500 kW inverters. As a result of the simulation, the plant's annual energy generation was calculated to be 48.3 GWh, and its performance ratio was estimated to be 83.05%. The results indicated that the total potential CO_2 saving of the plant is 0.953 megatons. The study determined that the project has a low environmental impact and provides clean energy to support the district's energy needs. According to the results, floating solar photovoltaic systems can reduce water evaporation, generate clean and reliable energy, and significantly reduce land costs.

Keywords Floating solar photovoltaic · On-grid · Water evaporation · Clean energy

1 Introduction

The need to generate affordable electricity and move toward clean energy is essential for combating climate change. Renewable energy sources play a significant role in the transition from fossil fuel-based energy to zero-carbon green energy. As an abundant and enormous energy source, solar energy is considered among the most promising renewable energy sources [1]. PV technology is most commonly used to produce solar energy, which is both eco-friendly and cost-effective. Solar panels require a considerable amount of space, which makes them expensive and uncompetitive to fossil fuels. It is also challenging to install large-scale land-mounted PV systems due to the lack of land availability. Therefore, relying solely on land-mounted and rooftop PV systems to achieve a nation's target solar energy production is quite challenging.

M. Salman (✉)
Heriot-Watt University, Dubai Knowledge Park, United Arab Emirates
e-mail: mmoh803@aucklanduni.ac.nz

S. Kumar et al. (eds.), *Control Applications in Modern Power Systems*, Lecture Notes in Electrical Engineering 1128, https://doi.org/10.1007/978-981-99-9054-2_4

The floatovoltaics system, also known as floating solar PV, is one alternative solution [2].

Floating solar panels can generate electricity from sizeable unused water bodies in the state, reducing land costs. Land shortages can be mitigated with floating solar PV (FSPV), and valuable land can be preserved for future use [1]. As a result of water's natural cooling effect, FSPV offers additional benefits in reducing overheating of panels. Covering the water surface with a floating structure can help reduce algal blooms and evaporation during the summer. Unlike land-based solar plants, FSPVs are usually hidden from view, which makes them unobtrusive to the public.

When PV panels are installed over water (lakes, ponds, reservoirs, etc.), they are naturally cooled, resulting in higher power output [3, 4]. As a result of the cooler environment, the system's lifespan is extended to a significant extent. There are no significant differences between FSPV systems and roofing or land-based single-axis tracking solar systems because the panels are the same as those available on commercial grounds. FSPVs are photovoltaic power plants installed on water, and they must maintain their ability to float without issue for long periods to produce power. Usually, the electricity produced by FSPV is routed to an inverter on land, which receives the power [5]. Like land-mounted or rooftop PV systems, it is possible to feed the energy into the grid or use it directly at the production site.

Numerous studies have been conducted on the design and feasibility of FSPV systems. Brillianto Apribowo et al. [6] designed a 1 MW FSPV system in three water bodies in Indonesia, indicating that it was environmentally friendly and clean. The performance of floating photovoltaic systems in an Indian reservoir was examined by Ravichandran et al. [2]. According to their findings, the system reduces water evaporation, and the majority of the construction costs are related to floating structures. Researchers have also studied floating photovoltaic plants' tracking and cooling advantages [5]. FSPV systems are highly efficient at using water resources, reducing evaporation losses, and allowing productive use of off-shore land while saving on-shore land [7, 8].

The project aims to design an FSPV plant for Vaigai Dam in Theni District, Tamil Nadu, India. There is an empty water space at the tail end of the dam that could be used as a site for the plant, which is away from the tourist area. This system will be connected to the local electricity grid and operate as an on-grid system. The energy generated can be used to power the dam, and surplus energy can be supplied to Andipatti Taluk. This study's novelty lies in its design and simulation of novel FSPV systems in one of southern India's major dams.

Five sections make up this paper, the first of which describes the research carried out and the implications it has. In the second section, resource, and demand data for the design of FSPV plant are presented. The third section discusses the methodology and design of the study. In the fourth section, the results and findings of this study are presented. As a conclusion, the fifth section summarizes the results of the study.

2 Resource and Demand Data

The following Fig. 1 represents the methodology of this study, firstly the resource and demand data for the project location was collected using online databases, following which the resource data is validated in the PVsyst software using the meteo data in software. The FSPV system is then designed in PVsyst taking the demand, size, and other factors into consideration. This design is then simulated to obtain the generation data and performance ratio of the plant. Lastly, the CO_2 emission savings is calculated. The design and simulation are carried out in PVsyst which is a widely used software for PV system design and simulation.

2.1 Resource Data

The necessary data is collected using online resources and simulation software to design the FSPV power plant. Using government databases, sun path, and daytime resources are gathered, and power generation capacity is determined for different timeframes over a year. Figure 2 depicts the selected location for installing the FSPV plant. With a height of 111 feet, the Vaigai Dam stretches across the River Vaigai near Andipatti in Theni, Tamil Nadu [9]. It is capable of storing 71 feet of water.

The location's solar resource data was accessed from Global Solar Atlas [10]. Figure 3 shows the data on the solar resources available at the Vaigai dam. The annual global horizontal irradiation at the location ranges from 5.38 to 5.55 kWh/m^3/day. According to the resource data, sunlight is available year-round but is relatively low during the monsoon season (June–November).

Throughout the region, there is a very low wind speed. The average wind speed in Theni is around 5 m/s throughout the year [12]. It is necessary to conduct a wind analysis before building an FSPV station. Figure 4 represents the monthly average wind speeds at the location.

A peculiarity of Theni is the presence of two rainy seasons, and heavy rain will cause a flood, which will impact the reservoir. This will be considered during the FSPV plant's design to withstand in-depth changes in water levels. The average temperature in the location ranges between 20 and 35 °C throughout the year. Figure 5 illustrates the rainfall and temperature data of the site.

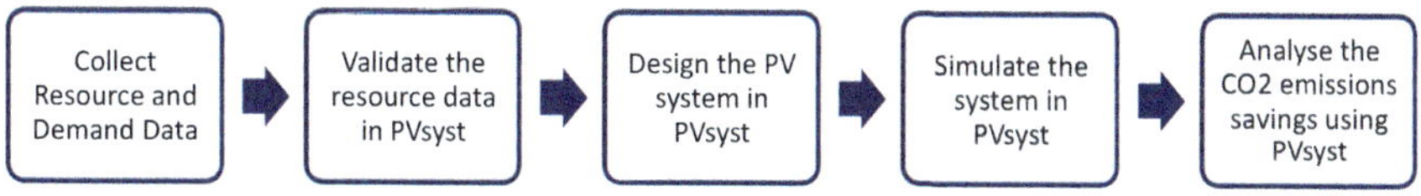

Fig. 1 Methodology flow chart

Fig. 2 Project location [11]

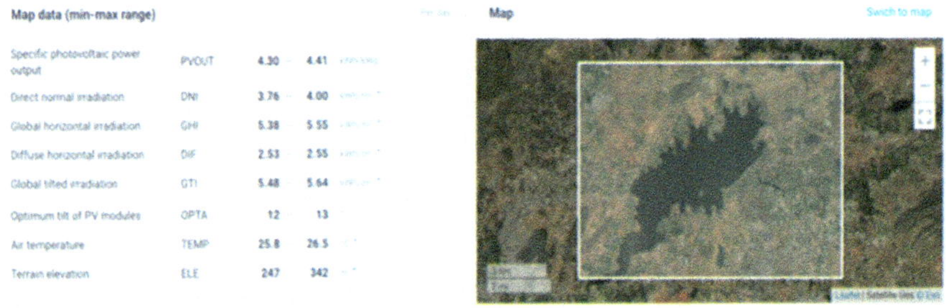

Fig. 3 Solar resource at location [10]

2.2 Demand Data

Tamil Nadu's average power demand ranges between 15,600 and 16,100 megawatts (MW) [13]. In recent years, the state has been able to meet the highest demand, 16,151 MW, on 3rd April 2019. Tamil Nadu's daily average consumption is expected to increase from 345 million units (MU) in 2019 to 390 in 2022 [14]. The state reports a power surplus because it has added 13,995 MW to the grid since 2011. The per capita consumption in Tamil Nadu from 2010 to 2019 is shown in Table 1.

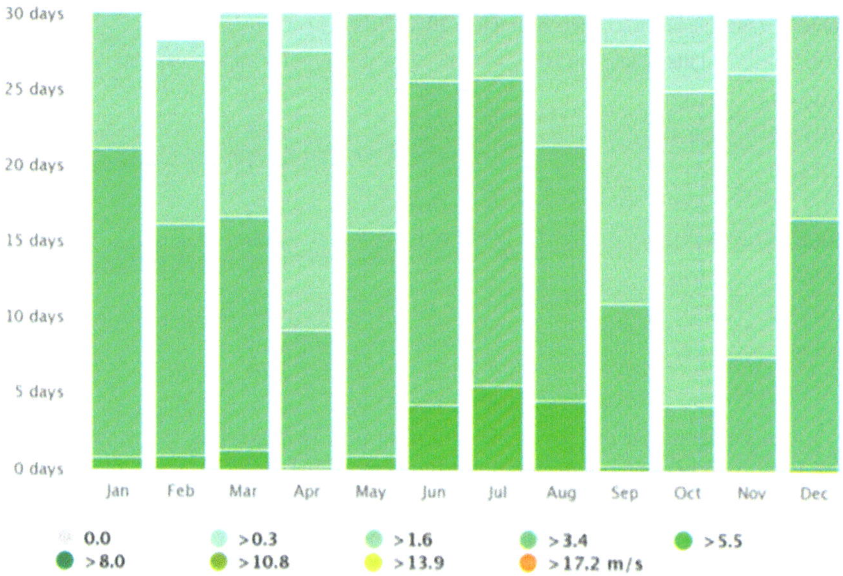

Fig. 4 Wind speeds at location [12]

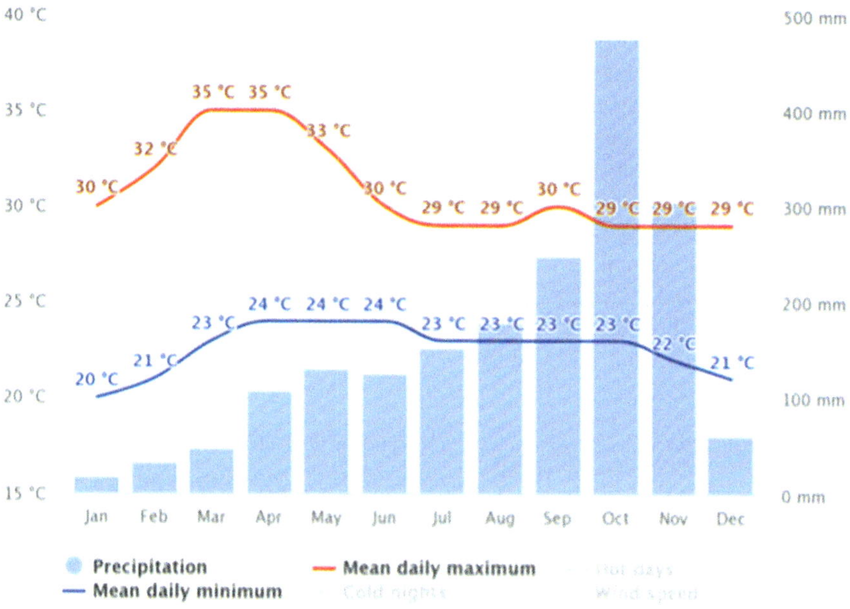

Fig. 5 Rainfall and temperature data of location [12]

Table 1 Tamil Nadu per capita consumption 2010–2019 [15]

Year	Installed capacity (MW)	Generation (MU)	Per capita consumption (kWh)
2010–11	10,237	25,639	1040
2011–12	10,364	27,942	1065
2012–13	10,515	25,301	1011
2013–14	11,884	31,276	1196
2014–15	13,231	34,569	1228
2015–16	17,976	34,709	1280
2016–17	18,733	29,223	1340
2017–18	18,747	27,873	1389

It can be seen that the per capita consumption is increasing every year, which will lead to higher energy demand. As Sivakumar [14], peak energy demand during summer will exceed 17,000 MW in 2022, and the state's renewable energy generation plays a vital role in meeting this demand. Hence, the state government is investing in renewable energy, such as solar and wind, to ensure energy supply during peak demand. The total power consumption of Theni during the year 2018 was 996.23 MU [16]. Table 2 illustrates the sector-wise power consumption of the Theni district in 2018.

Table 2 deduced that domestic and agriculture are the two major power-consuming sectors in Theni. The instantaneous peak demand in Theni during 2018 was 195 MW [16]. Theni is home to three hydropower stations and one thermal and gas power station [16]. The proposed FSPV plant in the Vaigai dam can support the Theni district to meet its energy demand. The plant will be connected to the local electric grid and act as an on-grid system. The electricity generated from the plant can be used to meet the energy needs of the domestic and agriculture sectors, as these are the two most power-consuming sectors in Theni. In addition, the FSPV plant can reduce the need for diesel generators to meet the peak demand during summer. Based on the assumption that Theni's energy consumption will increase by 15% in 2024, its consumption will reach 1,145.66 MU.

Table 2 Theni district's sector wise power consumption in 2018 [16]

Sector	Consumption (MU)	Percentage consumption (%)
Industries	172.2	17.28
Agriculture and huts	310	31.11
Domestic	340.03	34.13
Commercial	69.6	6.99
Public lighting and public works	37.55	3.77
Miscellaneous	66.85	6.72
Total	**996.23**	**100**

3 System Design

Before installing PV systems at any location, it is imperative to model, simulate and analyse the behaviour and characteristics of the generation under actual climatic conditions [17]. PVsyst, photovoltaic simulation software was used to design and simulate the proposed FSPV plant at Vaigai dam. The meteorological (meteo) data for the project location was loaded into PVsyst by inputting the latitude and longitude of the Vaigai dam. To validate the data, the site's meteo data from PVsyst SA [18] was compared to Meteoblue [12], and the data were in good agreement. Using these data, an FSPV plant for Vaigai dam was designed and simulated in PVsyst. The proposed FSPV plant in the Vaigai dam is illustrated in Fig. 6.

The panel title angle was optimized for the yearly irradiation yield of the site. The ideal tilt angle at which there are no losses was determined to be 13°. Figure 7 shows the title angle optimization for the project location.

A total area of 200,000 m^2 is available at the tail end of Vaigai Dam, where the proposed FSPV plant will be located. Based on this available area, the FSPV system is pre-sized, and the plant capacity was deduced to be 22 MW. Trina Solar's 300 Wp polycrystalline solar panels are selected for this project based on performance and value for money [19]. A total of 90,045 solar panel modules are required for the FSPV plant. A 2500 kW power inverter is selected after various optimizations and accounting for all possible losses. Using this, the direct current generated by the system is converted to alternating current. Sungrow's 2500 kW inverters were

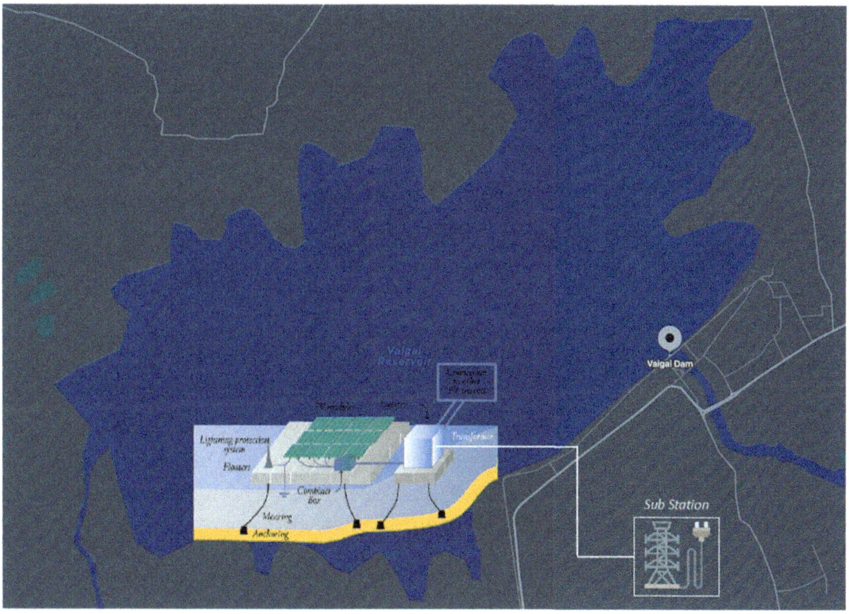

Fig. 6 Proposed FSPV plant for Vaigai dam

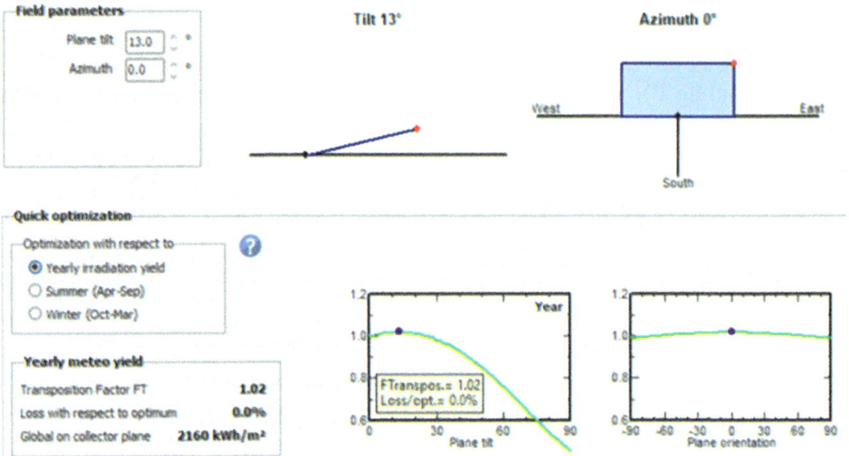

Fig. 7 Tilt angle optimization

selected for their reliability and cost-effectiveness. It is estimated that this 22 MW FSPV plant will require nine inverters.

Based on the specifications of solar panels and inverters selected for the project, the ideal number of modules in a string was determined to range between 26 and 29. Choosing the maximum number of modules possible in a string is critical to avoid Undervoltage. When we choose the smallest number of modules in a string, and one of the panels gets shaded, the voltage drops below the threshold. The inverter will shut down immediately. Hence, for this FSPV plant, one string will consist of 29 modules. An additional 20% inverter capacity during string sizing was considered to prevent generation loss during a mismatch. Even under non-ideal conditions, the string array will produce at the required capacity. Accordingly, the total number of strings for the 22 MW FSPV plant was 3105. The area needed to build this 22 MW FSPV plant in Vaigai Dam was estimated to be 176,684 m^2, within the available space. Figure 8 illustrates the horizon diagram of the location. It supports more accurate shading assessments and improved solar production estimates by automatically modelling the shade from the terrain surrounded by a site [20].

As part of the design process, all possible losses of the proposed FSPV plant were considered. The field thermal loss factor was modelled using the factor value of mounted panels with air circulation. As per the standards, the DC circuit ohmic losses were assumed to be 1.5%. The module efficiency losses were taken to be − 1.6% based on the data from panel manufacturer Trinasolar. The yearly soiling loss factor was considered to be 5% because the area is relatively windy during the winter season. This can lead to dirt accumulation on panels. To account for unplanned plant shutdowns, an unavailability time fraction of 1% was applied, around four days per year. Solar plants with a capacity over 20 MW typically have auxiliary losses of around 0.82% [21], so we assumed 1.8 MW of additional losses for this plant. The FSPV plant layout was designed in PVsyst to simulate the shading losses. The panels

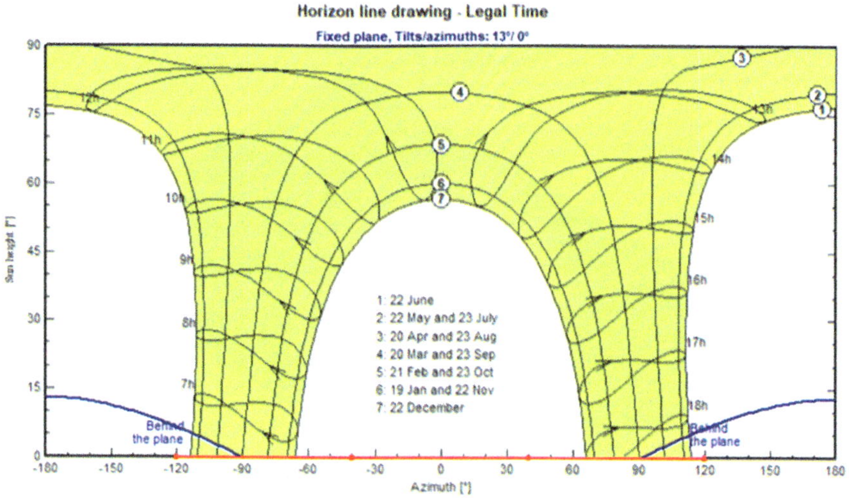

Fig. 8 Horizontal line diagram

were arranged in portrait orientation with 29 modules in series across the X direction. As a rule of thumb, the panels should be positioned 1.5 times the array width in the Y direction. Hence, the panels were placed at a pitch distance of 3 m in the Y direction. Figures 9 and 10 represent the proposed FSPV plant array layout and the shading loss diagram, respectively.

FSPV projects of a megawatt-scale typically take between one and three years to build [22]. A two-year timeframe is estimated for building the FSPV plant at Vaigai Dam. Therefore, it is expected that the plant will be operational by the year 2024. The lifespan of FSPV plants is typically 20–25 years [22]. It is assumed that the lifespan of the FSPV plant at Vaigai Dam is 25 years.

4 Results and Discussions

A simulation of the final design using PVsyst is performed after factoring in all the losses and potential challenges of the system. Floating structures and PV modules account for the majority of the costs, so the space on the float is utilized to the maximum. The FSPV plant consists of 3105 strings with 29 PV modules arranged in series, each string being arranged in parallel. Figure 11 shows 2024's monthly energy generation of the Vaigai dam FSPV plant.

The figure above shows that March has the highest energy generation (4707 MWh). Due to the monsoon season in the region, the generation in June–August and October–December is lower. Based on the design, it has been determined that the proposed FSPV plant in Vaigai Dam will generate 48.3 GWh per year. It is estimated that the plant will produce enough electricity to support 4.2% of the district's

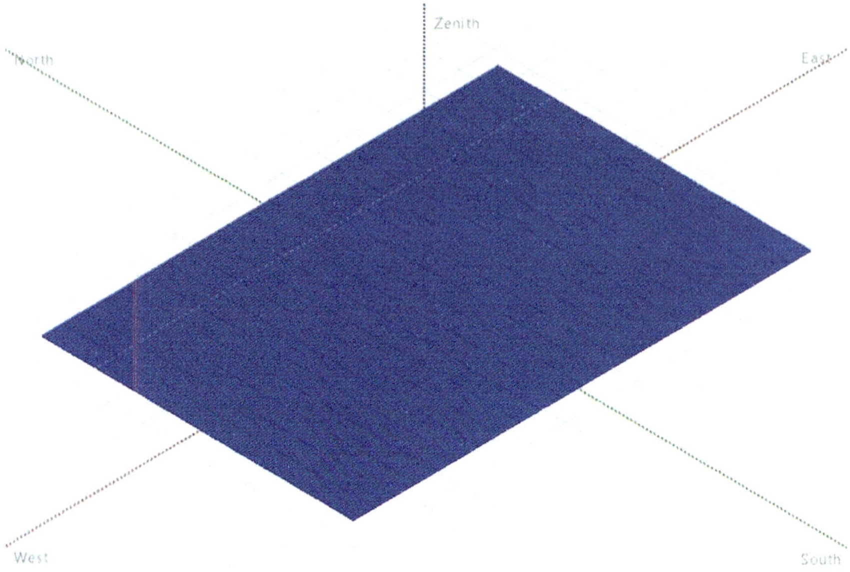

Fig. 9 Vaigai dam FSPV plant array layout

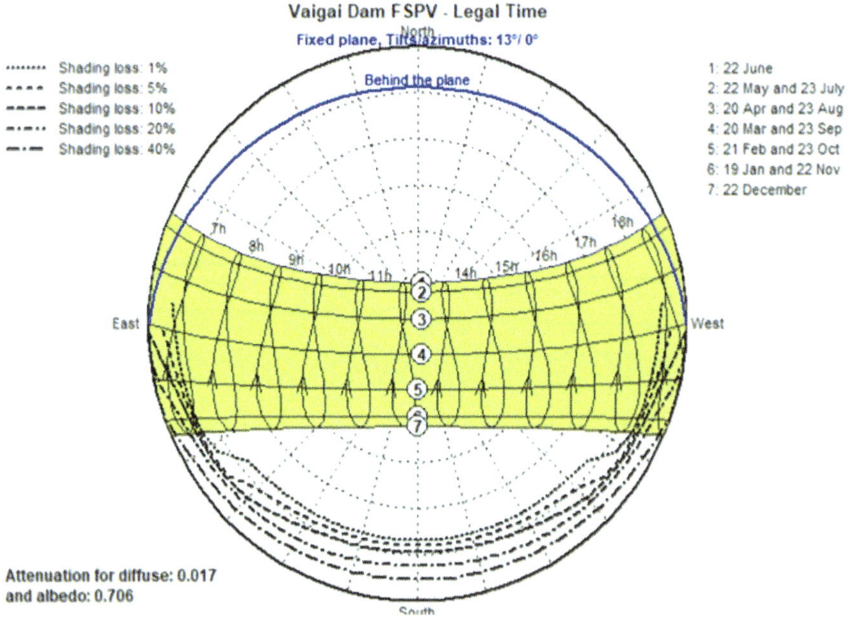

Fig. 10 Shading loss diagram

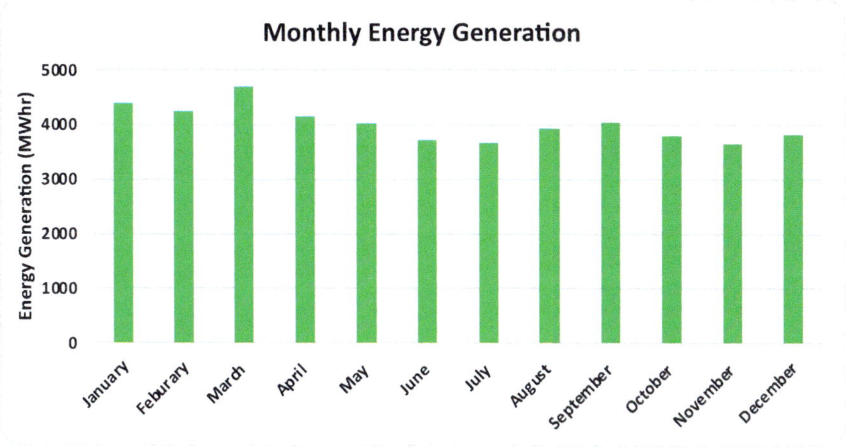

Fig. 11 Vaigai dam FSPV plant monthly energy generation profile for 2024

total energy needs. From the normalized production of the simulation, three fundamental parameters can be determined, namely the produced useful energy, the system, and the collection losses [23]. According to the results, both system and collection losses were found to be 0.9 kWh/kWp/day and 0.1 kWh/kWp/day, respectively, while inverter output or useful energy was found to be 4.9 kWh/kWp/day.

Often described as a quality factor, the performance ratio measures a PV plant's quality independent of location [24]. The PV plant's performance ratio shows how much energy is available for export to the grid after deducting thermal and conduction losses and the energy consumed for operation [24]. The monthly performance ratio of the Vaigai Dam FSPV plant is shown in Fig. 12.

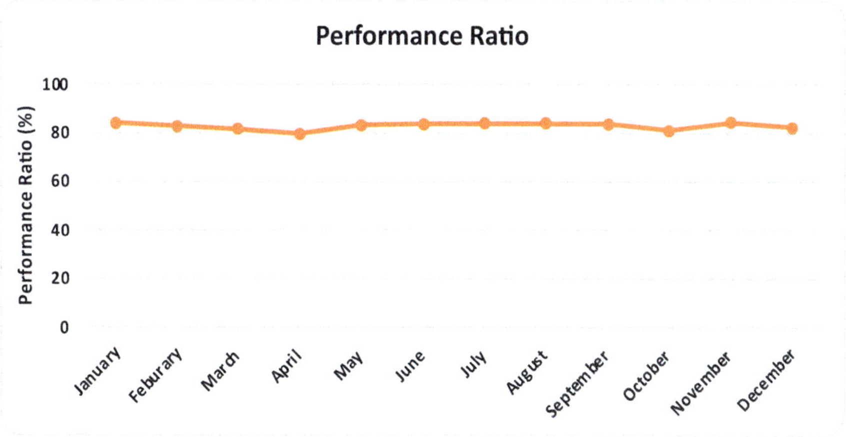

Fig. 12 Vaigai dam FSPV plant monthly performance ratio

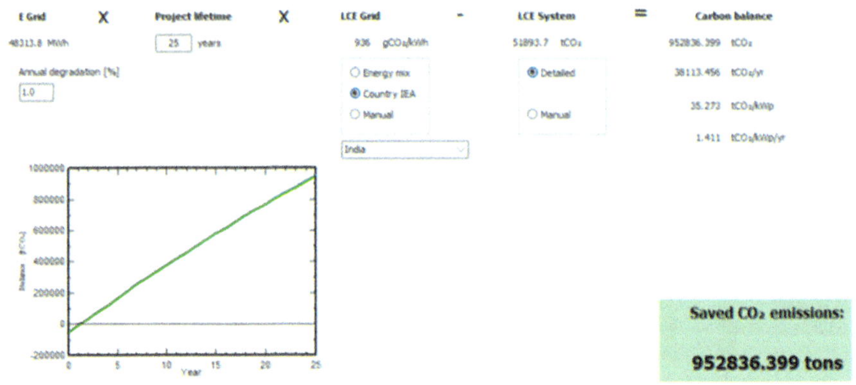

Fig. 13 Vaigai dam FSPV plant FSPV plant CO_2 savings

From the above graph, it is noticed that the performance ratio of the FSPV plant ranges from 79.9 to 84.5%. The overall performance ratio of the plant was estimated to be 83.05%. According to the study by Mittal et al. [25], performance ratios of FSPV plants are generally higher than that of land-based PV systems.

The carbon footprint is generally expressed in equivalent tons of carbon dioxide (CO_2), representing the number of greenhouse gases produced to, directly and indirectly, support human activities [26]. The carbon footprint is nothing but the sum of all CO_2 emissions resulting from a set of activities over a given timeframe. The use of renewable energy systems reduces pollution and carbon footprint. As a result, the construction of an FSPV plant at Vaigai Dam will ultimately result in a reduction of CO_2 emissions. Figure 13 illustrates the total reductions in CO_2 emissions from the Vaigai dam FSPV plant.

The above analysis shows that installing an FSPV plant at the Vaigai dam will reduce CO_2 emissions by 0.953 megatons over its lifetime, proving the system's feasibility. This is in addition to the fact that a renewable energy system such as an FSPV plant would significantly contribute to environmental conservation and enhance air quality, biodiversity, and ecosystems.

5 Conclusions

Numerous studies have been conducted on the feasibility of FSPV plants over the past few years, and the system is reliable and effective. Our study focuses on the design of an on-grid FSPV plant at Vaigai Dam in Tamil Nadu's Theni district. In the tail end area of the dam, an FSPV plant with a capacity of 22 MW was designed. A total of 90,045 Trina Solar crystalline solar panels and nine Sungrow 2500 kW inverters made up the system. A simulation was conducted using PVsyst, considering the plant's losses. The FSPV plant is expected to generate 48.3 GWh per year, which accounts for about 4.2% of the district's total energy consumption. Using this energy, the dam can be powered, and surplus electricity can be supplied to the taluk of Theni, which would be a great way to improve the rural area. According to the results, the plant had a performance ratio of 83.05%.

Over its lifetime, the FSPV plant will reduce CO_2 emissions by 0.953 megatons. The results of this study contribute significantly to the existing literature on FSPV systems. Additionally, similar projects can be deployed throughout the region to meet rural energy demands. Future research could assess the economics of FSPV plants.

References

1. Sahu A, Yadav N, Sudhakar K (2016) Floating photovoltaic power plant: a review. Renew Sustain Energy Rev 66:815–824
2. Ravichandran N, Ravichandran N, Panneerselvam B (2021) Performance analysis of a floating photovoltaic covering system in an Indian reservoir. Clean Energy 5(2):208–228
3. Patil Desai Sujay S, Wagh M, Shinde N (2017) A review on floating solar photovoltaic power plants. Int J Sci Eng Res 8:789–794
4. Manolache AI, Andrei G, Rusu L (2023) An evaluation of the efficiency of the floating solar panels in the western black sea and the Razim-Sinoe lagunar system. J Mar Sci Eng 11(1):203
5. Cazzaniga R, Cicu M, Rosa-Clot M, Rosa-Clot P, Tina G, Ventura C (2018) Floating photovoltaic plants: performance analysis and design solutions. Renew Sustain Energy Rev 81:1730–1741
6. Brillianto Apribowo CH, Santoso B, Suyitno S, Wicaksono F (2019) Design of 1 MWp floating solar photovoltaic (FSPV) power plant in Indonesia, p 030013
7. Matheus MN, Musti KSS (2023) Design and simulation of a floating solar power plant for Goreagab dam, Namibia. In: Vasant P, Rodríguez-Aguilar R, Litvinchev I, Marmolejo-Saucedo JA (eds) Human agro-energy optimization for business and industry. IGI Global, Hershey, PA, USA, pp 1–26
8. Pouran HM, Padilha Campos Lopes M, Nogueira T, Alves Castelo Branco D, Sheng Y (2022) Environmental and technical impacts of floating photovoltaic plants as an emerging clean energy technology. iScience 25(11):105253. https://doi.org/10.1016/j.isci.2022.105253
9. Centre NI (2022) Vagai Dam. Government of India. https://theni.nic.in/tourist-place/vaigai dam/. Accessed 25th Jun 2022
10. Solargis (2022) Global Solar Atlas. https://globalsolaratlas.info/map?c=11.523088,8.261 719,3. Accessed 25th Jun 2022
11. Google (2022) Google maps
12. Meteoblue (2022) Simulated historical climate & weather data for Theni. https://www.meteob lue.com/en/weather/historyclimate/climatemodelled/theni_india_1254745. Accessed 25th Jun 2022
13. Thangamani TP (2020) Energy department policy note. https://cms.tn.gov.in/sites/default/files/ documents/energy_e_pn_2019_2020.pdf. Accessed 2022 Jun 25
14. Sivakumar B (2022) TN's power consumption to cross 17,000 MW by April. In TheFedral
15. T. N. G. and D. C. Ltd. (2020) Electricity report 2020. https://www.tn.gov.in/deptst/electricity. pdf. Accessed 25th Jun 2022
16. Centre NI (2023) Theni district hand book 2017–18: electricity. https://tn.data.gov.in/catalog/ theni-district-hand-book-2017-18-electricity
17. Vinod RK, Singh SK (2018) Solar photovoltaic modeling and simulation: as a renewable energy solution. Energy Rep 4:701–712. https://doi.org/10.1016/j.egyr.2018.09.008

18. SA P (2022) PVsyst
19. Trinasolar (2022) Reliable and smart solar. https://www.trinasolar.com/us/product. Accessed 22nd Jun 2022
20. Harvey B (2019) What is horizon shading and why is it important for solar? Solar Power World. https://www.solarpowerworldonline.com/2019/12/what-is-horizon-shading-and-why-is-it-important-for-solar/
21. Pandey S, Kumar R (2017) Analysis of auxiliary energy consumption in utility scale solar PV power plant, 10th Jan 2017
22. Singapore SERIO (2019) Where sun meets water, floating solar handbook for practitioners. The World Bank. https://documents1.worldbank.org/curated/en/418961572293438109/pdf/Where-Sun-Meets-Water-Floating-Solar-Handbook-for-Practitioners.pdf. Accessed 25th Jun 2022
23. Baqir M, Channi HK (2022) Analysis and design of solar PV system using PVsyst software. Mater Today: Proc 48:1332–1338. https://doi.org/10.1016/j.matpr.2021.09.029
24. AG SST (2021) Performance ratio—quality factor for the PV plant. https://files.sma.de/downloads/Perfratio-TI-en-11.pdf. Accessed 26th Jun 2022
25. Mittal D, Saxena BK, Rao KVS (2017) Comparison of floating photovoltaic plant with solar photovoltaic plant for energy generation at Jodhpur in India. In: 2017 international conference on technological advancements in power and energy (TAP energy), 21–23 Dec 2017, pp 1–6. https://doi.org/10.1109/TAPENERGY.2017.8397348
26. Sharma S, Kurian CP, Paragond LS (2018) Solar PV system design using PVsyst: a case study of an academic institute. In: 2018 international conference on control, power, communication and computing technologies (ICCPCCT), 23–24 Mar 2018, pp 123–128. https://doi.org/10.1109/ICCPCCT.2018.8574334

Selection and Design of Energy Harvesting Circuit for Microbial Fuel Cell for Low Power Applications

Kushagra Bhatia, A. Lavanya, and Divya Navamani

Abstract The world's energy consumption has increased enormously due to increased energy demand. Therefore, this research focuses on alternate eco-friendly energy sources. A microbial fuel cell (MFC) is a bio-electrochemical source that converts the chemical energy in organic and inorganic compounds into electricity by using microorganisms as biocatalysts. MFCs attract substantial consideration as they can harvest renewable energy from organic matter. MFC is a low voltage source. An energy harvester is required to extract useable energy from MFC that can accumulate energy in a capacitor and increase the output potential of the MFC. In this paper, an overview of various energy harvesting circuits is discussed. The selection criteria for suitable energy harvesting circuit for a specified application is presented in this paper. Key parameters like cost, power, size of the MFC structure, and energy harvesting circuit (EHC) are discussed in the selection. This paper thoroughly analyses the design of energy harvesting system at the same time the literature works published in the past decade are presented. This review article categorizes the used topologies and comparatively presents their advantages and disadvantages.

Keywords Microbial Fuel Cell · Energy harvesting · Power management · Converter · Alternate source

1 Introduction

The energy consumption has increased enormously worldwide due to increase in energy demand. Therefore, in this research paper, we focus on the alternate eco-friendly energy sources. Microbial fuel cells generate electricity from organic wastes including sludge, green waste, and food waste by using electro-active pathogens. The

K. Bhatia (✉) · A. Lavanya · D. Navamani
Department of Electrical and Electronics Engineering, SRM Institute of Science and Technology, Kattankulathur, Tamil Nadu 603203, India
e-mail: kushagrabhatia0303@gmail.com; kb1607@srmist.edu.in

A. Lavanya
e-mail: lavanyaa@srmist.edu.in

S. Kumar et al. (eds.), *Control Applications in Modern Power Systems*, Lecture Notes in Electrical Engineering 1128, https://doi.org/10.1007/978-981-99-9054-2_5

power density and output voltage of these cells are usually low; therefore, we need energy harvesting circuit for real-time applications as we know that energy harvesting is an efficient method for converting low energy into electricity.

Now the energy produced from MFC is low hence we prefer using stacked MFCs[1, 2] to design an energy harvesting system that has the ability to acquire, store, and boost the power out of MFCs. Hence, we can derive useful powers from the fuel cells, and need to use an energy harvesting circuit or EHC. Where EHCs are designed circuits to harness the power generation by energy sources and to boost the voltage and current to a practical level. There are many ongoing as well as published studies on MFC and its power management system and EHC [3, 4–6, 7] which have conclusive results for powering low voltage devices and sensors.

Our aim of this research is to amend low currents and to maintain high efficiency for the same. The goal is to optimize the power harvested and maximize the energy conversion efficiency.

1.1 Microbial Fuel Cells-An Overview

Microbial Fuel cells or MFCs are an alternative and attractive technology producing renewable electrical energy from chemical catalytic reactions of organic compounds. MFC is an excellent alternative for electricity generation from wastewater or sludge, where bacteria in wastewater directly convert organic matter to electrical energy. The electricity generation is utilized by the electrons produced from anaerobic oxidation of organic/carbon substrates. MFC has two chambers, a cathode and an anode like in a battery. A membrane separates them, often known as an exchange membrane. The bacteria used to achieve this reaction are classified as Bio-Electrochemically active bacteria.

To understand the working of the MFC [8], comprehend the reactions that take place in anode and cathode chambers. MFC is based on the principles of redox reaction. The bacteria oxidize organic matter to produce carbon dioxide (CO_2), electrons, and protons. The anode chamber is anaerobic, and the cathode is aerobic. A membrane separates the two chambers known as the proton exchange membrane (PEM). Microbe's oxides the fuel/substrate at the anode to generate protons, electrons, and CO_2. Electrons are transferred to the cathode chamber by using an electrical circuit connection. Protons from the anode move over the membrane and are consumed by the electrons, combined with oxygen (O_2) to form water.

$$CH_3COO^- + H_2O \rightarrow 2CO_2 + 2H + 8e^-(\text{anodic}) \tag{1}$$

$$O_2 + 4H + 4e^- \rightarrow 2H_2O(\text{cathodic}) \tag{2}$$

1.2 Energy Harvesting Circuit

Energy harvesting is the term defined as capturing or harnessing of energy that is available to us through either conventional or non-conventional sources, and efficiently transforming it into usable electrical power. This power is adapted for direct use or storage for later. Small energy sources such as fuel cells or piezoelectric sources are pretty much useless on their own, but by implementing a harvesting technique can make this otherwise wasted power into usable electrical energy [5, 9, 10, 11].

Driving force for an energy harvesting circuit will be the ability to create a self-sustaining source of power with adequate circuit to provide uninterruptable power, which requires minimal to no oversite, unlike a battery powered device. Microenergy harvesting or energy scavenging is especially interesting for IoT applications such as remote sensing, node network, etc., as well as other applications in the lower power spectrum segment. Here are some existing power management systems[12, 13, 14, 15] that are incorporated with MFCs (see Fig. 1).

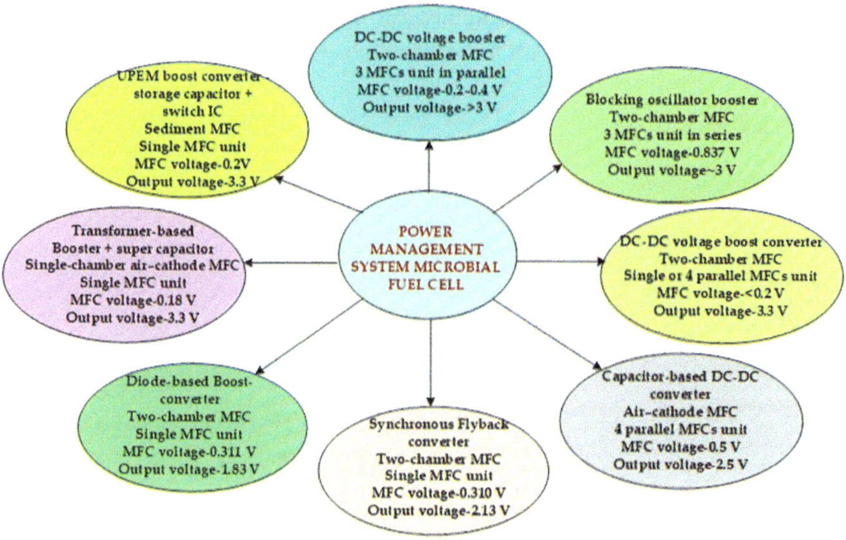

Fig. 1 Existing power management system for microbial fuel cell

2 Existing Systems of EHC

2.1 Charge Pump-Based System

A charge pump system uses a capacitor to achieve the higher voltage levels desired. In more simple terms, it is a voltage doubler. There are two states of this system between which it continuously toggles. The first state (see Fig. 2), is the charging state. Here C1, also referred as flying capacitor is charged to a voltage V_{in}.

The second state is the charge transfer state (see Fig. 3). This has two switches S1 and S2, which are in opposite configurations i.e., left switch up and right switch down. We assume that the capacitor C2 is at an initial voltage of 0 V. C1, which has V_{in} across its terminal prior to switching, transfers some charge in the second state of the systems to C2. This leads to a rise in voltage level across C2 and a fall of levels in C1 and the rise in value for C2 is between V_{in} and 2 V_{in}.

Therefore, we assume that at no load condition, V_{out} is at the ideal state to provide a supply voltage to some load. This setup is relatively light and maintains near to double the input voltage. These circuits can be built to provide higher voltages, usually with integral multiples of the input side voltage like 2*x*, 3*x*, 4*x*, and so on.

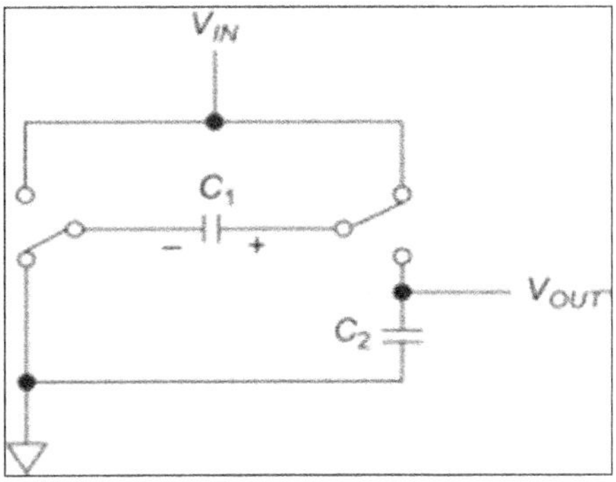

Fig. 2 Capacitor charging state [16]

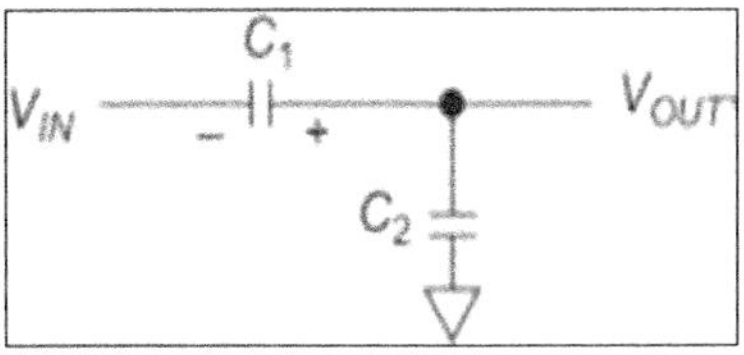

Fig. 3 Charge transfer state [16]

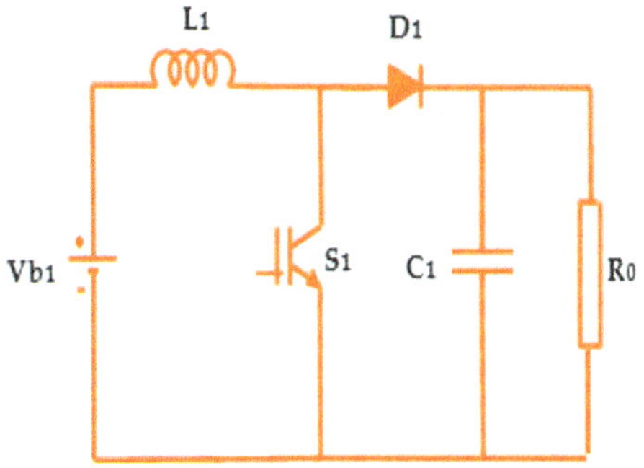

Fig. 4 Circuit for boost converter

2.2 *Boost Converter-Based System*

A boost converter is a type of switch mode, DC-DC converter that steps up voltage from its supply input to its load output. It consists of a switch (usually a MOSFET), an inductor, a diode, and a capacitor (see Fig. 4). Suitable DC sources power the boost converter such as batteries, rectifiers, solar panel, and DC generators. The switch requires a pulse signal, for its operation. The designing of the passive components is based on the desired output, acceptable ripple, and the control system they follow. There are 2 modes of operation. Mode 1, where the switch is ON, providing a short circuit (ideal) and charges the inductor coil. Mode 2, where switch is OFF, where the polarity of inductor reverses and energy is dissipated.

Use of boost converters in research of MFC is widely accepted. Since the MFC output is very low, a boost converter helps it reach a level which is usable. For example, $V_{in} = 0.8$ V from an MFC can be boosted to 3.3 V output, a level where small devices and sensors can start operation. In the literature published by F. Ma and his team, they have used a P-channel FET to replace the diode for their boost converter design [3]. The analysis of the energy harvesting circuit where a charged capacitor at 1.2 V, boosted the output at 3.3 V.

2.3 *Dynamic Maximum Power Point Tracking*

An MFC is a system whose internal resistance and power density change when pH, microbial biomass, and temperature change, implying that this static management system is unable to collect real-time peak energy and hence has a lower efficiency. MFCs are voltage sources with variable internal resistance and capacitance. Furthermore, because the MFC output is DC, the internal capacitance can be regarded as an open circuit, which is usually ignored.

Maximum power point refers to the voltage and current that produce the most power. It has been discovered that if the external resistance is not equal to the MFC internal resistance, there would be a power loss of more than 50%. Furthermore, by adjusting the load size, dynamic maximum power point tracking (DMPPT) technology may regulate the o/p efficiency. It can track inreal-time, minimizes start-up time, and enhances power generation, so boosting Coulomb and wastewater treatment efficiency. As a result, this technology can be employed on MFCs without the need for voltage inversion.

To match the load, DMPPT can be done by altering the resistance or by using the boost converter's switching frequency. The conductance increment approach and the gradient method are frequently used in PV systems. The stable time is more than 18 h, and the energy derived is 76 times that of the charge pump operation, with a 21-fold increase in coulomb efficiency.

As a result, DMPPT is the most efficient way to get energy from an MFC.

3 Proposed System

The discussed system is based on a four-block architecture, which typically consists of a power source, a switching circuit, an amplifying system, and finally a load element (see Fig. 5).

A representation of switching circuit coupled with the boost converter to produce desired output (see Fig. 6). The charge is first built up and stored within the supercapacitor and discharged into the boost converter which amplifies the received voltage respectfully.

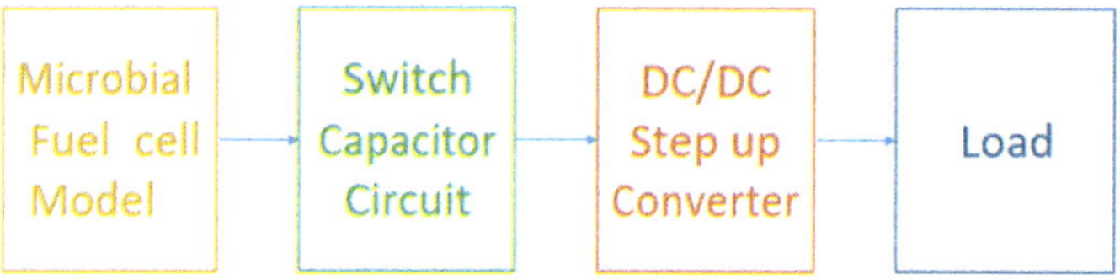

Fig. 5 Block diagram of designed system

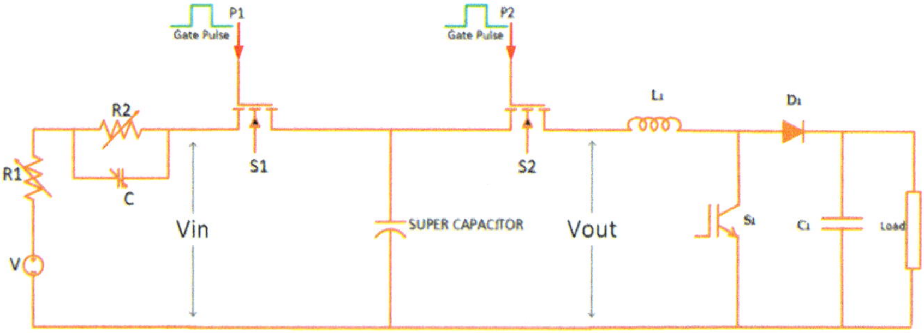

Fig. 6 Proposed circuit diagram

The proposed circuit works in a dual-operation method. The first mode (Mode 1) is namely the charging mode where the circuit charges the supercapacitor to the required value within the said time period. The second mode (Mode 2) is the discharging mode, wherein the switching circuit is activated and the secondary part of the circuit which is connected to the load starts discharging the Supercapacitor. Each mode has been thoroughly explained as follows.

3.1 Mode 1-Charging Mode

At pulse input P1 = 0, the MOSFET S1 is active. The pulse passes through the NOT gate and provides an inverse input, i.e., P2 = 1 to MOSFET S2. Hence charging takes place within the initial part of the circuit and discharging circuit connected to the load is open and not active. Therefore, the charging circuit is closed at P1 = 0 while the discharging circuit is open when it receives P2 = 1 as input signal. The low energy input is passed to the Supercapacitor which charges through an extensive time period since it charges gradually at $I_{\text{in}} = 1$ mA. In this mode, voltage across supercapacitor is initially $V_c = 0$ V at time period $T = 0$ s. Charging follows the capacitor charging equation

$$V_c = V_{\text{in}}\left(1 - e^{\frac{-t}{RC}}\right) \tag{3}$$

The final voltage achieved by the supercapacitor is $V_c = 1.7$ V and time taken by charging circuit to attain this is $T = 5016$ s (1 h 23.6 min) theoretically.

3.2 Mode 2-Discharging Mode

At pulse input P1 = 1, the MOSFET S1 is deactivated, and second circuit is in open condition. An inverse signal, i.e., P2 = 0 is provided to MOSFET S2 and discharging of supercapacitor occurs since circuit is connected to load. Therefore, the charging circuit is open at P1 = 1 while the discharging circuit is closed when it receives P2 = 0 as input signal. The supercapacitor has a charge of $V_c = 1.7$ V at this instant. When discharging circuit activates, the DHT11 sensor acts as a load for the system and discharging of capacitor is initiated. This process is instantaneous as compared to charging mode since charging is done at minimum conditions but discharging takes place at rated conditions for load. The current drawn by the sensor is $I_{\text{out}} = 13$ mA. In this mode the boost converter shall amplify the voltage received to the required output voltage of $V_{\text{out}} = 5$ V. A minimum threshold voltage of 1 V is considered so that the supercapacitor recharges at a faster rate in the next cycle.

4 Simulations

The proposed circuit simulation is divided into two parts. The first is the switch capacitor circuit with a source that mimics the characteristic of an MFC. Next at ideal charging condition is the entire circuit. Since the time of charging the super capacitor using an MFC is significant, there were limitations while performing these simulations using MATLAB Simulink. As follows the results of the simulation were conclusive of the calculations performed before setting the simulation.

4.1 Switch Capacitor Circuit Simulation

To understand the operation of the circuit, this simulation is the first iteration that is based on ideal conditions. To charge and discharge the capacitor (see Fig. 7), we need timed switches that open and close for their respective loops. S1 and S2 are charging and discharging switches respectively. They need to operate in complement to each other, so a basic inverter logic is given for the pulse generator for S2. As seen in the previous slide, pulses are in complement to each other, that is when S1 is *closed*, S2 is *open* and when S1 is *open*, S2 is *closed*. The input source used is a voltage source element initialized to provide a constant DC voltage of 1.8 V. The supercapacitor characteristics are mimicked using a capacitor element set at 1 F. The simulation works the same as explained previously, but instead of using a voltage source with no limits, an MFC is the source input. The simulation time sample is 2e4 s as capacitor charging and discharging time are calculated. The Duty cycle for the pulse is 96.7% set into the pulse generator.

The characteristic curves and pulse for the switches are scoped (see Fig. 8). As observed in Fig. 8(1) and (2), are the pulse signals P1 and P2 for the switches S1 and S2 respectively. P1 and P2 are complements of each other, hence a not logic is used to derive the P2 signal. Figure 8(3) and (4) depict the capacitor discharging and

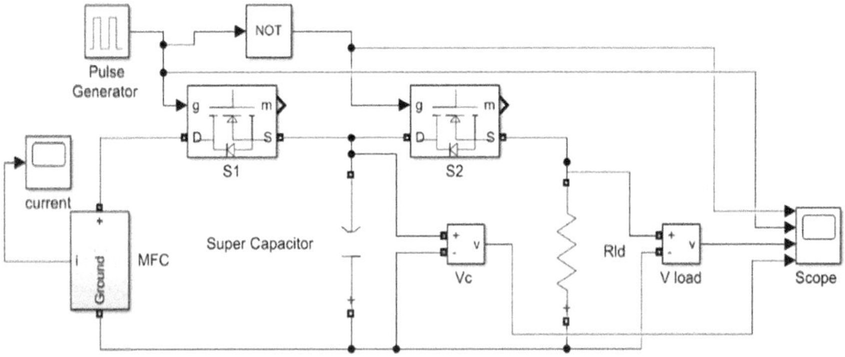

Fig. 7 Switch capacitor simulation; MFC = 1.8 V; Super Capacitor = 1F; $R_{ld} = 668\ \Omega$

charging curve. The voltage across the supercapacitor will increase with time (see Eq. 3). The current will decrease with increase in voltage as an inverse relation (see Fig. 9).

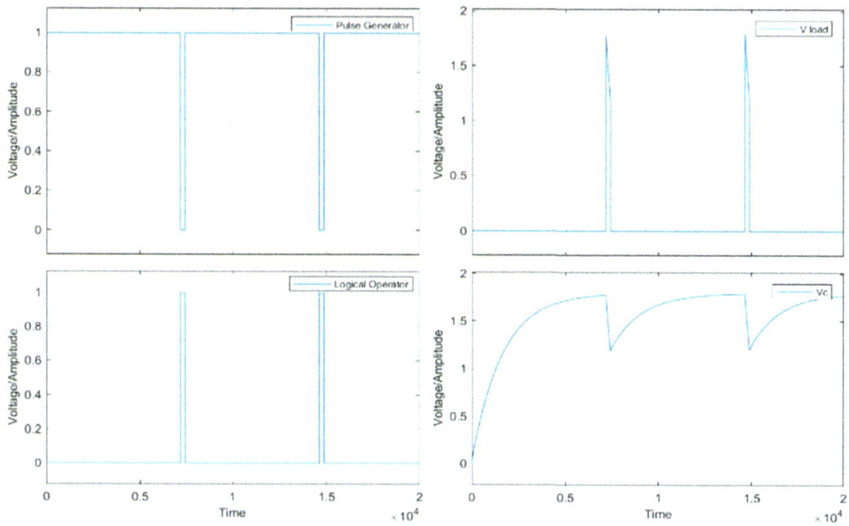

Fig. 8 Scope output graphs; (1) Pulse wave P1 (top left); (2) Pulse wave P2 (bottom left); (3) Voltage curve at load (top right); (4) Super Capacitor charging (bottom right)

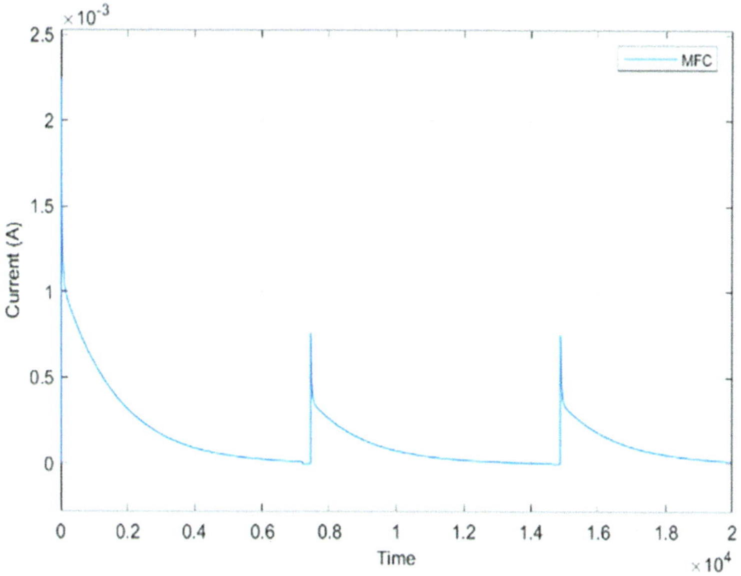

Fig. 9 Current from MFC

4.2 Proposed Circuit Simulation

The system is further linked to a boost converter which is initialized to function in such a manner that it produces a desired output voltage of ~5 V (see Fig. 10). The operational frequency of the boost converter for this specified circuit is set to 50 kHz, and the duty cycle is set at 64%. This setting is initialized through the pulse generator block connected to MOSFET block 3 of the boost converters. The load is set at 5kΩ to produce desired output. The switch capacitor is at ideal condition where its charges and discharges at 1/10 s with switching duty cycle of 50%. As the time sample for the realistic setup is very large, simulation for the complete circuit became tedious. Output scope gave a steady result of 4.909 V across the load (see Fig. 11).

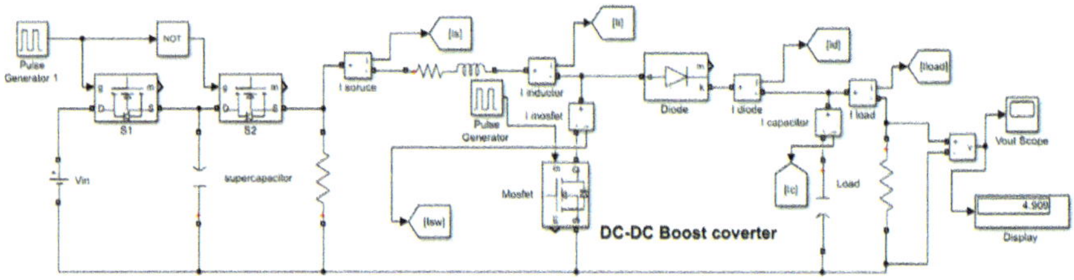

Fig. 10 Proposed circuit Simulation; V_{in} = 1.8v, Super Capacitor = 1F, Load = 5 kΩ

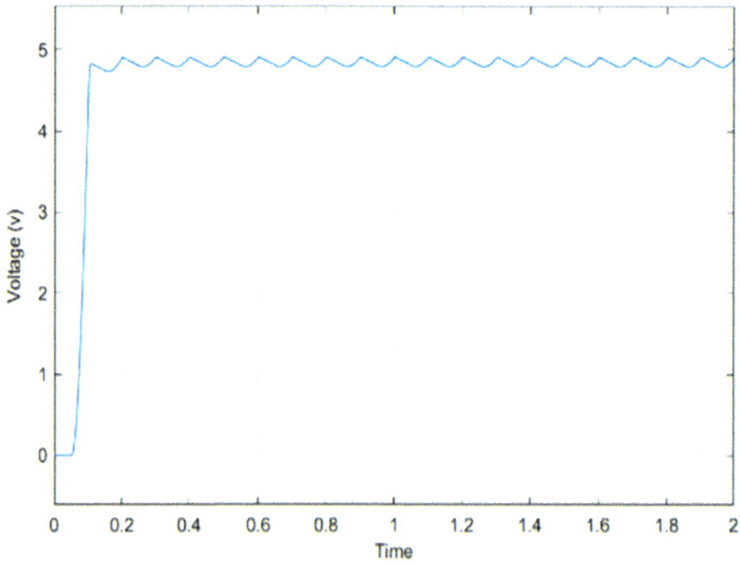

Fig. 11 Output waveform across load

5 Hardware Setup

To verify the theoretical calculation and simulations, the proposed circuit is realized into a working prototype. The setup consists of a supercapacitor of 1F, IRFZ44 MOSFET for switching, and an Arduino microcontroller for generating gate pulse for switches. For the boost converters, there are a few options on the shelf that have been used. These are IC based boost converter circuits readily available in the market. These boost converters' technical specifications are mentioned in Table 1.

ME2108A50 boost converter is pulse frequency modulation or PFM Step-up DC/DC converter IC which has a very low supply current by the CMOS process. A DC/DC up converter can be constructed with lower ripple and high efficiency with the use of ME2108 chip with only three external components. The operating voltage range is 0.9~6.5 V.

PSM-205 is a 5 V USB board that can provide a stable 5 V output from a low input voltage of 0.9 V. This circuit has a maximum efficiency of 96%. It can provide a maximum output current of 600 mA. Provides higher output current.

HT7750 boost converter is also an IC based, pulse frequency modulation step-up converter with a lower ripple and a higher efficiency criterion. A range of output voltages of 2.7 V, 3.0 V, 3.3 V, 3.7 V, or 5.0 V can be achieved with only three external components. Integrated CMOS technology makes sure for an extremely low current supply, making it an ideal choice for applications requiring battery-operated devices from one or more cells. The system employs a very low start-up voltage of 0.6 V, along ultra-low supply current of 4 μA and a low shutdown current of 0.5 μA. This integrated circuit employs pulse frequency modulation, or PFM, to ensure low input current and ripple voltage at low output loading.

For the load, a DHT11 sensor (see Fig. 12) which is a digital temperature and humidity sensor. Calibrated digital exclusive digital-signal-acquisition technique and the temperature and humidity sensor in one. It makes use of a proprietary digital

Table 1 Electrical characteristic comparison for boost converters

Sr. no	Parameter	ME2018A50	PSM 205	HT7750
1	Output voltage V_{out}	4.93 V	5.21 V	5.01 V
2	Current drawn from super cap. I_{in}	27 mA	17 mA	13 mA
3	Current drawn from boost conv. I_{out}	0.55 mA	1.08 mA	0.57 mA
4	Time taken to discharge circuit T	30 s	22 s	86 s
5	Efficiency of boost converter η	85%	96%	85%
6	Operating voltage range	0.8–3.3 V	0.9~5 V	0.6~5 V
7	Max. output current I_{max}	400 mA	600 mA	80 mA
8	Output voltage accuracy	±2.5%	±0.5%	±2.5%
9	Max. oscillator frequency f_{osc}	180 kHz	500 kHz	115 kHz
10	Oscillator duty cycle D_{osc}	84%	N/A	75%

Measuring conditions: At No load, $T = 32$ °C, $V_{in} = 1.8$ V, $I_{in} = 1$ mA

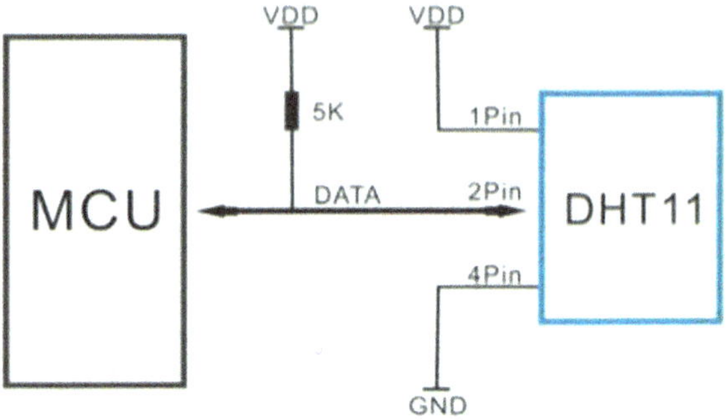

Fig. 12 Pin configuration of DHT11 sensor [17]

module capture tech as well as temperature and humidity sensor tech that is extremely reliable and stable over time. The sensor consists of a resistive element humidity sensor and an NTC temperature sensor, which are connected to a high performance 8-bit microcontroller and provide excellent quality, fast response, uninterruptable capability, and cost-effectiveness.

5.1 *Super Capacitor Charging Setup*

A simple charging set up with 1F super capacitor is in place (see Fig. 13). Since the EHC is designed for MFC, we mimic its power characteristics by using current limiting resistors of 1400 ohms and bench supply of 1.8 V. The current and voltage measurements are monitored with an interval of five minutes, and graph is plotted to be compared with the simulation findings. V_c at $t = 83$ min was 1.72 V which is close to the theoretical calculated voltage for capacitor.

Fig. 13 Supercapacitor charging-testing circuit

Fig. 14 Switching circuit with boost converter

5.2 *Switching Circuit Setup*

The charging and discharging blocks are separated on two bread boards (see Fig. 14), and connected with a bridge that has ground and V_{in} for the discharging block. First the charging for super capacitor is done with a pulse signal for 1,800,000 μs. Then the pulses are changed to activate the discharging circuit, which feeds the super capacitor voltage to the boost converter, which has a constant output voltage of 5 V. The super capacitor can power this circuit for a few seconds before the threshold voltage is reached and the boost converter stops operating.

The pulse can be programmed to charge the capacitor to 1.75 V and with a small continuous load of a white LED, the circuit works for roughly 8 min.

DHT11 temperature sensor is used for final testing and the programmer for sensor is on another microcontroller.

5.3 *Complete Hardware Kit*

Complete hardware kit (see Fig. 15) is fabricated with the flexibility to choose the boost converter to be used for the load. IRFZZ44 MOSFETS are used as digital switches. Comparative analysis is taken for each boost converter to find the best suit for this circuit.

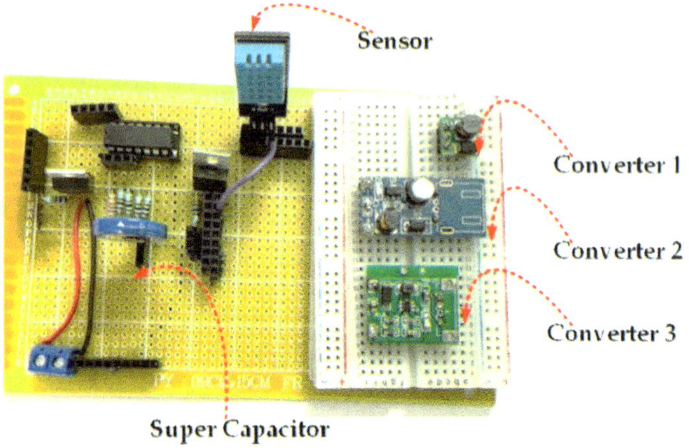

Fig. 15 Hardware kit

6 Results and Discussion

Charging time and characteristics are important to determine the switching duty cycle for the switch capacitor circuit. With the use of bench supply and current limiting resistors, the curve is determined. Figure 16 follows the findings in the simulation model also.

As per theoretical calculation, to achieve the capacitor voltage of $V_c = 1.75$ V, time taken $T_{\text{t} = 83\text{ min}}$. With the setup, at $T_{t = 83\text{ min}}$, the $V_c = 1.72$ V. The voltage/current graph matches the simulation model.

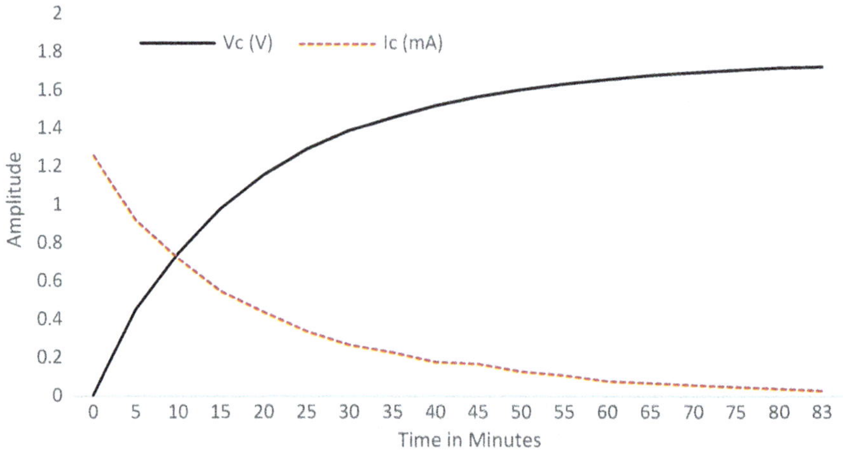

Fig. 16 Charging curve of supercapacitor; $V_{\text{in}} = 1.8$ V

Table 2 No load analysis

Energy harvesting circuit	I_c (mA)	V_{out} (V)	$V_{c\ drop}$ (V)
Converter 1	0.56	4.95	0.5
Converter 2	0.09	5.05	0.15
Converter 3	8.9	5.2	>0.9

Converter 1: ME2108A50, Converter 2: HT7750, Converter 3: PSM 205

For no load analysis, the energy harvesting circuit is divided into 3 different boost converter technologies. The no load analysis is carried out with a charged super capacitor with $V_c = 1.7$ V.

As shown in Table 2, a voltage drop $V_{c\ drop}$, which is a voltage drop that occurred as soon as the super capacitor is connected to the boost converter. Figure 17 shows that no load condition output wave forms were observed with the use of DSO to understand the ripple of the output voltage from the respective boost converters.

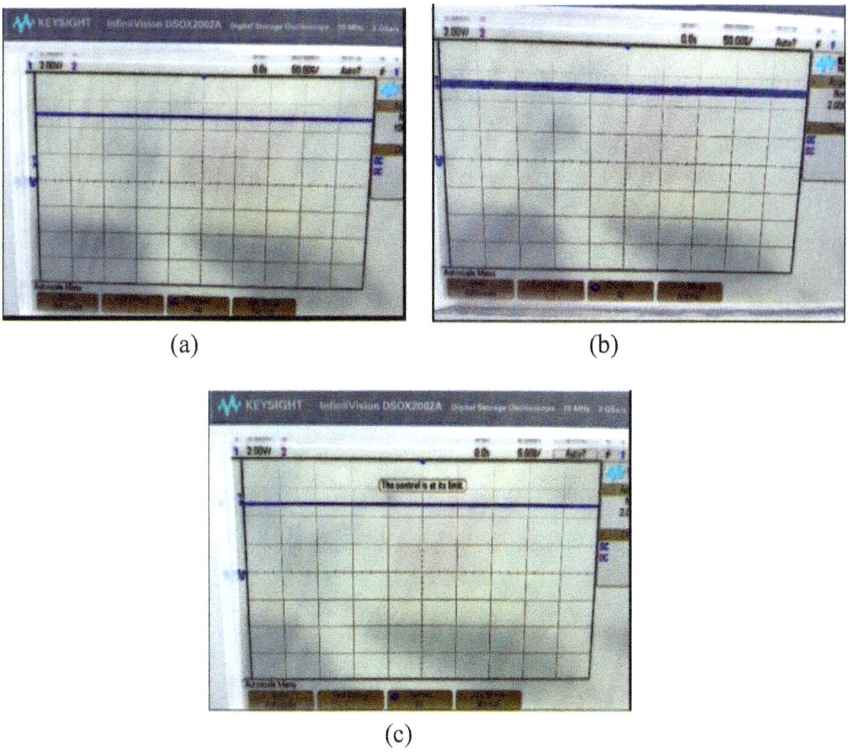

(a) (b) (c)

Fig. 17 DSO readings from the respective converters, labeled (a) is converter 1, (b) is converter 3, (c) is converter 2, refer Table 2

For load analysis, the proposed circuit is tested with the load on the output side. Load used is DHT11 sensor. This sensor provides temperature and humidity values on the serial monitor of the IDE used for the Arduino Microcontroller. Parameters for the analysis are taken, $V_{\text{safety}} = 1$ V, $V_{\text{in}} = V_c$ where V safety is the safety threshold chosen. At this voltage, the super capacitor with stop discharging and the charging circuit will come into action.

As observed through Table 3, converters 1 and 2 perform better than converter 3. Leakage current for circuits 1 and 2 respectively is −0.01 mA. Whereas, for converter 3, there is a standby leakage current for when the sensor is not measuring 2 mA.

Because of this, converter 3 performance is comparatively worse than 1 and 2. Hence no useful output could be measured with the help of converter 3. Discharging of the supercapacitor and the current measurement are plotted to understand and compare with the simulation.

Figure 18 represents the discharge characteristic for the supercapacitor while using converter 1. As the sensor is at measurement state after every 2 s, the voltage drops and bounces back to a reduced level. The trendline for the given values matches the simulation outputs. Each spike represents the measurement cycle for the sensor (see Fig. 19). The sensor on average draws 0.57 mA from the output side of the boost converter.

As a side-by-side comparison, Figs. 20 and 21 depict that, converter 2 performs twice as much better than converter 1. The time of discharge is more than 2 times. The drop in voltage while the measurement cycle is also less in comparison to converter 1. The current drawn from the capacitor when in operation is 48.3% less when compared to circuit 1. This leads to the converter used in circuit 2 to be the most efficient for our application.

Table 3 On-load analysis output readings

Energy harvesting circuit	V_c (V)	V_{out} (V)	V_{safety} (V)	I_c avg (mA)	I_{load} avg (mA)	Power (mW)	Time of discharge (sec)
Converter 1	1.69	4.93	1.06	27.7	0.57	25.5	30.4
Converter 2	1.77	5	1.11	14.32	0.57	13.24	86
Converter 3	1.72	5	1	17	-		12

Converter 1: ME2108A50, Converter 2: HT7750, Converter 3: PSM 205

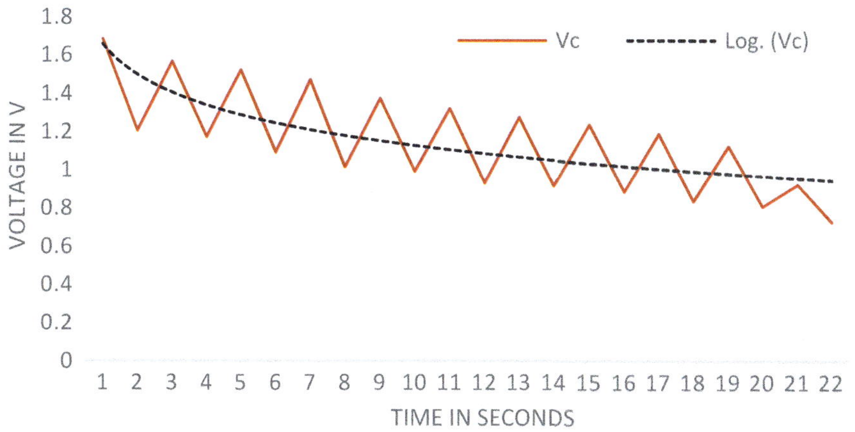

Fig. 18 Discharge curve for converter 1

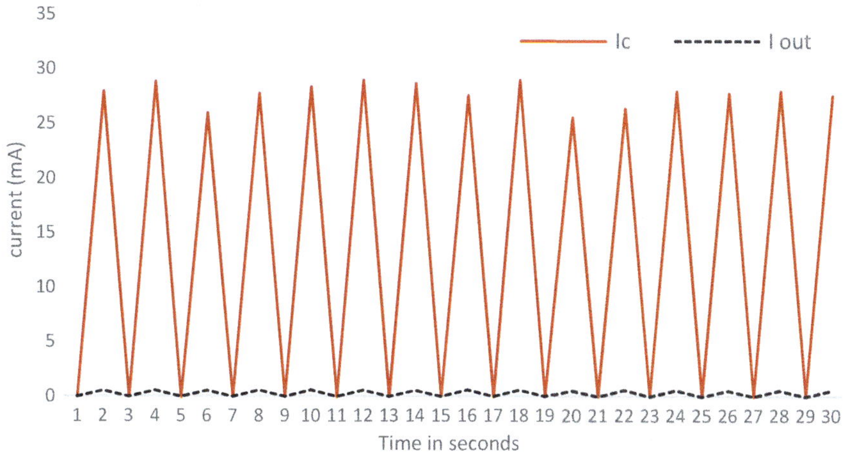

Fig. 19 Current measurement graph for converter 1

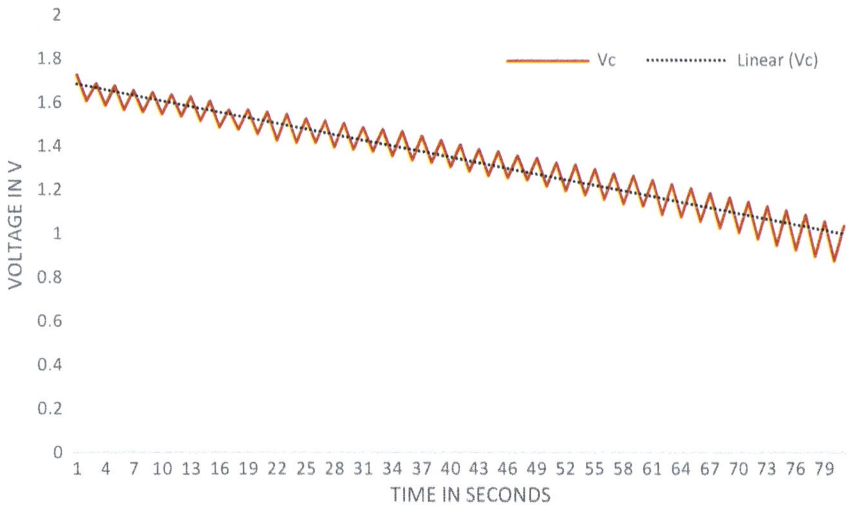

Fig. 20 Discharging graph for converter 2

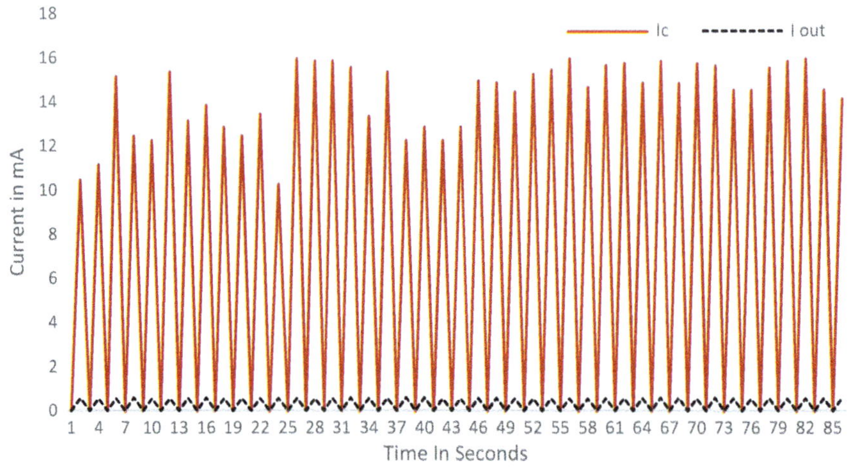

Fig. 21 Current measurement for converter 2

7 Conclusion

Energy harvesting is a crucial part of any system which deals with the extraction of energy from source in an efficient manner. MFC on its own is not useful and the energy if not tapped is wasted. But with the ongoing developments in the low power energy harvesting technology, we are getting close by the day to cracking the code to an efficient system.

Table 4 Summary of previous studies that are based on low power application using MFC

Main electronic component	V_{in}	V_{out}	References
Capacitor, voltage comparator, boost converter	500 mV	3.3 V	[18]
Boost converter (BQ25504)	>1 V	3-5 V	[19]
Capacitor, boost converter (BQ25504)	>1 V	3.6 V	[20]
Transformer, capacitor, boost converter	1.16 V	2.658 V	[6]
Capacitor, voltage comparator, boost converter	0.52 V	3-5 V	[21]
Custom IC	0.25–0.3 V	3.3 V	[22]
Supercapacitor, boost converter (ME2108A50, HT7750, PSM 205)	1.8 V	5 V	This study

Through this analysis, we are able to make a comparative study on the readily available solution for boost converter and with the help of the switch capacitor circuit, we are able to analyze the performance difference in all these circuits. Table 4 compares the results of other studies conducted in this area and the different techniques used to harvest energy from MFC, and convert it into usable and stable energy for low voltage devices.

Although, there are many systems which are available, the feasibility of securing and then testing these systems are sophisticated and given the resources currently available with us, it was not possible. But many more research papers publicly available have used these technologies to very their findings. Nevertheless, our approach for dealing with low power source energy harvesting has opened a gateway for comparative analysis of available systems and with the right harnessing circuitry, we can get promising results.

8 Challenges and Limitations

There are several available step-up converter IC in the existing market, but since these ICs are small outline packages (SOP), they aren't available over the counter and require additional cost (customs duty, taxes) to acquire. Not to mention, since these are SOP based, testing is not possible on breadboard.

Designing a boost converter that works efficiently for low voltage applications is time consuming and complex. Each passive element is to be of precise value based on the math and it's difficult to arrange these elements over the self. Moreover, for a properly functioning boost converter, there needs to be Feedback analysis from the output to efficiently vary its duty cycle. That adds to having more components and more computation power as a dedicated Microcontroller is needed for this operation. To compensate, there are Low power boost converter ICs available in the market which have defined and fixed output voltage with respect to their inputs, but since these ICs are fabricated over Small Outline Packages (SOP), they are extremely difficult to get over the counter and furthermore are difficult to implement over a breadboard setup. Hence, they are not cost effective.

9 Future Scope

MFCs are an exciting new technology that has the potential to significantly contribute to sustainable energy production and energy savings in wastewater treatment. However, the low power and voltage output of a single MFC are insufficient to meet the demands of actual devices. While numerous MFCs can be connected directly in series, the overall voltage of these cascaded MFCs cannot be increased sustainably due to voltage reversal, which reduces the stack voltage.

With more research in energy harvesting from low power sources and efficient design for these circuits, the energy race can see its next revolution. Entering the smart age of IoT and Industry 4.0, nodes-based systems will become second nature and power requirements can be met with a self-sustaining resource which is environmentally friendly.

References

1. Khaled F, Ondel O, Allard B, Degrenne N (2013) Voltage balancing circuit for energy harvesting from a stack of serially-connected microbial fuel cells. In: Proceedings of the 2013 IEEE ECCE Asia Downunder (2013), pp 392–397. https://doi.org/10.1109/ECCE-Asia.2013.6579126
2. Kim T, Yeo J, Yang Y, Kang S, Paek Y, Kwon JK, Jang JK (2019) Boosting voltage without electrochemical degradation using energy harvesting circuits and power management system-coupled multiple microbial fuel cells. J Power Sources 410–411:171–178
3. Ma F, Fu C, Rong X, Wang H (2017) Advanced energy harvesting in microbial fuel cells for environmental monitoring. Chin Autom Congress (CAC) 2017:7357–7361. https://doi.org/10.1109/CAC.2017.8244107
4. Nguyen C-L, Tartakovsky B, Woodward L (2019) Harvesting energy from multiple microbial fuel cells with a high-conversion efficiency power management system. ACS Omega 40. https://doi.org/10.1021/acsomega.9b01854
5. Tungpimolrut K, Hatti N, Phontip J, Komoljindakul K, Pechrach K, Manooonpong P (2011) Design of energy harvester circuit for a MFC piezoelectric based on electrical circuit modeling. In: 2011 international symposium on applications of ferroelectrics (ISAF/PFM) and 2011 international symposium on piezoresponse force microscopy and nanoscale phenomena in polar materials, pp 1–4. https://doi.org/10.1109/ISAF.2011.6014136
6. Prasad J, Tripathi R (2019) Energy harvesting from sediment microbial fuel cell to supply uninterruptible regulated power for small devices. Int J Energy Res 43:2821–2831. https://doi.org/10.1002/er.4370
7. Mukherjee A, Patel V, Shah MT, Jadhav DA, Munshi NS, Chendake AD, Pant D (2022) Effective power management system in stacked microbial fuel cells for onsite applications. J Power Sources 517
8. Santoro C, Arbizzani C, Erable B, Ieropoulos I (2017) Microbial fuel cells: from fundamentals to applications. A review. J Power Sources 356:225–244
9. Drew J (2015) Energy harvester produces power from local environment, eliminating batteries in wireless sensors. https://doi.org/10.1016/B978-0-12-800001-4.00226-X

10. Meehan A, Gao H, Lewandowski Z (2009) Energy harvest with microbial fuel cell and power management system. In: IEEE energy conversion congress and exposition
11. Shabani F, Philamore H, Matsuno F (2021) An energy-autonomous chemical oxygen demand sensor using a microbial fuel cell and embedded machine learning. IEEE Access 9:108689–108701. https://doi.org/10.1109/ACCESS.2021.3101496
12. Dutta A, Jacob CA, Das P, Corton E, Stom D, Barbora L, Goswami P (2022) A review on power management systems: an electronic tool to enable microbial fuel cells for powering range of electronic appliances. J Power Sources 517
13. Koffi NJ, Okabe S (2022) High voltage generation from wastewater by microbial fuel cells equipped with a newly designed low voltage booster multiplier (LVBM). Sci Rep 10
14. Nguyen CL, Tartakovsky B, Woodward L (2019) Harvesting energy from multiple microbial fuel cells with a high-conversion efficiency power management system. ACS Omega 4:18978–18986
15. Divya Navamani J, Mrinal R, Nair A, Chouhan S (2021) Extendable PSLC network-based high-gain DC–DC converter with switched-capacitor circuit for microbial fuel cell. J Circ Syst Comput 30(13):2150239
16. Harres D (2013) Chapter 20—creating high voltage. In: Harres D (ed) MSP430-based robot applications. Newnes, pp 299–312. ISBN 9780123970121. https://doi.org/10.1016/B978-0-12-397012-1.00020-5
17. Zheng Q, Xiong L, Mo B, Lu W, Kim S, Wang Z (2015) Temperature and humidity sensor powered by an individual microbial fuel cell in a power management system. Sensors (Basel) 15(9):23126–23144. https://doi.org/10.3390/s150923126.PMID:26378546;PMCID:PMC4610421
18. Donovan C, Dewan A, Heo D, Lewandowski Z, Beyenal H (2013) Sediment microbial fuel cell powering a submersible ultrasonic receiver: new approach to remote monitoring. J Power Sources 233:79–85. ISSN 0378-7753. https://doi.org/10.1016/j.jpowsour.2012.12.112
19. Ge Z, Wu L, Zhang F, He Z (2015) Energy extraction from a large-scale microbial fuel cell system treating municipal wastewater. J Power Sources 297:260–264. ISSN 0378-7753. https://doi.org/10.1016/j.jpowsour.2015.07.105
20. Khaled F, Ondel O, Allard B (2016) Microbial fuel cells as power supply of a lowpower temperature sensor. J Power Sources 306:354–360. ISSN 0378-7753. https://doi.org/10.1016/j.jpowsour.2015.12.040
21. Park J-D, Ren Z (2012) Hysteresis controller based maximum power point tracking energy harvesting system for microbial fuel cells. J Power Sources 205:151–156. https://doi.org/10.1016/j.jpowsour.2012.01.053
22. Yamashita T, Hayashi T, Iwasaki H, Awatsu M, Yokoyama H (2019) Ultra-low-power energy harvester for microbial fuel cells and its application to environmental sensing and long-range wireless data transmission. J Power Sources 430:1–11. ISSN 0378-7753. https://doi.org/10.1016/j.jpowsour.2019.04.120

Comparison of SMES and RFB Performance in Combined Voltage and Frequency Regulation of Multi Source Power System

M. Nagendra, Ch. Naga Sai Kalyan, and R. Srinivasa Rao

Abstract The analysis comported in this paper has been directed towards a novel control strategy in achieving the simultaneous regulation of voltage and frequency of a two-area multi-source power system (TAMSPS). The black widow optimization technique (BWOT) based integral-double derivative plus filter (IDDF) controller is suggested for combined voltage and frequency control. The investigation on TAMSPS is carried out by injecting the disturbance in a step load of 10% (SLD) in area-1. However, the efficacy of BWOT-based IDDF is validated with other controllers. Further, the investigation is stretched out with the integration of energy storage devices (ESDs) like redox flow batteries (RFB) and superconducting magnetic energy storage (SMES) with TAMSPS in both areas. Simulation results reveal the efficacy of RFB over the SMES device in attaining better dynamical behavior of the TAMSPS system under disturbed conditions.

Keywords Interconnected power system stability · Black widow optimization technique · IDDF controller · 10%SLD · Energy storage devices

1 Introduction

The IPS network in the modern-day environment is accommodated with various control areas for delivering electricity supply to the customer end. However, the electric supply delivering to the customers must deal with quality and reliability. The quality of the electric power is evaluated based on the indicators of frequency and

M. Nagendra (✉) · R. S. Rao
EEE Department, JNTUK College of Engineering, Kakinada 533 003, India
e-mail: nagendra.muppiri@becbapatla.ac.in

R. S. Rao
e-mail: drsrinivas@jntucek.ac.in

Ch. N. S. Kalyan
EEE Department, Vasireddy Venkatadri Institute of Technology, Guntur 522 508, India
e-mail: kalyanchallapalli@vvit.net

S. Kumar et al. (eds.), *Control Applications in Modern Power Systems*, Lecture Notes in Electrical Engineering 1128, https://doi.org/10.1007/978-981-99-9054-2_6

voltage. The electric supply with the fluctuating frequency might alter the efficiency and life expectancy of the IPS apparatus, measuring devices as well as the electrical apparatus of the consumers. Thus, appropriate control measures have to be adopted in delivering quality electric power to the customers. The frequency of the control areas in IPS can be regulated by regulating the real power (RPI) by performing the automatic control action of load frequency control (LFC). Whereas the control of terminal voltage is associated with the flow of reactive power in the IPS and that can be regulated with automatic voltage regulator (AVR) action [1].

The literature report demonstrates that considerable quantities of control mechanisms are available on LFC and AVR studies independently. Most of the control mechanisms in individual LFC and AVR studies are focused on the development of conventional PI/PID controllers [2]. While some of the research works addressed the development of fuzzy logic controllers (FLC) and these FLC are not extensively dragged the attention of the research community, especially for the stability of IPS. Implementing FLC in a realistic environment is hard and skilled technical operators are needed [3]. Further, the FLC includes several appropriations in various design stages that might affect the IPS performance. However, the efficacious performance of the above-discussed FLC-aided regulators and conventional PI/PID controllers are lean on their parameters.

Literature implemented various soft computing algorithms like donkey and smuggle optimization (DSO), backtracking search algorithm (BSA) [4], atom search algorithm (ASA), krill herd optimizer (KHO), artificial electric field (AEFA) algorithm, wind driven optimization (WDO) [5], artificial flora algorithm (AFA), gravitational search algorithm (GSA) [6], path finder algorithm (PFA), symbiotic organisms search algorithm (SOS) [7], African vulture optimization (AVO) algorithm, seagull optimization algorithm (SOA), imperialist competitive algorithm (ICA) [8], wild horse algorithm (WHA), grasshopper optimization algorithm (GOA) [9], dragon fly algorithm (DFA), crow search technique (CST) [10], bull lion optimization (BLO), slap swarm algorithm (SSA), bibliography algorithm (BBA), stochastic fractal search algorithm (SFSA), differential evolution (DE) [11], bat inspired algorithm (BIA), fruit fly algorithm (FFA), chaotic black widow optimization (CBWO), spider optimization technique (SOT), Archimedes optimization approach (AOA), water cycle algorithm (WCA) [12] etc. for regulators optimal design.

However, most of the above-discussed controllers and the soft computing algorithms are tested on numerous IPD networks for LFC analysis and the impact of AVR in combined effect is not considered. Though limited research is available on combined LFC and AVR (CLFC-AVR) analysis and the most significant are discussed here. For isolated power systems control approaches like hybrid flower pollination and PFA-based adaptive PID [13], improved PSO-tuned power system stabilizer [14], FLC-aided PID [15], and neural network controllers are adopted for single-area IPS. Further, it is stretched out to multi-area conventional IPS networks with and without considering renewable integration such as wind, geothermal, solar photovoltaic, and solar thermal plants. Control methodologies adapted to CLFC-AVR analysis of multi-area IPS network includes the PID optimized with simulated annealing [16], hybrid

DE-AEFA [17], firefly algorithm [18], moth flame optimizer, etc. are reported. Moreover, the FLC-aided PI/PID [19] and cascade FLC and fractional order type regulators based on AFA are available in the literature.

The ESDs integration with the IPS network for stability enhancement is gaining momentum these days. The secondary controller of the IPS network is only capable enough in handling the system dynamical behavior under small perturbations [20]. The regulators alone can't handle the IPS network stability when large perturbations emerge within the system. Hence additional control mechanisms are necessary to assuring grid stability while the emergence of large load disturbances. Integration of ESD [21] with the IPS acts as the better spinning reserves so that the RPI in the IPS can be regulated continuously. The ESDs get charged up when the generation is more than the load demand. The stored energy can be sent back to the grid instantly during peak durations in order to balance the RPI and hence the stability of the IPS is ensured effectively. A literature survey disclosed the integration of various ESDs like flywheel energy storage, SMES [22], ultra-capacitors, battery energy storage [23], RFBs [24], etc. with several IPS models. This paper analyses the performance efficacy of the SMES and RFBs that can be best suitable for obtaining the CLFC-AVR.

The contributions of this paper are

(a) Developed the TAMSPS for combined frequency and voltage regulation in MATLAB/SIMULINK.
(b) Presented the BWOT-based IDDF for CLFC-AVR analysis which is the primary attempt.
(c) Necessity of considering AVR coupling for IPS dynamical analysis is demonstrated.
(d) Investigated the performance efficacy of ESDs like SMES and RFBs in CLFC-AVR study.

2 Power System Model

The investigation compassed in this current work is to achieve and stable the IPS network frequency and voltage during the load perturbations. For the investigative analysis, the TAMSPS shown in Fig. 1 is considered an IPS network that comprises two unique generation areas. Area-1 and area-2 have hydro-thermal-nuclear plants as generation units and with the intention of performing the analysis near to reality the thermal-hydro units are considered with generation rate constraints. To attain control over the voltage and frequency simultaneously the AVR shown in Fig. 2 is coupled with the multi-source system using the coefficients of K_1, K_2, K_3, and K_4. However, the control actions that are performed in LFC to regulate the RPM there by the control area frequency, might affect the performance of the AVR. Similarly, the actions undergone in AVR while changing the generator filed excitation in order to regulate the terminal voltage might influence the LFC performance. Thus, the two power full indicators in the IPS network that is the voltage and frequency are to be controlled in combined as given in Eq. (1).

$$P_g = \frac{|V||E|}{X_s}\sin\delta \tag{1}$$

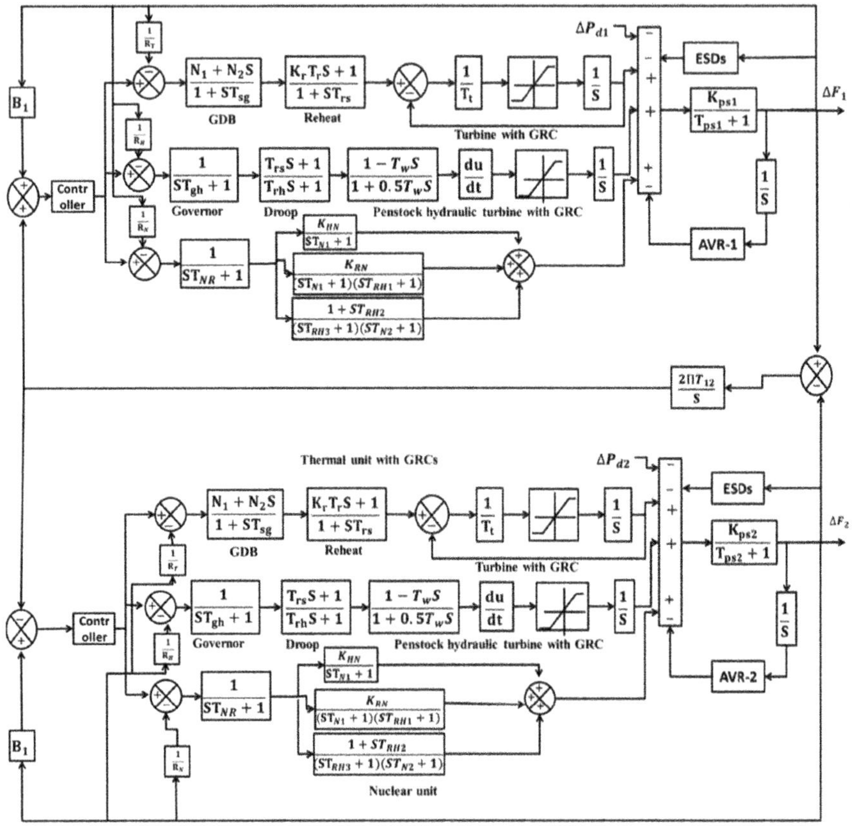

Fig. 1 Transfer function model of TAMSPS under investigation

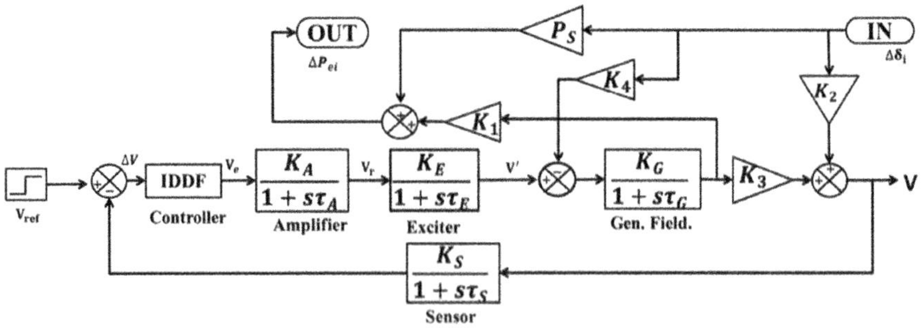

Fig. 2 Automatic voltage regulator with coupling coefficients

where the parameters 'P_g' represents the real power generation, 'E' indicates the voltage induced in the generator windings, 'V' is the terminal voltage, 'δ' is the load angle and 'X_s' is the synchronous reactance.

3 IDDF Controller

IDDF is the modified version of traditional controllers like PID. Moreover, the design of IDDF is as simple as that of PID and its implementation to realistic practice is easy. Further, it exhibits performance superiority with the DD parameter in achieving better dampening over the peak deviations. The structure of IDDF is shown in Fig. 3 and K_I, K_{DD}, and N are its parameters. The parameter 'N' facilitates the elimination of the disturbances in the reference control signal so that the chance of attaining the system responses to the stable position is enhanced. However, the efficacious of IDDF is depending up on the K_I, K_{DD}, and N parameters that need to be found using global optimization algorithms. The IDDF parameters optimization is performed in this work with regard to the integral square error (ISE) given in Eq. (2).

$$J_{\text{ISE}} = \int_0^{T_{\text{Sim}}} (\Delta f_1^2 + \Delta P_{\text{tie12}}^2 + \Delta f_2^2)dt \tag{2}$$

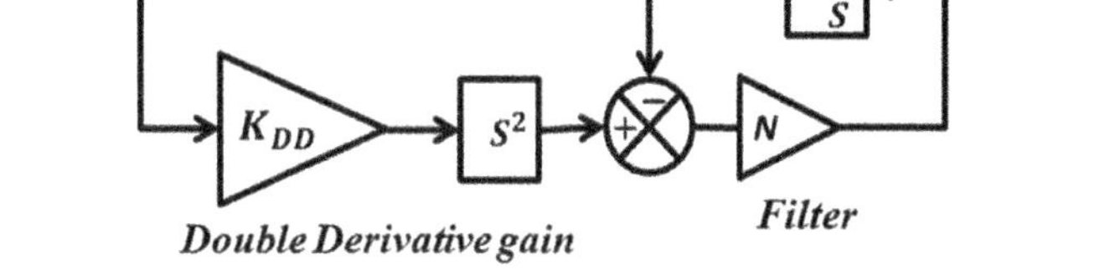

Fig. 3 Structure of IDDF controller

4 Black Widow Optimization Technique

The BWOT is the recently emerged population-based algorithm that mimics the socio-behavior of black widow spiders. The black widows usually can't sight during the daytime and at night they can spin their web. The BWOT incorporates several stages in the socio-behavior of the black widow spiders like reproduction and cannibalism. The female black widows made a special spot on the web to attract the male widows for mating. After mating the female widow lays eggs in the egg sac and hatches it for 8 or 11 days. The female widow might cannibalize the male widow after or before the mating. The female widow after hatching the web might cannibalize the baby widows. The baby widows are having the nature in cannibalizing the mother widow.

Like every optimization algorithm, the BWOT also starts with the initialization of the black widow spiders and each widow acts as the possible solution. In a problem of N-variables, the populations termed as widows (W) are initialized as given in Eq. (3) as an array.

$$W = [X_1, X_2, \ldots X_{N\text{var}}] \tag{3}$$

However, the position of each widow is a floating point and for each widow position the fitness values are evaluated as given in Eq. (4).

$$f(W) = f(X_1, X_2, \ldots X_{N\text{var}}) \tag{4}$$

The creation of off-springs (OSP) in BWOT is given in (5), as the widow pairs are independent and the mating procedure is done in each web. In realistic practice, approximately 1000 eggs are laid by the female widow after each mating and only the strongest baby widows would get survived.

$$\begin{aligned} \text{OSP}_1 &= \alpha * X_1 + (1-\alpha) * X_2 \\ \text{OSP}_2 &= \alpha * X_2 + (1-\alpha) * X_1 \end{aligned} \tag{5}$$

Where alpha is the random number and this process continues till the number of variables reaches half. Later, the process of cannibalism will begin and cannibalism in the socio-behavior of black widow spiders is of three kinds. The first one is female widow cannibalizing the male while mating and the second one is sibling cannibalism and the third one is baby widows cannibalizing the mother widow. After cannibalism, the mutation process would begin and the generation of new solutions is done by randomly mutating the chromosomes. After mutating, the new population can emerge and the update process will began and finally the best solutions are displayed. In this work, the BWOT is designed in Intel Core-i3 M370@2.4 GHz system MATLAB (R2017b) (.m file) for 100 trails with an initial population of 100. The detailed mathematical modeling of the BWOT is provided in [25], and the flow chart is displayed in Fig. 4.

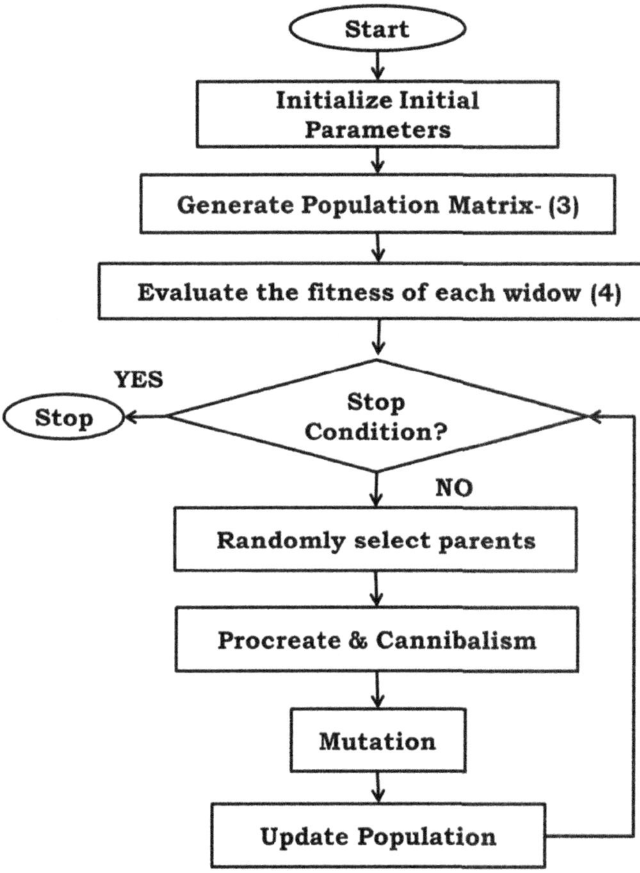

Fig. 4 BWOT flow chart

5 Energy Storage Devices

5.1 SMES

Considering the economic aspects and performance superiority the SMES devices will always be in the primary consideration for the electricity utility engineers as well as for the power system operators. The superconductivity and the zero loss of stored energy in the SMES devices made them a top priority consideration for IPS integrations. In SMES the energy is stored in DC form as the DC coil is coupled to the magnet and placed in cryogenic temperature. The SMES are always operated with the power conditioning units, and the energy can be inverted before injecting into the grid. The modeling of SMES is presented in Eq. (6), and the K_{SMES} and T_{SMES} are the gain and time constants that are considered as 0.967 and 00.713 s respectively in this work.

$$G_{\mathrm{SMES}} = \frac{K_{\mathrm{SMES}}}{1 + sT_{\mathrm{SMES}}} \tag{6}$$

5.2 RFBS

On the other hand, RFBs are also widely accepted ESDs for improving the performance of IPS networks. However, the working of the RFBs is based on oxidation and reduction chemical reactions. Compared to other ESDs, the RFB has test cycle time and noise-free operation. The modeling of RFBs is presented in [22] and is given in Eq. (7). The KRFB and TRFB are the gain and time constants and in this work, they are taken as 0.896 and 0.780.783 s respectively.

$$G_{\mathrm{RFBs}} = \frac{K_{\mathrm{RFB}}}{1 + sT_{\mathrm{RFB}}} \tag{7}$$

6 Simulation Results

6.1 Case-I: Analysis of TAMSPS Under Different Controllers

The dynamic nature of TAMSPS for 10%SLD in area-1 is analyzed under various BWOT regulators like IDDF, 2DOFPID, and TIDF individually in LFC-AVR control loops. The TAMSPS responses under different controllers are shown in Fig. 5 and are interpolated with reference to the settling time in seconds noted in Table 1. Noticing Table 1 and observing the TAMSPS dynamic behavior in Fig. 5, it is primarily concluded that the deviations are quickly damped out and quenched effectively under the BWOT-based IDDF controller. Moreover, the response's peak deviations are by far mitigated, and the ISE objective index is minimized effectively and is enhanced by 55.67% and 70.71% respectively with TIDF and 2DOFPID controllers. The optimal parameters of 2DOFPID, TIDF, and IDDF found with BWOT for TAMSPS are placed in Table 2.

Fig. 5 Case-I responses

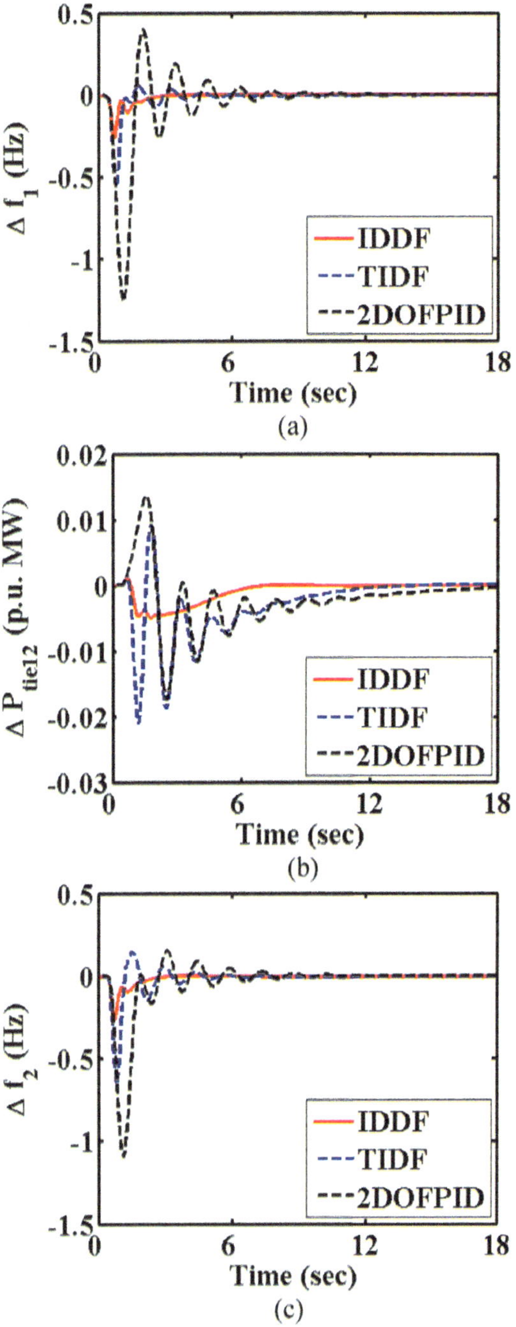

Fig. 5 (continued)

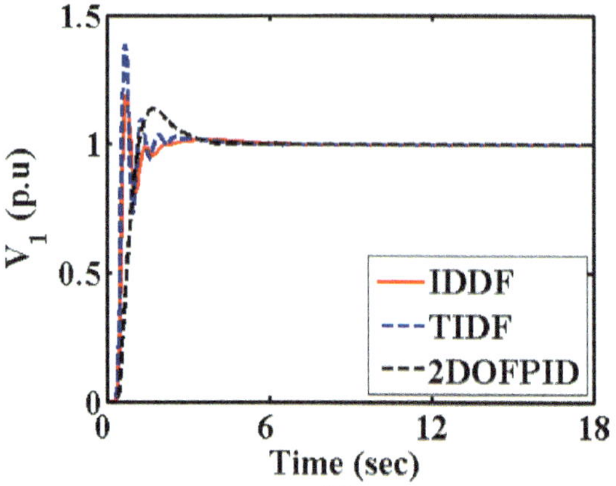

(d)

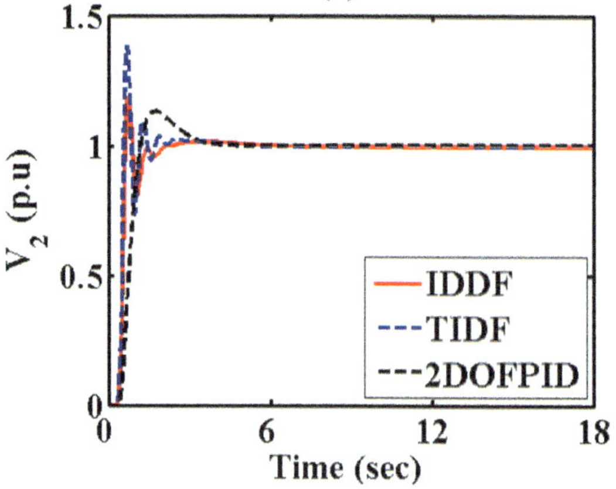

(e)

Table 1 Responses settling time

Settling time (s)	Δf_1	ΔP_{tie12}	Δf_2	V_1	V_2	ISE*10^{-3}
2DOFPID	11.74	18.75	10.97	6.58	6.45	127.34
TIDF	8.35	14.45	7.94	5.42	5.36	84.128
IDDF	5.77	9.66	5.67	4.54	4.25	37.286

Table 2 Controller optimal parameters

Parameters		Area-1			Area-2		
		IDDF	TIDF	2DOFPID	IDDF	TIDF	2DOFPID
LFC	K_P/K_T	–	1.165	0.893	–	1.211	1.217
	K_I	1.357	0.825	0.093	1.216	1.082	0.092
	K_D	–	0.367	0.167	–	0.764	0.653
	K_{DD}	0.893	–	–	0.672	–	–
	N	245.36	224.26	241.19	241.28	219.64	247.25
	K_1	–	–	0.014	–	–	0.043
	K_2	–	–	0.105	–	–	0.124
AVR	K_P/K_T	–	0.956	0.953	–	1.181	1.316
	K_I	0.827	0.321	0.116	0.763	1.167	0.166
	K_D	–	0.672	0.216	–	0.635	0.736
	K_{DD}	0.336	–	–	0.227	–	–
	N	212.45	219.35	239.97	231.67	211.24	228.36
	K_1	–	–	0.084	–	–	0.035
	K_2	–	–	0.093	–	–	0.103

6.2 Case-II: TAMSPS Behavior Under the Conditions of with and Without AVR

In contemplation of demonstrating the AVR on the LFC performance of TAMSPS, the investigation is needed to be undergone for the instances of taking and not taking the AVR coupling with the LFC. Analysis from the aforementioned case revealed that the IDDF is supreme in dealing with the stability of considered TAMSPS under load disturbances. Hence further analysis on TAMSPS is stretched out under the control of BWOT tuned IDDF regulator. In order to obtain the most comparative analysis, the TAMSPS under BWOT tuned IDDF is analyzed for 10%SLD in area-1 for the instances of without and with taking AVR coupling and the corresponding dynamic behavior are rendered in Fig. 6. Noticing Fig. 6, it is revealed that the TAMSPS dynamic behavior is more disturbed while considering the AVR coupling. It means, with AVR consideration the stability of the TAMSPS is assessed by considering the deviations in respective area frequency, tie-line power, and terminal voltage. Up on addressing the voltage and frequency of the TAMSPS in a collective manner, a compromise might have been taken place in altering the power plant operating point and changing the generator excitation. Moreover, the regulation of terminal voltage by the AVR in the TAMSPS had a considerable impact on regulating the real power and vice-versa as explained in Eq. (1). Hence, by addressing the voltage and frequency of IPS in a combined and simultaneous manner definitely, there will be an impact of one parameter in regulating the other parameter. Even though the LFC performance of TAMSPS is more fluctuated with considering the AVR, this work recommends considering stability analysis because in realistic practice the generation units are coupled with excitation systems for controlling the generator field excitation.

6.3 Case-III: Analysis of TAMSPS Under the Performance of SMES and RFBs

The ESDs like RFBs and SMES are integrated with the TAMSPS in both areas individually in order to further enhance the system dynamic behavior during load perturbations. The responses of TAMSPS under BWOT tuned IDDF controller with the ESDs are shown in Fig. 7. The responses settling time are indicated as a bar chart in Fig. 8 for easy identification of the ESDs with better performance. Noticing Figs. 7 and 8 it is concluded that the better enhancement in TAMSPS dynamic behavior is reported with integrating the SMES when compared to that of the RFBs. The capability of bulk storage and the quick response nature of SMES helps the TAMSPS dynamic behavior in attaining a stable position.

Fig. 6 Case-II responses

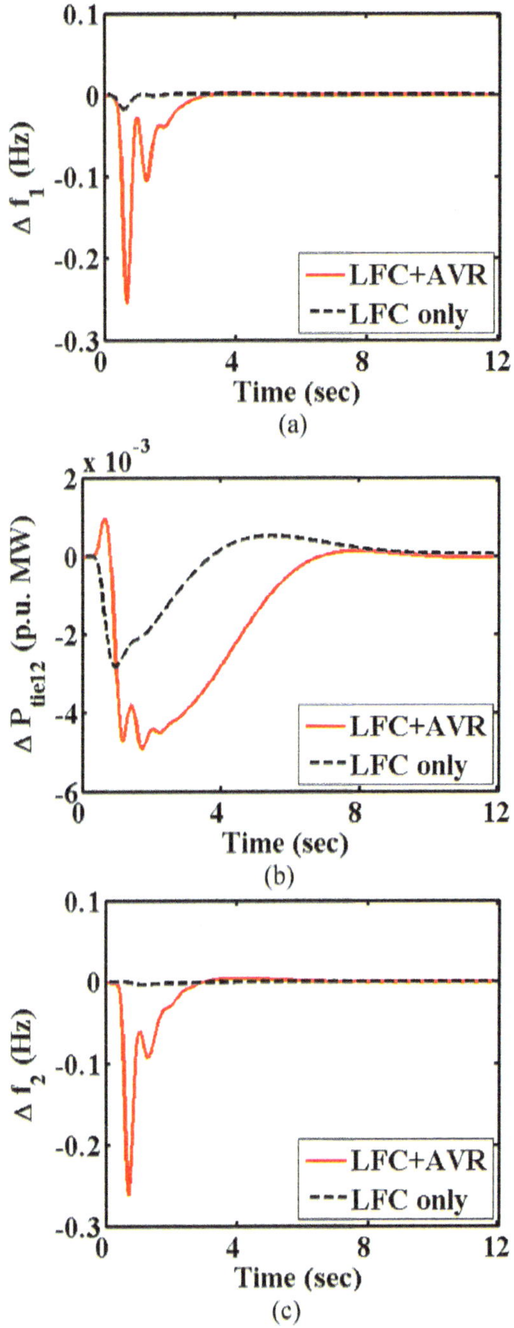

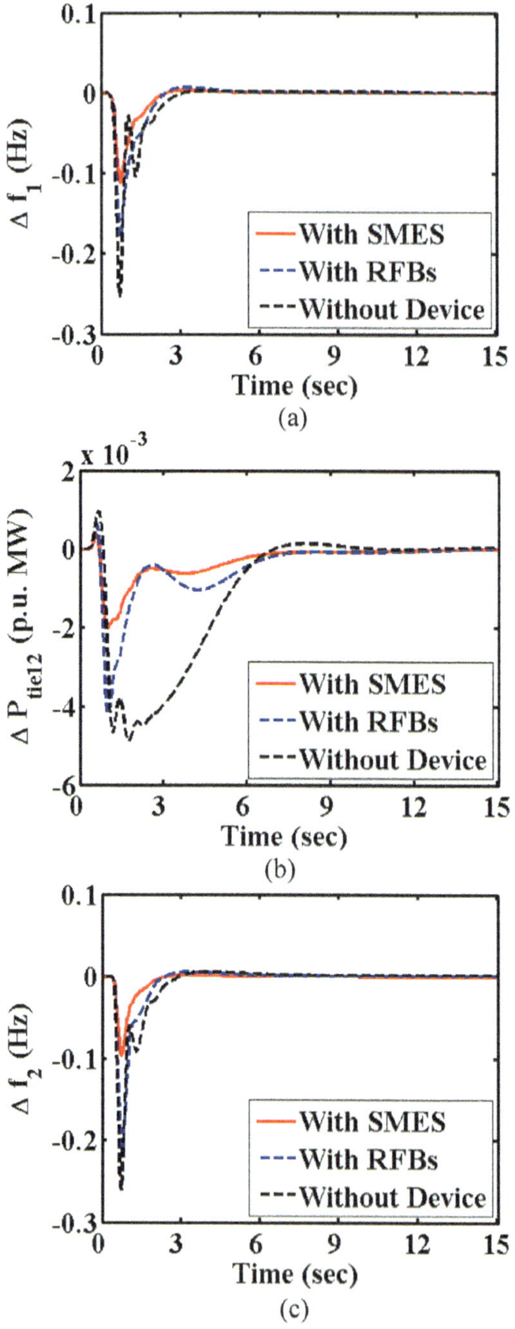

Fig. 7 Case-III responses

Fig. 7 (continued)

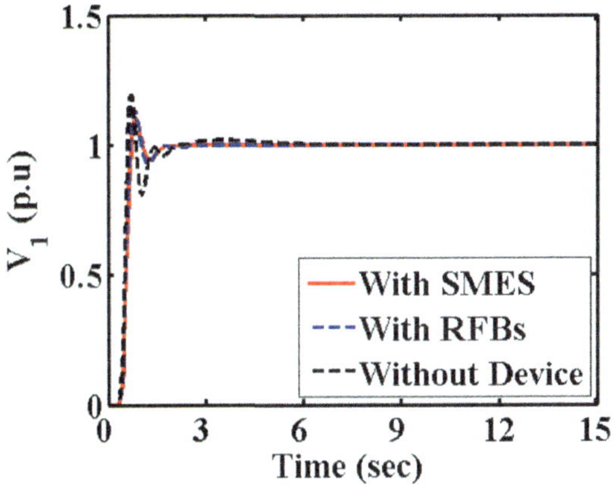

(d)

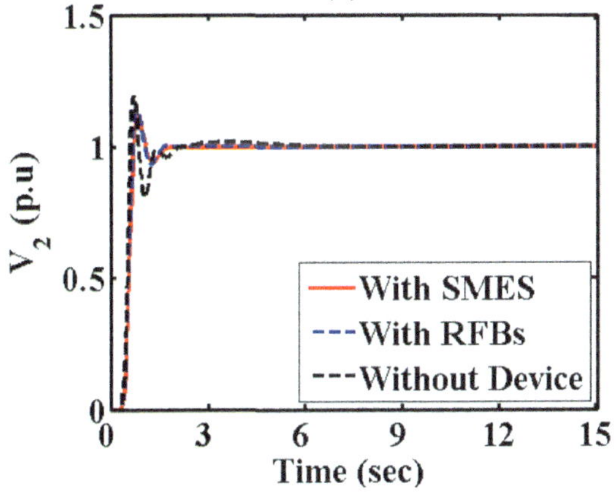

(e)

7 Conclusion

This work introduced the BWOT-based IDDF for the simultaneous control of voltage and frequency in the TAMSPS network. Initially, the analysis is performed subjected to the perturbation of 10%SLD in area-1 under the individual supervision of IDDF, TIDF, and 2DOFPID. The dynamic behavior of TAMSPS exhibits more efficacy in attaining simultaneous control over the voltage and frequency under the IDDF controller. Further, the analysis is performed on TAMSPS under the IDDF controller with the 10%SLD in area-1 for the instances of deliberating and omitting the AVR coupling. Analysis revealed the significance of AVR coupling on the IPS performance and is justified. Furthermore, the analysis of TAMSPS is extended with the integration of SMES and RFBs in area-1 and area-2 one after the other. The analysis reported the prominent enhancement in system frequency and voltage with the integration of RFB devices.

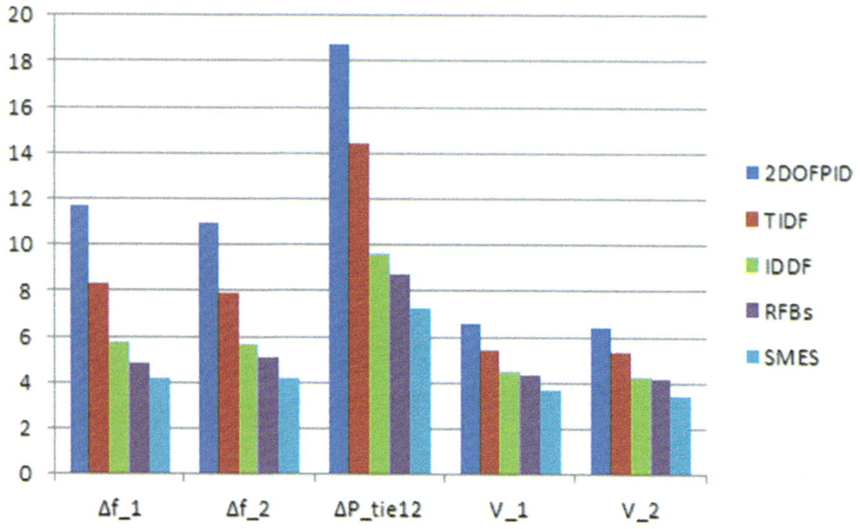

Fig. 8 Bar chart indicating the responses settling time

References

1. Kalyan CNS, Rao GS (2020) Frequency and voltage stabilization in combined load frequency control and automatic voltage regulation of multiarea system with hybrid generation utilities by AC/DC links. Int J Sustain Energ 39(10):1009–1029
2. Bai J, Zhao Y, Jiang H, Wei M, Yu S (2022) Load frequency control of power system with energy storage based on disturbance observer. Energy Rep 8(8):615–622
3. Oshnoei S, Oshnoei A, Mosllanejad A, Haghjoo F (2021) Novel load frequency control scheme for an interconnected two-area power system including wind turbine generation and redox flow battery. Electr Power Energy Syst 130:107033
4. Guha D, Roy PK, Banerjee S (2018) Application of backtracking search algorithm in load frequency control of multi-area interconnected power system. Ain Shams Eng J 9(2):257–276
5. Alhelou HH, Golshan MH, Fini MH (2018) Wind driven optimization algorithm application to load frequency control in interconnected power systems considering GRC and GDB nonlinearities. Electric Power Components Syst 46(11):1223–1238
6. Sahu RK, Panda S, Padhan S (2014) Optimal gravitational search algorithm, for automatic generation control of interconnected power systems. Ain Shams Eng J 5(3):721–733
7. Singh SP, Prakash T, Singh VP (2020) Coordinated tuning of controller parameters using symbiotic organisms search algorithm for frequency regulation of multi-area wind integrated power system. Eng Sci Technol 23(1):240–252
8. Zaheeruddin, Singh K (2022) Load frequency regulation by d-loaded tidal turbine power plant units using fractional fuzzy based PID droop controller. Appl Soft Comput 92:106338
9. Nosratabadi SM, Bornapour M, Gharaei MA (2019) Grasshopper optimization algorithm for optimal load frequency control considering predictive functional modified PID controller in restructured multi-resource multi-area power system with redox flow battery units. Control Eng Pract 89:204–227
10. Nayak JR, Shaw B, Sahu BK, Naidu KA (2022) Application of optimized adaptive crow search algorithm based two degree of freedom optimal fuzzy PID controller for AGC system. Eng Sci Technol 32:101061

11. Kalyan CNS, Rao GS (2020) Coordinated SMES and TCSC damping controller for load frequency control of multi area power system with diverse sources. Int J Electr Eng Inform 12(4):747–769
12. Kalyan CNS, Goud BS, Bajaj M, Kumar MK, Ahmed EM, Kamel S (2022) Water cycle algorithm tuned intelligent fuzzy controller for stability of multi-area multi-fuel power system with time delays. Mathematics 10(3):508
13. Oladipo S, Sun Y, Wang Z (2021) An effective hFPAPFA for a PIDA-based hybrid loop of load frequency and terminal voltage regulation system. IEEE PES/IAS Power Africa, Kenya, 23–27 Aug 2021
14. Morsali J, Esmaelli Z (2022) Proposing a new hybrid model for LFC and AVR loops to improve effectively frequency stability using coordinative CPSS. In: 28th Iranian conference on electrical engineering (ICEE), Tabriz, 04–06 Aug 2020
15. Sharma D, Kushwaha V, Pandey K, Rani N (2018) Intelligent AVR control of a single thermal area combined with LFC loop. Intell Commun Control Dev 624:779–789
16. Chandrakala KRMV, Balamurugan S (2016) Simulated annealing based optimal frequency and terminal voltage control of multi-source multi area system. Electric Power Energy Syst 78:823–829
17. Kalyan CNS, Rao GS (2022) Combined frequency and voltage stabilization of multi-area multi source system by DE-AEFA optimized PID controller with coordinated performance of IPFC and RFBs. Int J Ambient Energy 43(1):3815–3831
18. Anusha P, Patra S, Roy A, Saha D (2021) Combined frequency and voltage control of a deregulated hydro-thermal power system employing FA based industrial controller. In: International conference on computational performance evaluation (ComPE), Shillong, 01–03 Dec 2021
19. Ramoji SK, Saikia LC (2021) Optimal coordinated frequency and voltage control of CCGT-thermal plants with TIDF controller. IETE J Res. https://doi.org/10.1080/03772063.2021.1959420
20. Kalyan CNS, Rao GS (2022) Coordinated control strategy for simultaneous frequency and voltage stabilization of the multi-area interconnected system considering communication time delays. Int J Ambient Energy 43(1):5512–5524
21. Rajbongshi R, Saikia LC (2018) Coordinated performance of interline power flow controller and super conducting magnetic energy storage in combined ALFC and AVR system under deregulated environment. J Renew Sustain Energy 10:044102
22. Kalyan CNS, Rao GS (2020) Performance comparison of various energy storage devices in combined LFC and AVR of multi area system with renewable energy integration. Int J Renew Energy Res 10(2):933–944
23. Rajbongshi R, Saikia LC (2017) Performance of coordinated FACTS and energy storage devices in combined multi area ALFC and AVR system. J Renew Sustain Energy 9:064101
24. Kalyan CNS, Rao GS (2021) Impact of communication time delays on combined LFC and AVR of a multi-area hybrid system with IPFC-RFBs coordinated control strategy. Prot Control Modern Power Syst 6:1–20
25. Hayyolalam V, Kazem AP (2020) Black widow optimization algorithm: a novel meta-heuristic approach for solving engineering optimization problems. Eng Appl Artif Intell 87:103249

Hybrid Intelligent Algorithm Applied to Economic Dispatch of Grid-Connected Microgrid System Considering Static and Dynamic Load Demand

Rakesh Sahu, Pratap Kumar Panigrahi, Deepak Kumar Lal, and Bishwajit Dey

Abstract This paper presents a novel hybrid modified-Gray Wolf Optimization-Sine Cosine Algorithm-Cuckoo Search Algorithm (MGWOSCACSA algorithm) as an optimization tool to minimize the generation cost of a grid-connected microgrid system. Initially, the optimal placing of DGs is performed to benefit a 33-bus distribution system. Thereafter three different cases were studied to analyze the generation cost of the system. The cases include without DGs, with 2 DGs, and with 3 DGs. The DGs considered for the study are two fossil-fueled power plants and a wind farm. The generation cost was found minimum of 1–2% when 3 DGs are considered in the study. Also proposed hybrid algorithm yielded better quality results than other optimization tools used in the study.

Keywords Radial Distribution System (RDS) · Distributed Generation (DG) · Microgrid · Unit commitment · Grey Wolf optimization (GWO)

1 Introduction

There is a tremendous possible to generate clean and green energy sources such as hydro, wind and solar for smooth device utility. Unlike traditional fossil fuels which emit greenhouse gases and contribute to climate change, these renewable sources harness the natural energy available in the environment and convert it into electricity in a clean and sustainable manner. Solar energy is abundant and can be harvested using solar panels, which convert sunlight into electricity. Wind energy

R. Sahu (✉) · P. K. Panigrahi · B. Dey
Department of EEE, GIET University, Gunupur, Odisha 765022, India
e-mail: rakeshsahu@giet.edu

P. K. Panigrahi
e-mail: pkpanigrahi@giet.edu

D. K. Lal
Departments of EE, VSSUT, Burla, Odisha 768018, India

S. Kumar et al. (eds.), *Control Applications in Modern Power Systems*, Lecture Notes in Electrical Engineering 1128, https://doi.org/10.1007/978-981-99-9054-2_7

is generated by wind turbines that harness the power of the wind and convert it into electricity. Small-scale hydro power plants can be built in rivers and streams to generate electricity from the flowing water. The use of these renewable energy sources can help to reduce the dependence on fossil fuels and promote a more sustainable and cleaner energy system. In addition, these technologies have become increasingly affordable and accessible in recent years, making it easier for people without access to the electricity grid to generate their own power and improve their living standards.

In any case, there are still difficulties to survive, for example, the discontinuity of a few sustainable sources, and that implies that energy capacity and distribution frameworks should be created and improved to guarantee a dependable and stable supply of power. Nevertheless, the potential of these alternative energy sources is enormous and they have an important role to play in the transition towards a more sustainable and clean energy system. Distributed generators (DG) installed in different locations in distribution network are based on alternative/renewable energy sources which convert passive distribution network into active network. Integrating multiple renewable energy sources can help to address the issues of stability and voltage and power variation in isolated mini electric grid systems. Hybridizing renewable sources involves using the largest and most reliable renewable source(s) to provide the base load, while other intermittent sources can be used to augment the base load and meet peak demand [1]. For example, a hybrid renewable energy system might combine a large-scale solar array or wind farm to provide the base load, with smaller solar panels or wind turbines installed nearby to help meet peak demand. Hydroelectric or geothermal sources may also be used as base load providers in certain locations. Integrating multiple renewable sources in this way can help to ensure a more stable and reliable supply of electricity, while also increasing the overall efficiency of the system. Energy storage solutions, such as batteries, can also be used to store excess energy generated during periods of low demand and release it during peak demand, further improving system stability. Overall, hybridizing renewable energy sources is a promising approach to improving the stability and reliability of isolated mini electric grid systems, while also promoting the use of clean, sustainable energy sources.

A study explains the electrification of grid connected remote sites uses hybrid renewable energy sources [2, 3]. Many software tools are also available to help with the sizing and simulation of hybrid systems. The performance of a hybrid renewable energy system is highly dependent on the environmental conditions of the specific location where it is installed. Factors such as the availability of sunlight, wind speed, and water flow rates can all affect the performance of the system. To optimize the design and performance of a hybrid renewable energy system, it is important to conduct a thorough analysis of the given study area. This analysis should consider factors such as the local weather conditions, topography, and energy demand patterns. By analyzing these factors, it is possible to determine the most appropriate combination of renewable energy sources and energy storage solutions for the specific location. In addition to analyzing the performance of the system, it is also important to consider the associated costs and component sizes. This involves evaluating the capital and operational costs of the system, as well as the physical sizes of the various components such as solar panels, wind turbines, batteries, and inverters. By conducting a comprehensive analysis of the study area, it is possible to design a hybrid renewable energy system that is optimized for performance, cost, and component size. Electrification of faraway areas or grid connected systems promoted by the use of green and clean energy. Microgrid is composed of PV system, diesel and wind energy based sources. The author proposed to satisfy load demand and minimizing the fuel cost of generators. Two realistic cases are assumed by the author, first case includes one wind generator, one PV, three diesel generators and three rural consumers. Second case assumes larger DGs [4]. There are two modes of operation of microgrid i.e. grid connected mode and Islanded mode. In grid connected mode of operation utility grid act as a backup supply in case of failure of distributed energy sources or battery banks [5]. In [6] author concentrated on overall optimization of cost of the microgrid i.e. generation and transmission cost. Four types of metaheuristic algorithms are applied for overall reduction of microgrid cost.

Basak et al. applied three different types of techniques in order to minimize carbon emission, fuel, operation and maintenance cost. The author approached, Neighborhood based differential technique to minimize the cost of microgrid [7]. To solve optimal scheduling problems among the renewable energy sources, artificial fish swam algorithm is proposed [8]. Author suggested cuckoo search algorithm to solve the problems of power dispatch in microgrid [9]. Problems of economic/emission dispatch applied to 30-bus 6-unit systems solved by backtracking search algorithm which combines multi objective function into single objective [10]. Author proposed central controller of microgrid. The main advantages of central controller are that maximize the production of DGs and exchange the power within distribution grid [11]. The operational cost is minimized using proposed SIMBO-Q algorithm [12]. A modern hybrid approach is introduced for energy management of microgrid with renewable sources [13].

Overall, effective planning and control are essential for optimizing the performance of a power distribution system and achieving the benefits of reducing principal expenditure, decreasing loss of power, improving power factor, and increasing feeder capabilities. By leveraging advanced technologies and data analytics, it is possible to design and operate distribution systems that are highly efficient, reliable, and cost-effective. Golden Jackal Optimization (GJO) algorithm employed to determine the capacity and optimal location of the DG units [14]. For 33 buses RDN system, hybrid algorithm implemented and tested successfully. An IPSO algorithm improves voltage profile and minimizes power loss of distributed network considered four types of DG's. Proposed IPSO algorithm allocates DG's in proper bus and also calculates the size of the DG units [15]. A sine–cosine algorithm (SCA), a grey wolf optimizer (GWO), and a crow search algorithm (CSA) are built to execute EED, and their results are compared with basic GWOs and other hybrid algorithms. These current breakthroughs in improved optimization techniques are combined in this hybrid approach. [16–19].

Orientation of Paper

The remaining sections of the paper are presented as follows: Sect. 2 discusses fitness & limitation function; mathematically modeling's of hybrid algorithm; Sect. 3

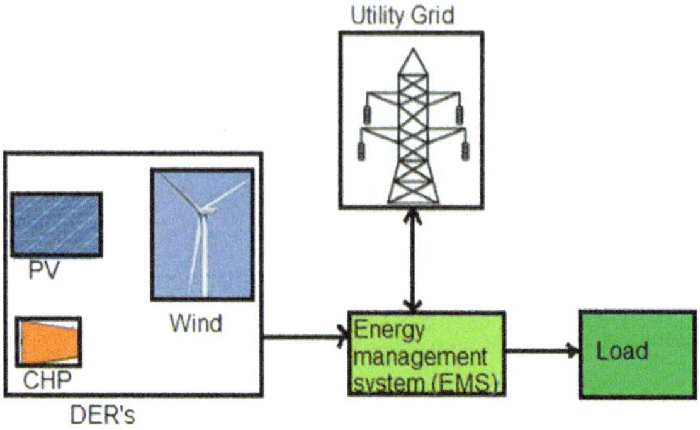

Fig. 1 Block diagram of Grid connected microgrid system

describes the case studies investigated and the statistical analysis used to support these conclusions. Section 4 lists most important findings (Fig. 1).

2 Methodology

2.1 Problem Formulation

Fuel cost of economic dispatch included utility grid can be mathematically represented in quadratic equation such as

$$F(g) = \sum_{t=1}^{24} \sum_{m=1}^{n} \left[\left(a_m P_m^2(t) + b_m P_m(t) + c_m \right) + d_{grid}^t P_{grid}(t) \right] \tag{1}$$

where t and m are representing operating period of microgrid and number of total DG sources. Power delivered by DG sources is denoted by $P_m(t)$ at tth hour and power delivered by grid is $P_{grid}(t)$ at tth hour. Fuel cost coefficient of quadratic equation is denoted by a, b and c of mth DG sources. Utility grid electricity market price is denoted by d_{grid}^t. Utilization percentage (UP) is defined as [8] the ratio of DG sources that can be supplied total rated power throughout its working hour to DG sources delivered total power. A positive UP indicates power supplied by DG sources and negative UP means DG sources withdraw power from the utility grid.

2.2 Generation Limit and Equality Constraint

$$P_{m/grid,\min}(t) \le P_{m/grid} \le P_{m/grid,\max} \tag{2}$$

$P_{m/grid,\min}$ and $P_{m/grid,\max}$ denoted by minimum and maximum output powers of DG sources at tth hour.

Equality constraint between power supplies to demand can be represented by,

$$\sum_{m=1}^{n} P_m(t) + P_{grid}(t) = P_{load}(t) \tag{3}$$

2.3 *Modified GWOSCACSA Algorithm*

Literature reviews suggest the proposed hybrid Modified GWO-SCA-CSA (MGWOSCACSA) is superior to other hybrids and GWO algorithms [16]. For mathematical constructing the MGWO algorithm, α is considered to be best solution. β represents second best solution and δ represents third best solution. Mathematically, others are represented by Ω. MGWOSCACSA algorithm is mathematical combination of GWO, SCA and CSA. Mathematically it is represented by,

$$\overrightarrow{D_\alpha} = \begin{cases} rn * \sin(rn) * \left|\overrightarrow{C_\alpha}\overrightarrow{X_\alpha} - \vec{X}\right| & \text{if, } rn > 0.5 \\ rn * \cos(rn) * \left|\overrightarrow{C_\alpha}\overrightarrow{X_\alpha} - \vec{X}\right| & \text{otherwise} \end{cases} \tag{4}$$

$$\overrightarrow{D_\beta} = \begin{cases} rn * \sin(rn) * \left|\overrightarrow{C_\beta}\overrightarrow{X_\beta} - \vec{X}\right| & \text{if, } rn > 0.5 \\ rn * \cos(rn) * \left|\overrightarrow{C_\beta}\overrightarrow{X_\beta} - \vec{X}\right| & \text{otherwise} \end{cases} \tag{5}$$

$$\overrightarrow{D_\delta} = \begin{cases} rn * \sin(rn) * \left|\overrightarrow{C_\delta}\overrightarrow{X_\delta} - \vec{X}\right| & \text{if, } rn > 0.5 \\ rn * \cos(rn) * \left|\overrightarrow{C_\delta}\overrightarrow{X_\delta} - \vec{X}\right| & \text{otherwise} \end{cases} \tag{6}$$

$$\overrightarrow{D_\Omega} = \begin{cases} rn * \sin(rn) * \left|\overrightarrow{C_\Omega}\overrightarrow{X_\Omega} - \vec{X}\right| & \text{if, } rn > 0.5 \\ rn * \cos(rn) * \left|\overrightarrow{C_\Omega}\overrightarrow{X_\Omega} - \vec{X}\right| & \text{otherwise} \end{cases} \tag{7}$$

Whether α, β, δ and Ω wolves will consider or only depending on α wolf only i.e. decided by AP. AP is calculated by formula as:

$$AP = 1 - \left(\frac{1.01 \times iter^3}{Max\_iter^3}\right) \tag{8}$$

3 Results and Discussions

3.1 System Description

A low voltage grid connected MG system was considered for economic analysis. The single line diagram of the 33-bus distribution network is shown in Fig. 2. The DGs are placed at optimal locations in a 33 bus distribution system [14]. The system was studied for different cases which focused on different number of DGs and load demand profiles. The DGs incorporated are, two numbers of fossil fueled generators and a wind farm. The parameters of the DG are displayed in Table 1 [17]. Initially a static load of 3715 kW was considered for the study. Thereafter, dynamic load demand was considered with 3715 kW as the peak load demand [18]. The dynamic load demand and the grid pricing are shown in Fig. 3. The hourly output of the wind farm is shown in Fig. 4. Algorithms such as GWO and WOA were used as the optimization tool for comparing the results obtained with the proposed MGWOSCACSA algorithm. For all the cases, the population size was considered as 80 and the maximum iteration was considered as 500.

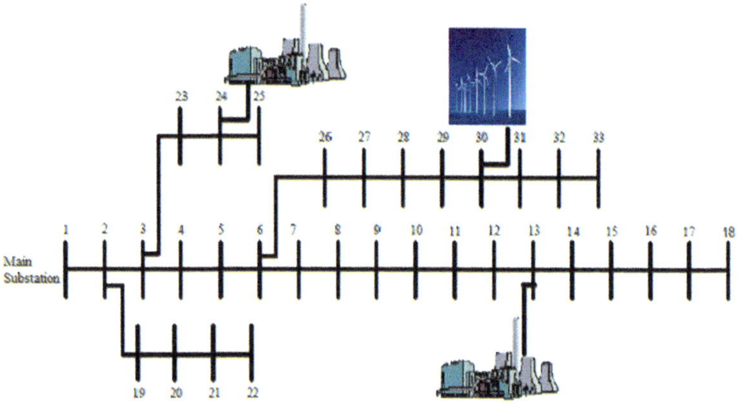

Fig. 2 Single line diagram of a standard 33bus radial distribution system

Table 1 Parameter of DG's [17]

DG's	G1	G2	Grid
Max power (KW)	3	3	5
Min power (KW)	0.8	0.8	−5
Bid ($/KW)	64.1	72.3	Appendix
Off hours	1,2,3	1,2	0

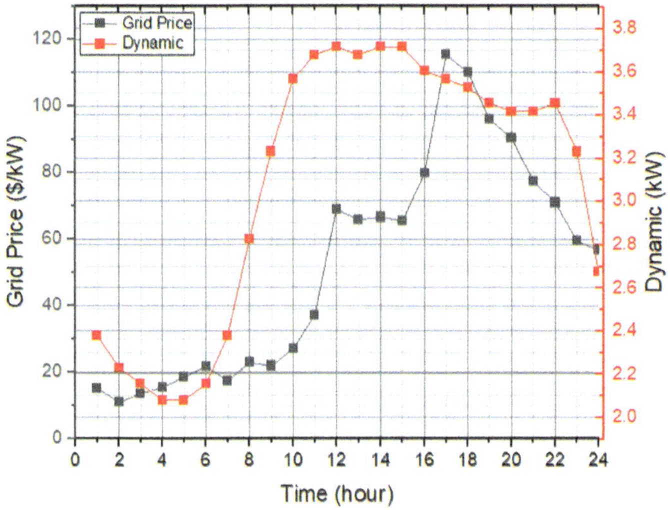

Fig. 3 Dynamic load demand and electricity market price

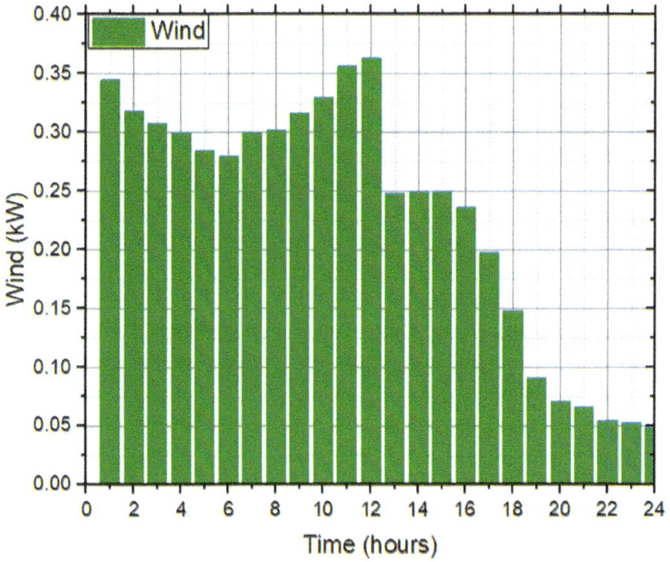

Fig. 4 Hourly output of the wind farm

3.2 *Descriptive Analysis of Results Obtained*

To start the analysis, the generation cost of the distribution system was calculated without DGs for static and dynamic load demand i.e. the entire load demand was to be supplied by the grid. It was seen that the generation cost was $4623 for static load of 3715 kW and $4187 for dynamic load demand.

Thereafter, the generation cost was minimized considering 2 DGs placed in bus 13 and 30 of the 33 bus distribution network. Both the DGs were fossil fueled in nature. The minimum generation cost was found to be $4293 and $3860 using proposed MGOWSCACSA algorithm for static and dynamic load demand, respectively. This result was lesser than the case when no DGs were participating and the entire load demand was met by the grid. Also, the proposed MGWOSCACSA yielded better quality result than GWO and WOA. Thereafter, the generation cost was minimized considering 3 DGs. The fossil fueled DGs were placed in bus 13 and 24 and a wind farm was place in bus 30 of the 33 bus distribution network. The proposed MGOWSCACSA algorithm yielded a minimum generation cost of $4063 and $3626 for static and dynamic load demand, respectively. This result was lower than in case where no DGs participated and the grid met the entire load demand. Furthermore, the observations show that the proposed MGWOSCACSA produces higher quality results than GWO and WOA. Table 2 gives a complete description of the costs obtained during various load demand profiles and considering different numbers of DGs. Figures 5 and 6 show the hourly output of DERs when 2 and 3 numbers of DGs were used to share the load demand. The active participation of grid and economic load sharing among DGs results in minimization of the generation cost of the MG system.

Figure 7 shows the comparative chart using different algorithms and different number of DGs. The minimum generation cost was obtained when 3 DGs were considered. Also, proposed MGWOSCACSA yielded the better quality results for all the cases. Figure 8 shows the convergence curve characteristics of all the algorithms

Table 2 Cost comparative analysis using various algorithms

Algorithms	Static load demand (3715 KW)			Dynamic load demand		
	Without DG's	With 2 DG's	With 3 DG's	Without DG's	With 2 DG's	With 3 DG's
GWO	4623.6147	4299	4067	4187.3481	3870	3630
WOA		4298	4063		3866	3627
MGWOSCACSA		4293	4063		3860	3626

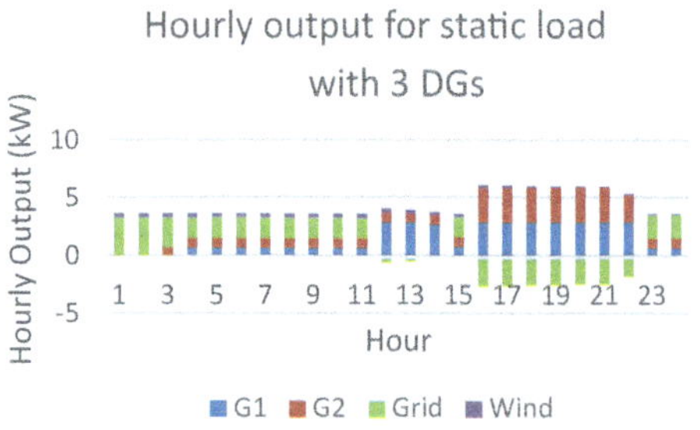

Fig. 5 Plot of Hourly outputs of DGs for static load with 3 DGs demand using MGWOSCACSA algorithm

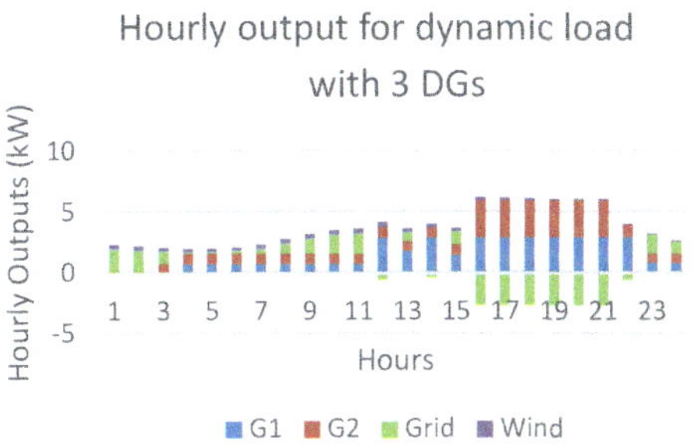

Fig. 6 Plot of Hourly outputs of DGs for dynamic load with 3 DGs demand using MGWOSCACSA algorithm

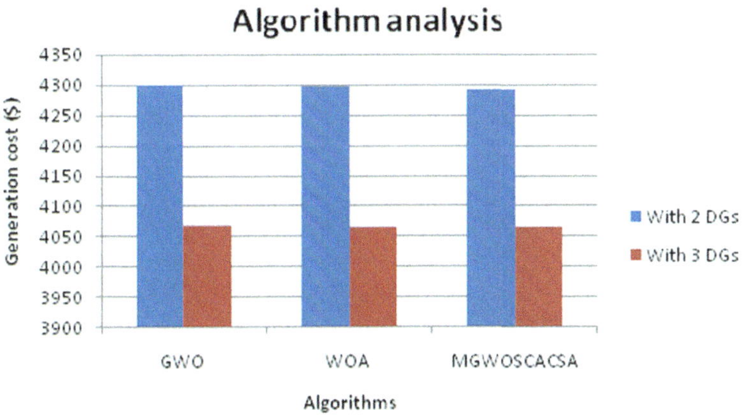

Fig. 7 Comparative analysis amongst generation costs using different algorithms

when the generation cost was minimized using 2 DGs. It can be seen that the proposed MGWOSCACSA converges smoothly and much faster than GWO and WOA.

4 Conclusions

The findings from MGWOSCACSA were of the highest caliber when compared to GWO and WOA. The thorough investigation and abuse ability of MGWO produces quick results of the highest caliber with the greatest robustness and accuracy during each individual iterations when coupled with the position updating techniques of SCA and CSA. Another important element in bringing down the microgrid cost to that level was the grid's active involvement in purchasing and selling power from the microgrid. When compared to the scenario where the load was provided solely by the grid, the increased number of DGs helped to reduction the cost of

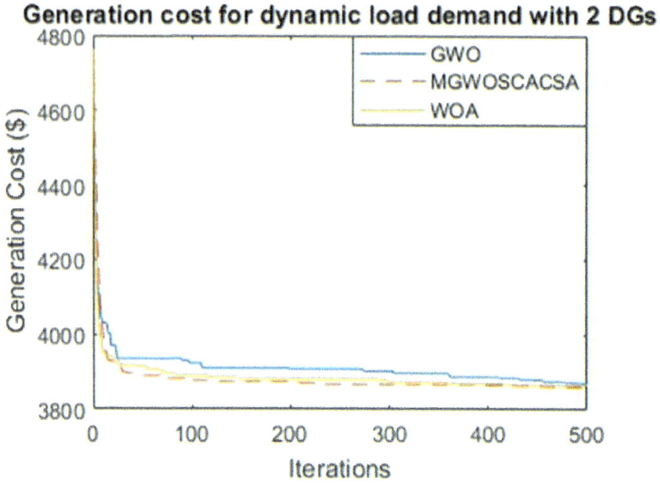

Fig. 8 Convergence curve characteristics

generation of the microgrid system. Because of its increased computational speed and exploitability, MGWOSCACSA can be regarded as one of the most powerful optimization instruments for resolving a variety of power system optimization issues.

In future more powerful algorithm can be developed and used for the study of distribution system planning and operation, which comprises renewable resources based hybrid power system.

References

1. Lal DK, Dash BB, Akella AK (2011) Optimization of PV/wind/micro-hydro/diesel hybrid power system in HOMER for the study area. Int J Electr Eng Inform 3(3):307–325
2. Rajoriya A, Fernandez E (2010) Sustainable energy generation using hybrid energy system for remote hilly rural area in India. Int J Sustain Eng 1–9
3. Gupta A, Saini RP, Sharma MP (2010) Steady-state modeling of hybrid energy system for off grid electrification of cluster of villages. Renew Energy 35:520–535
4. Nwulu NI, Xia X (2017) Optimal dispatch for a microgrid incorporating renewable and demand response. Renew Energy 101:16–28
5. Velik R, Nicolay P (2014) Grid-price-dependent energy management in microgrids using a modified simulated annealing triple-optimizer. Appl Energy 130:384–395
6. Dey B, Bhattacharyya B, Sharma S (2019) Robust economic dispatch of microgrid with highly penetrated renewables and energy storage system. Int J Energy Optim Eng 8(1):67–87
7. Basak S, Bhattacharyya B, Dey B (2022) Combined economic emission dispatch on dynamic systems using hybrid CSA-JAYA algorithm. Int J Syst Assur Eng Manage 13:2269–2290
8. Kumar KP, Saravanan B, Swarup KS (2016) Optimization of renewable energy sources in a microgrid using artificial fish swarm algorithm. Energy Procedia 90:107–113
9. Basu M, Chowdhury A (2013) Cuckoo search algorithm for economic dispatch. Energy 60:99–108

10. Modiri-Delshad M, Barati M, Abd Rahim N (2016) Economic power dispatch in microgrids through backtracking search algorithm. In: 4th IET clean energy and technology conference (CEAT 2016). IET, pp 1–8
11. Tsikalakis AG, Hatziargyriou ND (2011) Centralized control for optimizing microgrids operation. In: IEEE power and energy society general meeting, pp 1–8
12. Sharma S, Bhattacharjee S, Bhattacharya A (2018) Operation cost minimization of a microgrid using quasi-oppositional swine influenza model based optimization with quarantine. Ain Shams Eng J 9(1):45–63
13. Dey B, Sharma S (2015) Energy management of microgrids with renewables using soft computing techniques. In: Annual IEEE India conference (INDICON), pp 1–6
14. Manjhi R, Lal DK, Biswal S (2022) Optimal allocation of DGs in radial distribution network for power loss minimization based on LSF and GJO algorithm. In: 2022 IEEE international symposium on sustainable energy, signal processing and cyber security (iSSSC). IEEE
15. Purlu M, Turkay BE (2022) Optimal allocation of renewable distributed generations using heuristic methods to minimize annual energy losses and voltage deviation index. IEEE Access 10:21455–21474
16. Dey B, Basak S, Bhattacharyya B (2021) MGWOSCACSA: a novel hybrid algorithm for energy management of microgrid systems. In: Advances in smart grid automation and industry 4.0: select proceedings of ICETSGAI4 0, pp 669–678
17. Dey B, Biplab B (2019) Hybrid intelligence techniques for unit commitment of microgrids. In: 20th international conference on intelligent system application to power systems (ISAP). IEEE
18. Dey B, Dutta S, Pedro Garcia Marquez F (2023) Intelligent demand side management for exhaustive techno-economic analysis of microgrid system. Sustainability 15(3):1795
19. Dey B, Pedro García Márquez F, Basak S (2020) Smart energy management of residential microgrid system by a novel hybrid mgwoscacsa algorithm. Energies 13(13):3500

Design of Hybrid Energy Storage System Model with Multi-input Converter

Nabadeep Patra, Aditi Chatterjee, and Rakesh Kumar Sahoo

Abstract This paper aims to design and analyze the hybrid energy storage system (HESS) model with multiple input converter (MIC) configurations in simulation as well as real-time models. In the conventional design of HESS systems, each source has its local regulators for better power transfer. Still, in MIC configuration, a single bidirectional regulatory circuit is used to handle every source to get a better power output. In this paper, the HESS model has a renewable source (PV array), a battery unit and an ultracapacitor. All the sources are connected to the MIC converter, and the converter is connected to the AC grid. The purpose of the model is to synchronize the system output to the AC grid as well as use the grid energy for storing energy in the battery as a bidirectional system.

Keywords Hybrid Energy Storage System (HESS) · Multiple Input Converter (MIC) · PV array · Bi-directional regulator

1 Introduction

In the present scenario for generating and delivering electrical power, non-conventional sources and renewable sources are playing a vital role [1]. Different types of power generation sources are integrated through a single system and deliver the required electrical power. Hybrid Energy Storage System (HESS) has become a central concept for stable power management [2–10]. Due to this arrangement, the limitations of each source are eliminated and power transfer becomes smooth due to

N. Patra (✉) · A. Chatterjee · R. K. Sahoo
Electrical Engineering Department, CGU, Bhubaneswa, Odisha, India
e-mail: pnabadeep@cgu-odisha.ac.in

A. Chatterjee
e-mail: aditichatterjee@cgu-odisha.ac.in

R. K. Sahoo
e-mail: sahoorakeshkumar66@gmail.com

S. Kumar et al. (eds.), *Control Applications in Modern Power Systems*, Lecture Notes in Electrical Engineering 1128, https://doi.org/10.1007/978-981-99-9054-2_8

the power-sharing between the sources. Diverse types of storage units are present in this type of system due to which this system can handle transients more efficiently.

Solar, wind, fuel cells, biomass, geothermal, and tidal energy are some of the renewable energy sources accepted as an alternative sources of power generation. We cannot rely on a single RE source to generate power up to the required quantity and quality. Take an example for a solar-based HESS system, we could not get the required amount of generation through the PV array for every weather condition. The amount of shading, cloudy or rainy weather, and pollution in the environment all affect the generation of PV arrays. Hence the output from solar is non-linear so the output of the HESS may not be constant.

To deal with the varying output of the system, every HESS system uses a different topology to mitigate the non-linearity. Each topology consists of different kinds of regulators like DC-DC, DC-AC, and AC-DC individually or in groups. In many HESS, each source relates to a converter circuit that may be unidirectional or bidirectional to extract the necessary amount of energy from it. Due to this, the number of semiconductor devices used in HESS increases. The major problem arises due to an increase in semiconductor devices that increase the non-linearity of the system. For a better power profile as well as to maintain a linear generation, the structure of converters should be changed and semiconductor devices need to be reduced. Here multiple input converters come into the picture. These converters have fewer numbers of switches but function the same as conventional converters [11, 12].

In this paper, the proposed HESS system is designed by PV arrays, battery units and supercapacitors connected with the MIC. The proposed system is operated in bidirectional mode that means as per the requirement, converter acts as a regulator as well as inverter. Here, since MIC concept is introduced, energy can be transferred in between sources as well as to the output. The output load can be shared by PV arrays, battery and ultracapacitor individually or simultaneously. The switching properties like the switching frequency and duty cycle of each switch are controlled by a close loop mechanism that results in a stable output profile. Here the proposed model is designed and simulated in MATLAB, and the results are compared with the output of the prototyped hardware model.

2 Modeling of HESS Model with Multi-input Converter

This system aims to add electrical energy to the grid in a synchronous manner, derived from various sources. This system comprises a method to synchronize its output to the energy grid it is supplementing. The AC output of the system is derived through an inverting circuit, where the input is a DC bus. This bus is typically supplied through various Hybrid energy storage systems (Figs. 1 and 2).

It functions as a multiple-input single-output system (MISO). The main sources for this system are PV arrays and batteries. An ultracapacitor (UC) is also used momentarily for maintaining power quality [2].

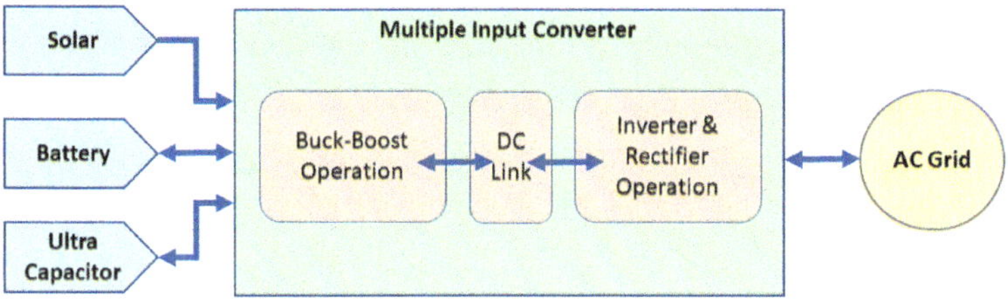

Fig. 1 Block diagram of proposed model

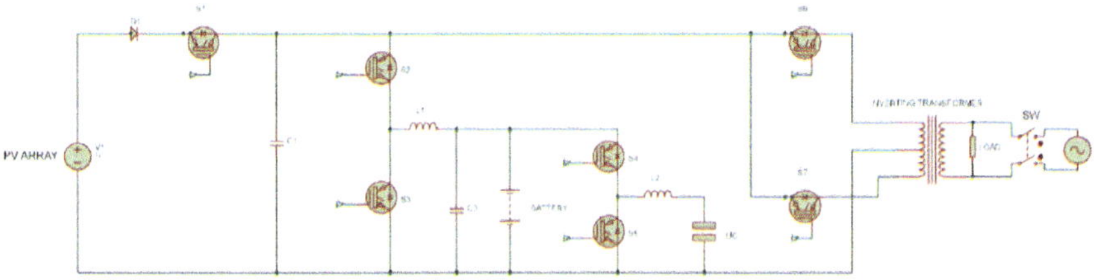

Fig. 2 Complete circuit diagram of HESS system with MIC

Simply, it can be said that this device operates in various conversion modes like inversion operation, and rectification mode depending upon the input conditions. When one input cannot provide the required power, another source will take over the place. During the change of inputs, a dip in the power graph may be experienced. To eliminate that situation, an ultracapacitor will deliver its stored power to the system [11].

For a stable and sustainable system, power quality maintenance should not be the only requirement. In the HESS system, anti-islanding of the grid, auto cut-off of battery and different protection schemes are also as important as power quality management. This project can deliver the requirements said above.

This proposed system is designed to maintain the power quality index mainly. DC sources, i.e., PV arrays and batteries, are used mainly to do so. To get a constant AC output, a controlled inversion is taken place with the help of an IGBT-based DC-AC converter. For a stable AC output, a constant DC source is compulsory. But, here voltage rating of the PV array taken for the experimental model and battery storage unit is different. So, both inputs are connected to the MISO system and deliver constant DC output, which is later fed to a common DC bus. The DC bus is used as input in the inverter circuit [11]. The model is described in the next section.

2.1 When Solar Energy is Fed to the System

Using solar energy directly to the system cannot maintain good power quality [9]. To get maximum power from PV arrays, a switch S_1 is used to transfer full power to

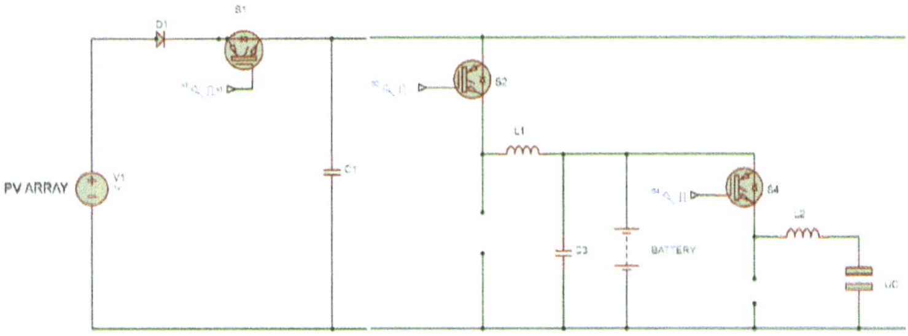

Fig. 3 Representation of switching sequence when PV arrays are active

the system. The triggering pulses are generated through the microcontroller, and the duty cycle is decided by perturb and observe-based MPPT algorithm.

When the PV array is connected as a source, its output voltage rating is maintained the same as the required DC bus voltage through the MPPT algorithm. The secondary operation is to charge the battery unit and ultracapacitor. Both battery and ultracapacitor voltage rating is different so regulation takes place. The PV array's model V_{oc} is 22 V, and the battery rating is 12 V. For buck regulation, IGBT S_2 gets the triggering pulse from the microcontroller and the duty cycle is defined accordingly (Fig. 3). The ultracapacitor used is of 5.5 V rating so the operation for the ultracapacitor is the same, and IGBT S_4 is used for this purpose [3].

2.2 *When the Battery Energy is Fed to the System*

When the battery is connected as an input to the system, the arrangement is such that it will lift the voltage level to the required DC link voltage. In this model, the battery voltage is 12 V while the DC bus is at 14 V. For this purpose, the system must act like a boost circuit and deliver the power to the inverting circuit. During drop conditions, the ultracapacitor must involve its power to the system to make power quality stable. Again, we need to boost this voltage to the upper level. Here it is boosted to the voltage level of the battery which reduces the drop at the battery terminal during load conditions (Fig. 4).

Here in the situation where the battery is connected, S1 and S3 become inactive and S2 and S4 become active. S4 is used to boost the ultracapacitor (UC) voltage to the battery voltage. S2 is responsible for boosting battery voltage to DC bus voltage. The duty cycle for each switch is decided through the closed-loop operation [2]. The complete process is represented in Fig. 5.

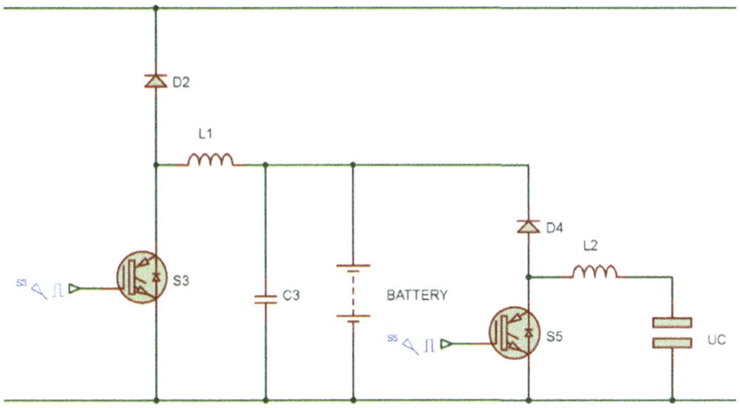

Fig. 4 Representation of switching sequence when battery and ultracapacitor are active

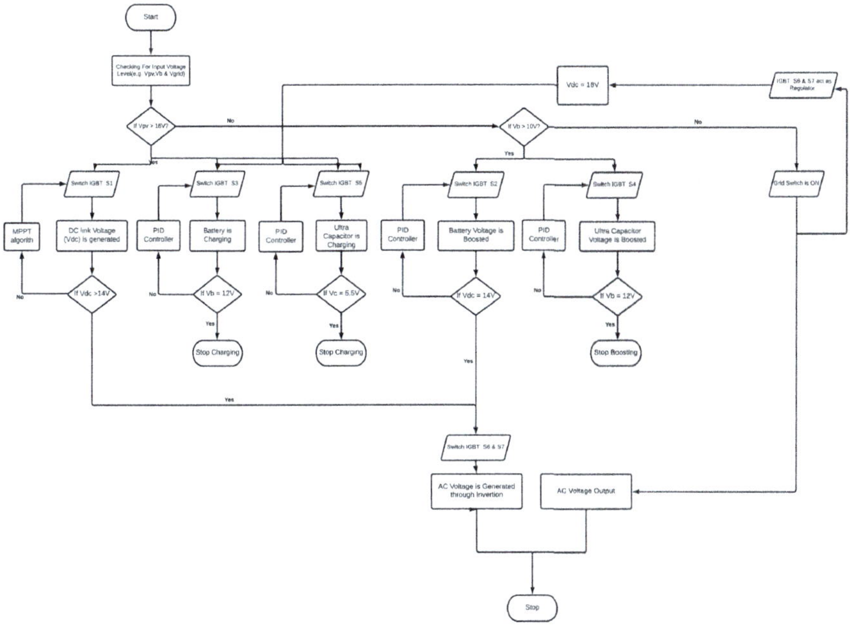

Fig. 5 Flowchart of complete process for proposed model

3 Prototyping of HESS Model with Multi-input Converter

During the operation of the model, switches are operated in a defined way and corresponding duty cycles are maintained through the closed-loop arrangement. In summary, when PV arrays are connected in the system, switches S_1, S_2 and S_4 are operated with the required duty cycle. During this time, the circuit behaves like a multilevel buck circuit where variable PV voltage is lowered to 14 V for the DC link by S_1. Again, it is reduced to 12 V for battery voltage level through S_2. Further, it is lowered to 5.5 V for the ultracapacitor by an S_4 switch. The necessary calculations for the operation are as follows [11].

- *Duty cycles of respective switches*

$$D_i = V_y / V_x \tag{1}$$

- *Inductor Ripple Current*

$$\Delta ILi = \left[V_x(\max) - V_y \right] / \left[fsi \times L_i \right] \tag{2}$$

- *Inductor Selection*

$$L_i = V_y \left[V_x - V_y \right] / \left[\Delta Ili \times fsi \times Vx \right] \tag{3}$$

where

V_x = Input Voltage (known)

V_y = Output Voltage for a switch (required)

fsi = Required Operative frequency of the switch

= Efficiency

When PV arrays are not operational in the system due to various reasons like weather, temperature or irradiation, switches S_1, S_2 and S_4 are deactivated as well as switches S_3 and S_5 are operated. During this time, the circuit behaves like a multilevel boost circuit where variable ultracapacitor voltage is raised to 12 V for battery by S_5. Again, it is raised to 14 V for the DC link through S_3. The necessary calculations for the operation are as follows [11]:

- *Duty cycles of respective switches*

$$D_i = 1 - \left(V_x \times \eta / V_y \right) \tag{4}$$

- *Inductor Ripple Current*

$$\Delta IL_i = \left[V_x(\min) \times D_i \right] / \left[fs \times L_i \right] \tag{5}$$

- *Inductor Selection*

$$L_i = V_y \times [V_x - V_y]/Ili \times fsi \times V_x \tag{6}$$

where

V_x = input voltage (Known)

V_y = output voltage (Required)

fsi = required operative frequency of the switch

η = Efficiency

DC bus, after receiving the energy from primary sources, feeds its energy to the inverter circuit where AC power develops [13]. For inverter circuits, IGBTs are used in half wave configuration. Both the switches are inverted to each other and run in a 50% duty cycle and at 100 Hz frequency. In this arrangement, collectors of both switches are connected to the positive terminal of the DC bus. The emitters are connected to positive leads of the 12-0-12/230 V transformer, and the centre tap point is connected to the negative terminal of the DC bus. Secondary terminals of the transformer are connected to the grid.

4 Simulation Results

The proposed model is simulated using MATLAB/SIMULINK software, and the simulation parameters used for the system are tabulated in Table 1.

Table 1 Model parameters are taken during simulations

Parameters	Values
PV array	V_{oc} = 22 V, 80W
Battery	12 V, 7.2 Ah
Ultracapacitor	5.5 V, 3F
Grid voltage	230 V (rms)
Switching frequency	13 kHz
Filter parameters	L_1 = 3 mH C_1 = 1200 μF L_2 = 1.6 mH
IGBT ratings	1200 V, 50 A
Transformer rating	230/12 V, 100 VA
Switch (SW)	230 V, 10 A

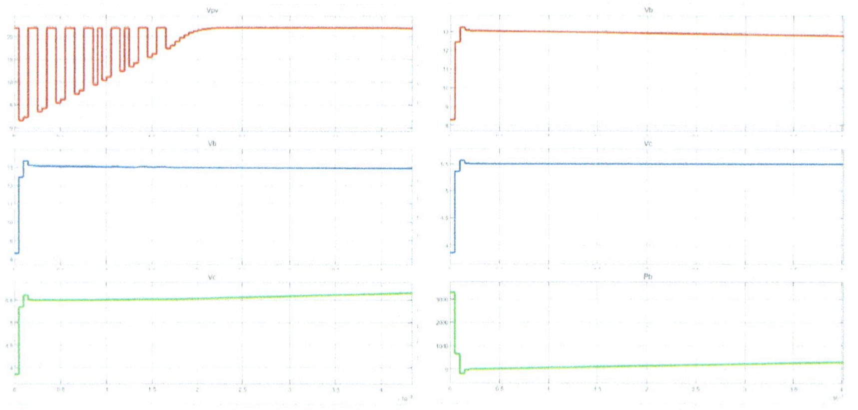

(A) (B)

Fig. 6 PV arrays ouput voltage (V_{pv}), Battery voltage (V_b) and Ultracapacitor voltage (V_c) when the PV array is active

In Fig. 6a, it is observed that the central control switches S_1, S_2 and S_4 when it senses the PV arrays are active (shown in the figure as V_{pv}). In this MATLAB simulation, that is represented here, switching of S1 is controlled by the P & O (perturb and observe) algorithm to extract maximum output from PV arrays. Switches S_2 & S_4 are being switched through a PID control process taking 12 V and 5.5 V (shown in the figure as S_2 switching & S_4 switching) as references respectively.

In Fig. 6b, it is observed that the central controller switches S_3 & S_5 when it senses that PV arrays are not providing the threshold power as per requirement. In this state, the regulator circuit is switched to boost mode. Now, the battery voltage (shown in the figure as V_b) of 12 V is boosted to 14 V for the inverting circuit through the S_3 switch. During the transition of the source from PV arrays to the battery storage, damping may occur in the power graph of the system. To overcome that, UC voltage (shown in the figure as V_c) is also boosted to 12 V during the transition through the S_5 switch. Both S_3 & S_5 switch is controlled by the PID controller. Since the No-load condition is demonstrated in this figure, it is observed that the battery and UC are not discharging.

Another objective of this proposed model is to synchronize the output AC voltage to the grid voltage (shown in the figure as V_{grid}) to fulfil the objective of the HESS model. For that, a switch SW is assigned to check whether the output of the model has the same voltage level and frequency or not. If the credentials match it should connect to the grid. In MATLAB simulations results shown in Fig. 7, it is observed that the proposed model can provide a continuous and uninterrupted sinusoidal voltage output (shown in the figure as V_o) despite the change in sources.

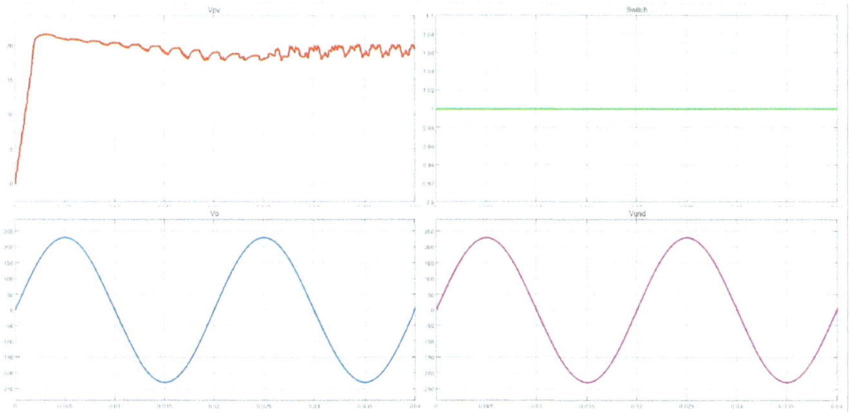

Fig. 7 PV array output voltage (V_{pv}), AC output voltage (V_o) and grid voltage (V_{grid})

5 Hardware Implementation

After verifying the proposed model by MATLAB simulation, the hardware model was constructed using the components described in Table 1. The snapshot of hardware model is shown in Fig. 8. It is shown that PV arrays are providing a voltage of 17.27 and 15.56 V is provided by the MPPT controller which is implemented by raspberry pi microcontroller. In the following inverter circuit, AC voltage is obtained at the output. The output voltage waveform is observed using a digital storage oscilloscope, which is around 200 V as shown in Fig. 9.

Fig. 8 Snapshot of hardware model operation

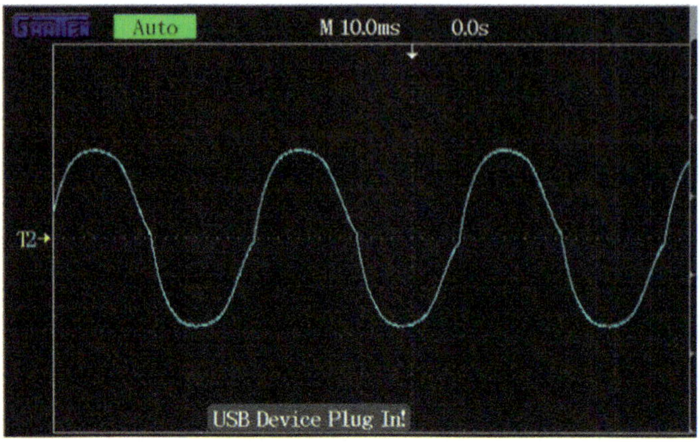

Fig. 9 Output voltage of inverter ch2: 20 V/div

6 Conclusion

In this paper a model of hybrid energy storage system (HESS) using multi-input converter (MIC) is proposed. The proposed model is simulated in MATLAB/SIMULINK and a hardware prototype of the model is also constructed in the laboratory.

The proposed MIC can integrate battery and ultracapacitor with the PV array, which is very significant to improve the output power quality. The MIC is a multipurpose converter which can act as both buck converter at one time and a boost converter at other as per requirement. The simulation results and the hardware implementation validates the efficacy of the proposed converter for its use in microgrids with renewable energy generation plants with hybrid energy storage system.

References

1. Chatterjee A, Mohanty K, Kommukuri VS, Thakre K (2018) Model predictive current controller for performance enhancement of grid-integrated single-phase photovoltaic distributed generation plants. Trans Inst Meas Control 40(3):762–775
2. Osama, Al Tamimi S, Amoura F (2022) A hybrid battery/Ultracapacitor energy storage solution for PV systems. In: 2020 6th IEEE international energy conference (ENERGYCon). IEEE
3. Saha R, Dey J (2019) PV, battery and ultra-capacitor-based hybrid energy storage system. In: 2019 IEEE 16th India council international conference (INDICON). IEEE
4. Mamen A, Supatti U (2017) A survey of hybrid energy storage systems applied for intermittent renewable energy systems. In: 2017 14th international conference on electrical engineering/electronics, computer, telecommunications, and information technology (ECTI-CON). IEEE
5. Adhikari S et al (2016) A battery/supercapacitor hybrid energy storage system for DC microgrids. In: 2016 IEEE 8th international power electronics and motion control conference (IPEMC-ECCE Asia). IEEE

6. Babu TS et al (2020) A comprehensive review of hybrid energy storage systems: converter topologies, control strategies and prospects. IEEE Access 8:148702–148721
7. Lakshmi JJ et al (2019) Improved bidirectional converter for PV system with battery storage. In: 2019 international conference on advances in computing and communication engineering (ICACCE). IEEE
8. Alam MJE, Muttaqi KM, Sutanto D (2013) Mitigation of rooftop solar PV impacts and evening peak support by managing the available capacity of distributed energy storage systems. IEEE Trans Power Syst 28(4):3874–3884
9. Wang G, Ciobotaru M, Agelidis VG (2014) Power smoothing of large solar PV plant using hybrid energy storage. IEEE Trans Sustain Energy 5(3):834–842
10. Zhou H et al (2010) Composite energy storage system involving battery and ultracapacitor with dynamic energy management in microgrid applications. IEEE Trans Power Electron 26(3):923–930
11. Sahoo RK, Jha V, Sen P (2022) Modeling of multiple inputs converter with renewable source. In: 2022 international conference on communication, computing and internet of things (IC3IoT). IEEE
12. Patra S, Chatterjee A, Patra KC (2022) A comparative analysis of multiport DC-DC converters in microgrid using small-signal modelling. In: 2022 IEEE 2nd international symposium on sustainable energy, signal processing and cyber security (iSSSC), Gunupur, Odisha, India, pp 1–6
13. Argyrou MC et al (2021) Modeling a residential grid-connected PV system with battery–supercapacitor storage: Control design and stability analysis. Energy Rep 7:4988–5002

Optimal Tuning of Single Input Power System Stabilizer Using Chaotic Quasi-Oppositional Differential Search Algorithm

Sourav Paul, Sneha Sultana, Provas Kumar Roy, Pravin Kumar Burnwal, Devjeet Sengupta, and Nirmalya Dey

Abstract In this research work, the concept of the Chaotic Quasi-Oppositional Differential Search Algorithm (CQODSA) has been successfully implemented on a single input power system stabilizer for the optimum tuning so as to damp low-frequency oscillations. The Heffron-Phillips model has been considered to evaluate the efficiency of the proposed algorithm by incorporating it in a device with a separate infinite bus. The enactment in terms of the prototype algorithm is tested for wide-loading scenarios to claim the applicability of CQODSA under flexible scenarios. By contrasting the findings with those of other well-known algorithms, the superiority of the established method has been proven.

Keywords Power system stabilizer · Single machine infinite bus · Low-frequency oscillation · Chaotic quasi-oppositional based learning · Differential search algorithm

1 Introduction

A power plant often has a lot of synchronous generators installed to control the energy at the contact spontaneously. The Power network component , which has the innate potential to increase power system cohesion limits while maintaining high electrical Power Strength (PS), is seen to be a useful tool in many Distributed Generation (DG) applications. The scale and complexity of modern power systems are constantly

http://www.bcrec.ac.in.

S. Paul (✉) · S. Sultana · P. K. Burnwal · D. Sengupta · N. Dey
Dr. B. C. Roy Engineering College, Durgapur, India
e-mail: sourav.p01@gmail.com

P. K. Roy
Kalyani Government Engineering College, West Bengal, India

S. Kumar et al. (eds.), *Control Applications in Modern Power Systems*, Lecture Notes in Electrical Engineering 1128, https://doi.org/10.1007/978-981-99-9054-2_9

growing, which increases the need for a reduced oscillations damping in the power network stabiliser. The low magnitude and low frequency often prolong in some scenarios even restricting the power transfer capacity. As a result, low-frequency oscillation or electro-mechanical oscillations (EMO) [1] conventionally takes place in large power systems and occasionally make the feeding system unstable. Also, the incorporation of AVR used to improve the transient stability during fault also contribute to the reduction in damping of the power system owing to its high gain and fast-acting effect. Power system stabiliser (PSS) is utilised in the auxiliary feedback to add additional braking to the system to dampen these a reduced oscillations on the rotor in order to mitigate the negative impacts of the AVR.

The lead-lag compensators, also known as the PSS or conventional PSS (CPSS), are made up of gain stage K, a high pass filter, and temporal constants T_1 through T_4. To enhance the system damping, these parameters must be fine-tuned at a specific set of operating circumstances, often nominal. The operational environment is continually changing because of how highly nonlinear the power system is. As a result, it may be necessary to fine-tune the CPSS's parameters if they do not give an appropriate performance. Ferber et al. [2] Graham [3] details the impact of excitation and PSS. Several academics have worked to improve the PSS parameter tweaking methods. Yet, because of the non-linearity and multi-modality features, modelling the PSS is a difficult endeavour.

After changing system operating conditions, the PSS may not provide satisfactory damping in an unstable system. Several academics have worked to improve the PSS parameter tuning methods. These solutions include robust control [4, 5], optimization strategies [6, 7], and artificial intelligence techniques like neuro-fuzzy [8, 9] and fuzzy logic [10, 11]. These methods demand a lot of calculation time because they are complicated and use numerous particles in the optimization process.

A number of intelligence control and evolutionary optimization strategies have been effectively used in the past few years to address challenging optimization issues. Power system planning, design, operation, and control issues have frequently been addressed by the employment of high-speed computational tools. Researchers have given up on investigating various heuristic methods like particle swarm optimization (PSO) [12], and cultural algorithm [13]. For the best tuning of the PSS template utilised in the single machines infinite-buses (SMIB) system,the chaotic-orthogonal based learning (COBL) and Gaussian bare-bones implementation of the mean rating of vector (INFO) optimizer (GBB) methods is known as INFO-GBB [14]. The proposed INFO-usefulness GBB's and efficacy have been confirmed by comparing the results with eighteen well-known algorithms. With an overall ranking for the standard issues of more than 61 percent, the proposed algorithms surpass the bulk of the algorithms that were taken into consideration. Klein et al. [15] demonstrated the influence of thePSS about local and cross variations in the linked power network. Using generic and multiband (MB) PSS [16], the performance ofA hybrid 9 MW hydroelectric power plant and wind complex have been evaluated. A 200 MVA, 25 KV energy system is also coupled to the planned system to show how accurate the suggested Simulink model is.[17] Features smart, fuzzy-based synchronized control for AVR & PSS, to avoid declining concurrently succeeding unforeseen failures & reaching a relevant control point. In [18, 19], ANN techniques were employed to design PSS for resolving various difficult The PSS parameters have been effectively tuned using a variety of well-known population-based optimization approaches, including particle swarm optimization (PSO) [20], honey bee mating optimization [21], and others, to get the best fitness functions. Many areas of science and technology have found success using the aforementioned strategies [22]. Gurung et al. [23] suggested changing the IEEE 68 bus system to implement the concept of a deterministic and probabilistic method by coordinating several PSS to raise the system's small signal stability (SSS). The problem was then finally resolved to utilise a novel directional BAT algorithm. Panda et al [24] introduced the In order to determine the ideal PSS parameters and static synchronous series compensator (SSSC) settings, BFO (Hybrid Bacteria Foraging Optimization) and PSO (HBFOA-PSO) programs were used. The transient stability of both the SMIB system and the multi-machine system was improved by the author using the aforementioned strategies. The experiment's outcome was applied to a grid sample of four two-zone machines to reduce low-frequency fluctuations of native & interregional methods & thus enhance the steadiness of the system. The Cuckoo Search (CS) algorithm is used by Elazim et al. [25] to tune PSS parameters to their ideal values.

This opens the door for contemporary researchers to incorporate novel algorithms, in particular free input control parameters, into all variations of optimisations issues with the possibilities to solve the desired targeted function. This is referred to as "no-free-lunch" as none of the single optimization techniques will be able to be implemented successfully for all types of optimization problems. As a result, it is still feasible to make further advancements and go past the aforementioned drawbacks. With reference to "no-free-lunch" as none of the single optimization techniques will be able to be implemented successfully for all variations of optimisation issues, This opens the door for modern researchers to include novel algorithms, in particular free input control parameters, into all sorts of optimization problems with the possibilities of solving the desired targeted function. As a result, it is still possible to continue to improve the situation and overcome the aforementioned disadvantages. The main topic of the task in this specified chapter is the dampening of rotor oscillations in machine feeding systems equipped with PSS. Also, the introduction of the fact that the transitional effectiveness of the double insert IEEE-PSS is superior to the other, utilising various optimization strategies that are described, in the designated section, among the classic phasing adjustment PSS only one input and IEEE-PSS. The remaining papers were arranged in the manner mentioned below. In Part 2, a number of PSS models, including the SMIB, are fully detailed. The problem formulation and the system restrictions are illustrated in section "Problem Formulation". The solution strategy is covered in sections "Differential Search Algorithm (DSA)", "Quasi-Oppositional Based Learning (Q-OBL)" and "Cqodsa Steps for PSS Problem", where the optimization method for PSS is described. The job is finally completed in section "Conclusion" after the simulation results and various loading situations have been thoughtfully considered and reflected to section "Results and Discussion".

2 Smib Model and Various PSS Models

The present study incorporates a SMIB system [26] as illustrated in Fig. 1. Figure 2 represents the Pictorial representation of AVR and PSS [10]. The objective of PSS is to establish a As a logic input to regulate generator stimulation, velocity variation is employed to dampen the torque element. The inclusion ofasynchronous generating with high output thyristor alternator, AVR, and PSS block in the SMIB system has been illustrated in Fig. 3 [10]. The two-axis and fourth-order modes can be used to represent the generation system equipped with an IEEE type ST1A thyristor excitation system and an AVR. The Heffron-Phillips model's CPSS, which is a single input conventional PSS, is used to illustrate the linearized model of the SMIB system in Fig. 1, where the input is represented by $\Delta\omega_r$ and the output by ΔV_{PSS}. The CPSS primarily consists of a gain phase (K_{PSS}) that determines how much damping the stabilisers will apply, a washout stage that is designed to only react to speed oscillations, and a block stage that blocks dc offsets. Lead-lag compensators ($T_1 - T_4$) are used in the final stage to account for the phase difference brought on by the AVR and the reactors' fields circuits. The lead-lag variables are adjusted to give the rotor a damping torque when speed oscillations occur. The literature survey also justifies that in the present scenario, the modern excitation system is also predominantly relying on this interaction owing to the fact that these tools produce comparatively excessive gain at high frequency [20]. Predominantly, for light loading requirement for synchronous generators wherever the intrinsic mechanical damping is comparatively less, this may result in stabilizer-torsional instability with an excessive-responsive excitation system.

Therefore, it is essential to mention that stabilizers don't induce torsional instabilities. An extra blocks with a fixed latency time T_5 and lead time fixed T_{10} may be utilized for modelling stabilizers that may assimilate a more composite phase lead function (Fig. 4).

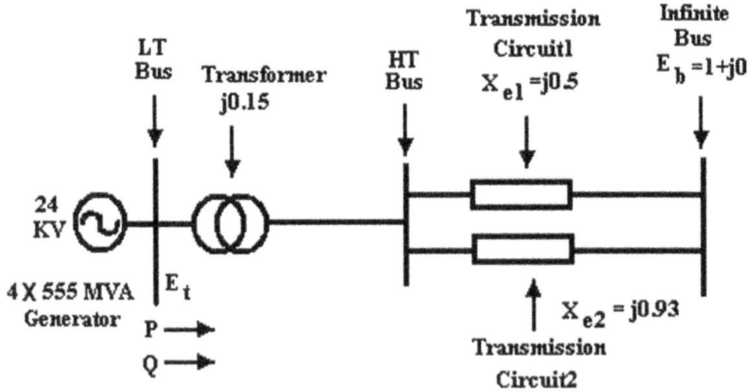

Fig. 1 Single-machine infinite bus test system

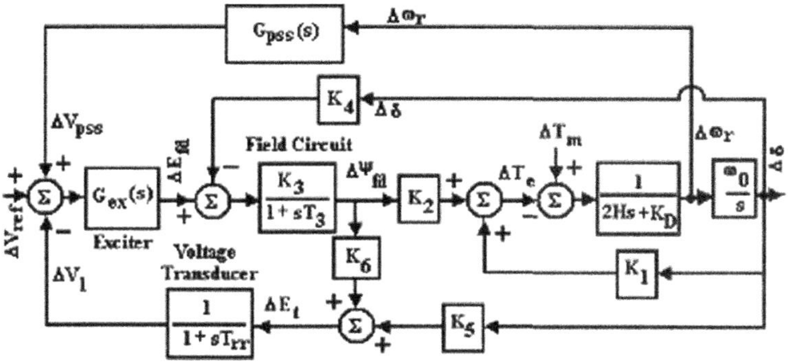

Fig. 2 Block diagram representation with AVR and PSS

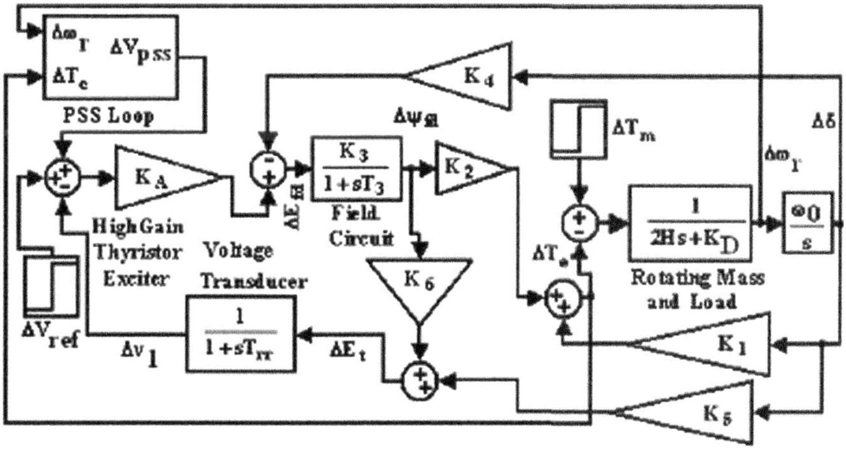

Fig. 3 Block diagram of the SMIB system showing the thyristor high gain exciter, synchronous generator, PSS and AVR

The four linear differential problems of first level serve as the synchronous machine model representation Eqs. (1–4). The fourth-order generator model is represented by these equations. Utilizing saturated mutual inductance values, magnetic saturation is either disregarded or taken into account based on the various levels of complexity as:

$$\frac{\mathrm{d}\delta}{\mathrm{d}t} = \omega - \omega_0 \tag{1}$$

$$\frac{\mathrm{d}\omega}{\mathrm{d}t} = \frac{1}{2H}(-D\omega + T_M + -T_e) \tag{2}$$

$$\frac{\mathrm{dE}'_{\mathrm{q}}}{\mathrm{d}t} = \frac{1}{T'_{d_0}}\left(-E'_q + \left(X_d - X'_d\right)i_d + E_{fd}\right) \tag{3}$$

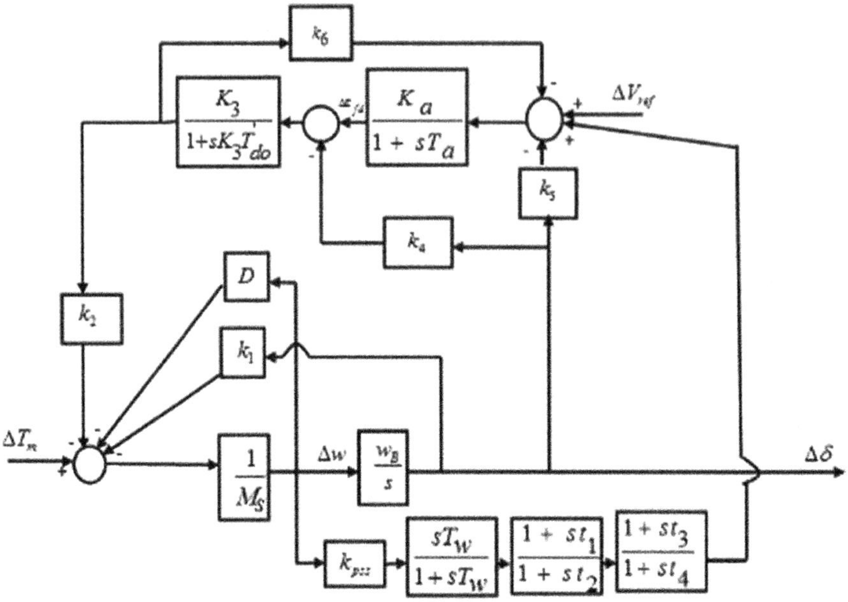

Fig. 4 Heffron-Phillips model

$$\frac{dE'_d}{dt} = \frac{1}{T'_{q0}}\left(-E'_d + \left(X_q - X'_q\right) i_q\right) \tag{4}$$

where, E_{fd} is the voltage proportional to field voltage; E_d' and E_q' are the voltages are the axis transient time constants are the elements i_d and i_q of the armature currents; and are equal to the dampers winding and fields flux.

3 Problem Formulation

In this research work, a single objective function has been reviewed to discover the efficacy of the presented CQODSA algorithm.

3.1 *Case-I: Eigenvalue Minimization*

By changing the position of the eigenvalues, the correlative stability of the system is obtained. In the case of real eigenvalues, the system corresponds to a non-oscillatory mode. Therefore, the system seems to behave unstably if the eigenvalue is positive and likewise the system goes into the decaying mode if the eigenvalue is negative. The system response shall be faster with farther the eigenvalues are in the negative-s plane. Like the complex form, as shown in Eq. 5, the eigenvalues happen to be in conjoined pairs. On the dependence of the real component 'σ', which either is negative or positive, it enlarges the magnitude of oscillations to either total instability or die out the oscillation. Here the eigenvalue-based objective function (OF_1) has been used as depicted in Eq. 6

$$s = \sigma \pm j\omega \tag{5}$$

$$OF_1 = \sum_i (\sigma_0 - \sigma_i) \tag{6}$$

The real part of the eigenvalue σ_i is the number of states for which the optimization is carried out, represented by n. σ_0 determines the relative stability with respect to the damping factor margin assuming for compelling the situation of the eigenvalues throughout the optimization process. The value σ_i shall depend on accordance with the given problem.

3.2 Case-II: Minimization of Damping Ratio

The damping ratio in Eq. 10, shows how rapidly the fluctuations are damped.

$$\zeta = \frac{\sigma}{\sqrt{\sigma^2 + \omega^2}} \tag{7}$$

Smallest Damping Ratio Taken Into Account, $\zeta = 0.3$. Therefore, reduction of the objective expression will reduce the undershoot, overshoot and settling time as depicted in Eq. 8

$$OF_2 = \sum_i (\sigma_0 - \sigma_i) \tag{8}$$

4 Differential Search Algorithm (DSA)

4.1 Inspiration of DSA

This part addresses a novel meta-heuristic optimization technique namely the differential search algorithm (DSA) established by Civicioglu [68]. The migration behaviour of living organisms in nature inspires DSA. The availability and efficiency of food areas existing in nature vary due to periodical changes during the year. For this reason, living organisms show migration behaviour throughout the year. This allows living organisms to move from one habitat to another where the availability and efficiency of natural resources are higher. This is the primary rationale used in the literature to derive DSA. Civicioglu [27].

4.2 DSA Mathematical Modelling

4.2.1 Initialization

In DSA, the initial population, referred to as the $artificialorganism$, is created through an arbitrary procedure that resembles to the optimisation problem. The parameters for artificial organism initialization in DSA are Eq. (2.8) [27].

$$X(i, j) = rand \times (ub(j) - lb(j)) + lb(j) \tag{9}$$

where,$i = 1, 2, 3, ..., n_p$, $j = 1, 2, 3, ..., dim$, $X(i, j)$ is the initial population of the jth variable of ith agent, $rand$ is an arbitrary number initiated between [0, 1], n_p and dim are referred to the number of search agents (superorganism) and a number of control variables of the optimization problem, respectively, $ub(j)$ and $ib(j)$ are the higher and lower limits of jth variable, respectively.

4.2.2 Random Shuffling of Superorganism

The need to determine the $stopoversite$, which is crucial for an effective migration, arbitrarily selects a $superorganism$ from the initial population and shifted towards the intended $donor$ by random shuffling. In order to regulate the positions of the artificial $superorganism$ members, a $scale$ value is generated by a gamma random generator by using Eq. (2.9) [27].

$$scale = randg(2 \times r_1) \times (r_2 - r_3) \tag{10}$$

where, $randg$ is the arbitrary value selected from a gamma distribution and r_1, r_2, and r_3 are the arbitrary numbers picked out between (0,1). In DSA, the $stopoversite$ of $superorganism$ is considered as [27].

4.2.3 Boundary Condition

There are only two control parameters in DSA, i.e., p_1 and p_2, those are randomly selected members of $superorganism$ to participate in $stopoversite$. In general, the values of and are generated between [0, 0.3]. If any element of the $stopoversite$ exceeds the habitat's limits, a boundary condition is put on the elements of the $stopoversite$ to compel them into the habitat. Civicioglu [27].

5 Quasi-Oppositional Based Learning (Q-OBL)

OBL was first introduced by Tizhoosh [28] as a cutting-edge notion for soft computing or intelligence-based techniques to improve different optimization methods. It appears to be one of the most successful theories in computational intelligence for tackling nonlinear optimization problems, enhancing the searching capabilities of traditional population-based optimization strategies. The OBL algorithm starts with the initialization of an initial estimate that is based on some prior information about the solution or randomly. The optimal solution could be in any direction or at least in opposite direction. For convergence, an opposite set of estimates are considered, which iteratively replaces the initial estimates for a better solution in the direction of optimality.

5.1 Opposite Number

Let, the real number be denoted by $P \in [y, z]$ and P^0 be its corresponding opposite number is defined real number by:

$$P^0 = y + z - P \tag{11}$$

5.2 Opposite Point

Say $R = (X_1, X_2, \ldots\ldots, X_n)$ is a point in n-dimensional space, where, $P_r \in [y_r, z_r]$, $r \in 1, 2, \ldots, n$. The $R^0 = ({X_1}^0, {X_2}^0, \ldots\ldots {X_n}^0)$ opposite point is defined by its components:

$$ {P_r}^0 = y_r + z_r - P_0 \tag{12}$$

5.3 Quasi-Opposite Number and Quasi-Opposite Point

Let P be a real number between $[y, z]$. The quasi-opposite number is defined as:

$$P^{Q_0} = rand(C, \tilde{P}) \tag{13}$$

where C is given by: $C = \frac{y+z}{2}$ Let P be a real number between $[a, b]$. The quasi-opposite point ${P_r}^{Q_0}$ is defined as:

$$ {P_r}^{Q_0} = rand(C_r, \tilde{P}_r) \tag{14}$$

$$\text{where} C_r = \frac{y_r + z_r}{2}$$

5.4 Chaotic DSA

Chaotic DSA (CDSA), which combines chaotic behaviour with Differential search optimization (DSA), was created to lessen the shortcomings of DSA. Chaos, which is stochastic and non-repeating by nature, conducts general searches at higher speeds, which is essential for quickening a metaheuristic algorithm's convergence rate. A chaotic set is composed of a total of 10 chaotic maps, each of which exhibits different behaviour. This set's starting point has been determined to be 0.7 between 0 and 1. The many chaotic map behaviours help to solve issues with local optimum and convergence speed.

6 Cqodsa Steps for PSS Problem

Different steps of the CQODSA for solving the PSS problem are presented below:

Step 1: Initialize $superorganism$ and create the first wave of people (P)of each individual at random. Also, generate the PSS control parameters with the defined solution space and calculate the fitness function.

Step 2: Generate chaotic quasi opposite population and calculate the fitness function.

Step 3: According to value, only a select handful of the best answers are privileged status-preserved solutions.

Step 4: The not belonging to the elite solutions are once more subjected to the varied chaotic map conduct of the CDSA in order to change the independent variables. In CDSA, the whales target the focuses particularly and the location of the fixates is regarded as the best location.

Step 5: Randon shuffling of $superorganism$ takes place and evaluate the scale using the gamma distribution.

Step 6: The method is finished and the outcomes can be seen if the stopping requirement is met; otherwise, proceed to **Step 4**.

7 Results and Discussion

The applicability and validity of the different optimization techniques have been tested on SMIB system by considering two types of objective functions, namely single objective function and multi-objective function. In all the cases, the synchronous machine model has been considered. All the simulations are carried out using MATLAB R2022b and implemented on a personal computer with Intel Core i7 2.20 GHz processor with 8GB-RAM.100 iterations are chosen as the terminating. For SMIB system, inertia constant, $H = 5$, $M = 2H$, nominal frequency, $f_0 = 50Hz$, $0.995 \leq \bar{E}_b \leq 1.0$; the angle of $\bar{E}_b = 0$; $0.2 \leq P \leq 1.2$; $-0.2 \leq Q \leq 1.0$; $0.4752 \leq X_e \leq 1.08$; $0.5 \leq E_t \leq 1.1$. To look into the proposed method, two new methods *i.e.*, minimization of eigenvalue and minimization of damping ratio are measured independently to review the strength of the suggested method. Well-known optimization technique, namely CQODSA is analysed and compared with the results of other well-known algorithms to verify the supremacy of the said algorithms.

7.1 Case-I: Minimization of Eigen Value

The Heffron-Phillips model is a component of the linearized SMIB system. To dampen out the low-frequency fluctuations affecting the power system, PSSs are installed at respective locations. The state space matrix eigenvalue is presented as:

$$det\,|A - sI| \tag{15}$$

where s is the eigenvalue; I is the identity matrix and is the state matrix. On a wide variety of operational conditions, the algorithms are tested as heavy ($P = 1.45$; $Q = 0.70$), nominal ($P = 1.25$; $Q = 0.50$) and light loading ($P = 0.65$; $Q = 0.55$) conditions. The comparative results of the fitness functions along with the optimal PSS parameters have been depicted in Tables 1, 2 and 3. It can be clearly shown from the corresponding tables that there is an improvement in the fitness function values in addition to improved overshoot and undershoot in all three-loading environments. By taking into account the suggested CQODSA method, the overshoot (O_s), undershoot (U_s), and fitness function values are improved.

Table 1 Eigen value based fitness function with light loading condition ($P = 0.65$; $Q = 0.55$)

Algorithm	Eigen value	Overshoot	Undershoot	Fitness value
CQODSA	**-1.5000±6.1976i**	**7.0065**	**13.4532**	**0.0053**
QODSA	-1.4999±6.8762i	7.6218	14.2342	0.0087
OCRO[29]	-1.5000±6.2387i	7.1250	13.9311	0.0099
CRO[30]	-1.5000±6.1432i	7.0782	14.0325	0.0106
OCOA[31]	-1.500±6.1901i	2.3179	13.2901	0.0246
COA[31]	-1.5000±6.2645	4.7879	14.1270	0.0877
ODSA[32]	-1.5000±6.1177i	5.6053	12.0614	0.0037
DSA[32]	-1.5000±6.1002i	7.1343	13.9602	0.0240
OGSA[33]	-1.5000±6.3855i	4.6761	13.0352	0.0319
GSA[33]	-1.5000±6.7786i	4.8443	13.4608	0.0859

Table 2 Eigen value based fitness function with light loading condition($P = 1.25$; $Q = 0.50$)

Algorithm	Eigen value	Overshoot	Undershoot	Fitness value
CQODSA	**−1.4999±5.1346i**	**8.7653**	**15.6512**	**0.0009**
QODSA	−1.5000±5.2546i	8.8976	16.0000	0.0010
OCRO[29]	−1.4999±6.7610i	8.9697	15.7167	0.0011
CRO[30]	−1.5000±6.7786i	9.0115	15.6368	0.0860
OCOA[31]	−1.5000±6.7034i	6.5138	13.7391	0.0015
COA[31]	−1.5000±5.4852	8.4237	14.4600	0.0074
ODSA[32]	−1.4999±6.8766i	5.7082	13.0012	0.0031
DSA[32]	−1.5000±6.6214i	7.8280	14.2250	0.0044
OGSA[33]	−1.5000±6. 2018i	7.1298	12.2386	0.0030
GSA[33]	−1.5000±4.2791i	4.6761	13.1397	0.0738

Table 3 Eigen value based fitness function with heavy loading condition ($P = 1.45$; $Q = 0.70$)

Algorithm	Eigen value	Overshoot	Undershoot	Fitness value
CQODSA	**−1.5000±3.0165i**	**3.1954**	**9.4117**	**0.0010**
QODSA	−1.5000±3.2397i	3.3181	9.4817	0.0014
OCRO[29]	−1.4999±3.1313i	3.3183	9.4795	0.0016
CRO[30]	−1.5000±3.1067i	3.3131	9.4713	0.0381
OCOA[31]	−1.5000±3.1913i	2.7106	7.8137	0.0024
COA[31]	−1.5000±3.1912i	3.6121	9.5125	0.0045
ODSA[32]	−1.4999±3.1357i	3.3173	9.4803	0.0012
DSA[32]	−1.5000±3.1366i	3.3138	9.4800	0.0029
OGSA[33]	−1.5000±3.1602i	3.3131	9.4865	0.0074
GSA[33]	−1.5000#3.1067i	3.3131	9.4713	0.0381

7.2 Case-II: Damping Ratio Optimization

To determine the applicability of the suggested CQODSA algorithm and to manage the stability of the power system, damping ratio minimizations are taken into account. The oscillation is muted according to the rate of damping. The algorithms are tested under the aforementioned loading circumstances. The overshoot and undershoot, together with the fitness function, are depicted in Tables 4, 5 and 6 on the basis of final optimization values.

Table 4 Results of damping ratio minimization compared between various techniques for light a loads conditions ($P = 0.65$; $Q = 0.55$)

Algorithm	Eigen value	Overshoot	Undershoot	Fitness value
CQODSA	**−1.1513±6.3480**	**5.3289**	**13.4380**	**0.0032**
QODSA	−1.2768±6.3289	6.9358	14.3201	0.0055
OCRO[29]	−1.2190±5.9323i	8.1975	14.2575	0.0250
CRO[30]	−0.8754±4.2883i	5.7747	12.3212	0.0572
OCOA[31]	−1.2685±6.2145i	7.9363	14.0000	0.0170
COA[31]	−1.2515±6.3242i	7.9436	14.0000	0.0169
ODSA[32]	−1.2624±6.1843i	6.6200	12.1270	0.0010
DSA[32]	−1.2419±6.0836i	7.9445	13.5749	0.0153
OGSA[33]	−1.2506±6.1267i	7.9589	13.7845	0.0346
GSA[33]	−1.2323±6.0383i	7.9949	14.0674	0.4315

Table 5 Results of damping ratio minimization compared between various techniques for a nominal loads conditions ($P = 1.25$; $Q = 0.50$)

Algorithm	Eigen value	Overshoot	Undershoot	Fitness value
CQODSA	**−1.3981±6.2347i**	**8.4398**	**13.6534**	**0.0023**
QODSA	−1.3951±6.8347i	9.3277	15.6768	0.0032
OCRO[29]	−1.3839±6.7798i	9.3656	15.7411	0.0048
CRO[30]	−1.3568±6.6466i	9.4227	15.8934	0.0712
OCOA[31]	−1.3981±6.8494i	7.3162	13.6500	0.0222
COA[31]	−1.3879±6.8154i	7.9271	13.9846	0.0563
ODSA[32]	−1.3993±6.8551i	8.7143	13.6627	0.0155
DSA[32]	−1.3818±6.7694i	9.3140	14.7523	0.02807
OGSA[33]	−1.3892±6.8057i	9.3425	15.6839	0.0172
GSA[33]	−1.3672±6.6995i	9.4051	16.1162	0.5264

Table 6 Results of damping ratio minimization compared between various techniques for a heavy loads condition ($P = 1.45$; $Q = 0.70$)

Algorithm	Eigen value	Overshoot	Undershoot	Fitness value
CQODSA	**−0.6223±3.2134i**	**6.5687**	**10.6540**	**0.0033**
QODSA	−0.6954±3.4200i	6.8094	10.9845	0.0065
OCRO[29]	−0.6853±3.3577i	6.8998	10.8801	0.0075
CRO[30]	−0.6841±3.3512i	6.8994	10.8786	0.0811
OCOA[31]	−0.4365±2.1385	4.0389	8.3912	0.0357
COA[31]	−0.6028±3.4543i	4.9815	8.8371	0.0395
ODSA[32]	−0.4985±2.4421i	1.9867	9.1208	0.0111
DSA[32]	−0.7107±3.4815i	1.9951	10.8733	0.0395
OGSA[33]	−0.6967±3.4130i	2.5969	6.8726	0.0543
GSA[33]	−0.6841±3.3512i	4.80008	9.6810	0.0811

8 Conclusion

For this survey, the CQODSA access is used in an effort to increase the SMIB's adaptable stability when including PSS. Initially, the unique DSA is used to catch out an optimum range of PSS variables and then to speed up the convergence and increase the stability and accuracy of the solution, the Chaotic Q-OBL theory was also included. The two separate objective functions known as minimization of the values were normalized and minimization of impact factor are employed to test the superiority of the expected CQODSA technique. Furthermore, to demonstrate its superiority, is compared to other state-of-the-art algorithms, and it is discovered that it performs much better than all other algorithms addressed in this study in terms of solution quality, speed, and final solution stability. The proposed algorithm also has a high degree of convergence. Finally, on the basis of originality, the suggested algorithm is a very promising evolutionary optimisation approach for global optimisation of any PSS problem.The capacity to remove the drawbacks of premature convergence and create a global or near-the-globally ideal answer may be summed up as the best convergence qualities of the aforementioned CQODSA.

References

1. Leonard LG (2006) Electric power engineering handbook. CRC Press LLC, London
2. Ferber RS, Harvey DH, Glenn EM, Ellis EH (1968) Excitation control to improve powerline stability. IEEE Trans Power Apparatus Syst 6:1426–1434
3. Graham R (2012) Power system oscillations. Springer Science & Business Media
4. Rajeev G, Sambariya DK, Reena G (2006) Fuzzy logic based robust power system stabilizer for multi-machine power system. In:2006 IEEE international conference on industrial technology. IEEE, pp 1037–1042
5. Sambariya DK, Rajendra P (2013) Design of pss for smib system using robust fast output sampling feedback technique. In: 2013 7th international conference on intelligent systems and control (ISCO). IEEE, pp 166–171
6. Muawia AM, Nordin BS, Rosdiazli BI (2014) Power system stabiliser for single machine in infinite bus based on optimal control methods. In: 2014 IEEE 8th international power engineering and optimization conference (PEOCO2014). IEEE, pp 313–317
7. Mariano SJPS, Pombo JAN, Calado MRA, Felippe de Souza JAM (2016) Damping of power system oscillations with optimal regulator. In: Electricity distribution: intelligent solutions for electricity transmission and distribution networks, pp 173–198
8. Eke I, Taplamacıoğlu MC, Kwang YL (2015) Robust tuning of power system stabilizer by using orthogonal learning artificial bee colony. IFAC-PapersOnLine 48(30):149–154
9. Ravi S, Avdhesh S, Kothari ML (2004) A self-tuning power system stabilizer based on artificial neural network. Int J Electr Power Energy Syst 26(6):423–430
10. Bhati PS, Rajeev G (2013) Robust fuzzy logic power system stabilizer based on evolution and learning. Int J Electr Power Energy Syst 53:357–366
11. Hiyama T, Kita T, Miyake T, Andou H (1999) Experimental studies of three dimensional fuzzy logic power system stabilizer on damping of low-frequency global mode of oscillation. Fuzzy Sets Syst 102(1):103–111
12. Butti D, Mangipudi SK, Rayapudi SR (2019) Design of robust modified power system stabilizer for dynamic stability improvement using particle swarm optimization technique. Ain Shams Eng J 10(4):769–783

13. Khodabakhshian A, Hemmati R (2013) Multi-machine power system stabilizer design by using cultural algorithms. Int J Electr Power Energy Syst 44(1):571–580
14. Snášel V, Rizk-Allah RM, Izci D, Ekinci S (2023) Weighted mean of vectors optimization algorithm and its application in designing the power system stabilizer. Appl Soft Comput 110085
15. Klein M, Rogers GJ, Moorty S, Kundur P (1992) Analytical investigation of factors influencing power system stabilizers performance. IEEE Trans Energy Conv 7(3):382–390
16. Kamal KS, Akhil G, Gagandeep K, Raman K, Jasgurpreet SC, Shubham S, Jujhar S, Nima K, Alibek I (2021) Power quality and transient analysis for a utility-tied interfaced distributed hybrid wind-hydro controls renewable energy generation system using generic and multiband power system stabilizers. Energy Report 7:5034–5044
17. Khezri R, Bevrani H (2015) Stability enhancement in multi-machine power systems by fuzzy-based coordinated avr-pss. Int J Energy Optim Eng (IJEOE) 4(2):36–50
18. Zhang Y, Chen GP, Malik OP, Hope GS (1993) An artificial neural network based adaptive power system stabilizer. IEEE Trans Energy Conv 8(1):71–77
19. Mahabuba A, Abdullah Khan M (2009) Small signal stability enhancement of a multi-machine power system using robust and adaptive fuzzy neural network-based power system stabilizer. Eur Trans Electr Power 19(7):978–1001
20. Peres W, Júnior ICS, Filho João Alberto P (2018) Gradient based hybrid metaheuristics for robust tuning of power system stabilizers. Int J Electr Power Energy Syst 95:47–72
21. Taher N, Hasan DM, Hamed ZM, Bahman BF (2011) A new honey bee mating optimization algorithm for non-smooth economic dispatch. Energy 36(2):896–908
22. Marmolejo JA, Jonás V, Héctor JS (2017) An adaptive random search for short term generation scheduling with network constraints. PloS One 12(2):e0172459
23. Gurung S, Jurado F, Naetiladdanon S, Sangswang A (2020) Comparative analysis of probabilistic and deterministic approach to tune the power system stabilizers using the directional bat algorithm to improve system small-signal stability. Electr Power Syst Res 181:106176
24. Sidhartha P, Narendra KY, Sangram KM (2013) Hybrid bfoa-pso approach for coordinated design of pss and sssc-based controller considering time delays. Int J Electr Power Energy Syst 49:221–233
25. Abd Elazim SM, Ali ES (2016) Optimal power system stabilizers design via cuckoo search algorithm. Int J Electr Power Energy Syst 75:99–107
26. Prabha SK, Om PM (2022) Power system stability and control. McGraw-Hill Education
27. Civicioglu P (2012) Transforming geocentric cartesian coordinates to geodetic coordinates by using differential search algorithm. Comput Geosci 46:229–247
28. Hamid RT (2005) Opposition-based learning: a new scheme for machine intelligence. In: International conference on computational intelligence for modelling, control and automation and international conference on intelligent agents, web technologies and internet commerce (CIMCA-IAWTIC'06), vol 1. IEEE, pp 695–701
29. Paul S, Maji A, Roy PK (2016) The oppositional chemical reaction optimization algorithm for the optimal tuning of the power system stabilizer. In: Foundations and frontiers in computer, communication and electrical engineering: proceedings of the 3rd international conference on foundations and frontiers in computer, communication and electrical engineering (C2E2-2016). Taylor & Francis Books Ltd, pp 101–106
30. Sourav P, Roy PK (2015) Optimal design of single machine power system stabilizer using chemical reaction optimization technique. Int J Energy Optim Eng (IJEOE) 4(2):51–69
31. Paul S, Roy PK (2015) Oppositional cuckoo optimization algorithm for optimal tuning of power system stabilizers. In: Michael faraday IET international summit 2015. IET, pp 176–181
32. Paul S, Roy P (2018) Optimal design of power system stabilizer using a novel evolutionary algorithm. Int J Energy Optim Eng (IJEOE) 7(3):24–46
33. Paul S, Roy PK (2014) Optimal design of power system stabilizer using oppositional gravitational search algorithm. In: 2014 1st international conference on non conventional energy (ICONCE 2014). IEEE, pp 282–287

Electrical and Optical Properties Optimization for Efficiency Improvement of CZTSSe Solar Cell

Ambrish Yadav, Rajan Mishra, and Alok Kumar Patel

Abstract Cu_2ZnSnS_4(CZTS) is a thin film compound structure, considered an excellent photovoltaic absorber material for Photovoltaic (PV) cell. The proposed structure of solar cell is front-contact/ZnO/CdS/CZTS/NiO/back-contact. One of the layers that significantly affects the conductivity of the carriers and improves the power conversion efficiency of PV cell is the hole transport layer (HTL). The goal of this examination is to determine the impact of NiO hole transport layer on the performance of proposed CZTS absorber material based solar cell. For this solar cell structure, the open—circuit voltage (V_{oc}), short—circuit current density (J_{sc}), fill factor (FF), and power conversion efficiency (PCE) are calculated at various thickness of the absorber material, operating temperature, and series resistance of the device. At optimized thickness of 2 μm, efficiency is achieved 23.13% while V_{oc} is 0.9813 V, J_{sc} is 27.8 mA/cm^2, and FF is 84.59%. This work provides guidelines for the fabrication of thin film CZTS solar cell.

Keywords CZTS · SCAPS 1D · Solar Cell

1 Introduction

Because of the serious worldwide temperature alteration peculiarity brought about by the utilization of petroleum derivatives, as an elective energy asset and for manageable turn of events, is requesting to further develop the PCE and decrease the general creation cost of sunlight based PV modules [1]. PV semiconductor devices are widely employed in power generation and play a vital role in commercial applications. Sunlight radiation is absorbed by PV cells, which then transform it into electrical energy [2]. PV cells can be crystalline or non-crystalline, and they are built of both elemental and synthetic materials. Crystalline materials have greater PCE, but their production costs are higher than those of poly- and micro-crystalline materials. The poly- and micro-crystalline families of materials excellent performance, affordable

A. Yadav (✉) · R. Mishra · A. K. Patel
Madan Mohan Malaviya University of Technology, Gorakhpur, India
e-mail: ambrishyadav2012@gmail.com

S. Kumar et al. (eds.), *Control Applications in Modern Power Systems*, Lecture Notes in Electrical Engineering 1128, https://doi.org/10.1007/978-981-99-9054-2_10

manufacture, dependability, and stability attracted increased research interest [3]. Economical PV devices with high power conversion efficiency (PCE) that don't degrade the environment are currently being developed for the purpose of generating electricity [4]. Thin film PV cells are becoming more significant as a result of increased production capabilities after many years of market dominance by silicon-based solar cells [5]. The first vacuum—deposited CZTS solar cell with a 0.66% PCE was announced in 1997 [6]. For PV solar cells built on the CZTS platform, a conversion efficiency of 6.7% was stated in 2009 [7]. The highest PCE ever remained 8.4% in 2011 [8]. For CZTSSe in 2013, a conversion Effectiveness of 12.6% was observed [9]. The benchmark for new PV technologies is the Shockley Queisser (SQ) limit. It specifies the highest rate of solar energy conversion attainable for a given substance [10]. The Shockley Queisser limit states that by adjusting the bandgap, a CZTS based PV cell has the highest theoretical conversion efficiency of around 28%. However, because of a lack of knowledge about the properties of the materials, it is still not possible to experimentally produce a solar cell with a PCE of 28%. The performance of cells and a better understanding of material properties both benefit greatly from numerical analysis. According to numerical study the solar cell with CZTS/CdS based structure measured PCE in 2010 is approximately 7.55% [11] and in 2016, it has a PCE of 15% [12]. A solar cell with the CZTS/CdS structure has PCE of 18.05% was presented in 2017 [13]. In order to analyze and improve the various PV performance parameters of the suggested device, the SCAPS-1D simulation of solar cell specific (NiO/CZTS/CdS/ZnO) structure was used throughout the project. The electrical properties of CdTe, CZTS, and CIGS PV cells, including current in form of DC and AC, have been simulated using SCAPS-1D simulation software. Additionally, it makes it possible to describe solar cells using their quantum efficiency (QE) and J-V properties. The QE and J-V curve also make it possible to identify current losses caused by reflection and recombination in various layers. This is especially beneficial for enhancing the effectiveness of PV modules [14].

2 Device Structure and Numerical Simulation

Figure 1 shows the suggested solar schematic diagram of ZnO/CdS/CZTS/NiO, where NiO is represented as HTL, CdS is represented as Buffer layer, and CZTS is represented as Absorber layer. A vital way for PV cell characterization from numerical simulation. Controlling the overall device and material qualities is one of its features. The model input parameters can be added via numerical simulation and change in the J-V or QE can be quantified. One-dimensional simulation software called SCAPS-1D is used to model devices [15].

A lot of people use SCAPS-1D to simulate CIGS and CdTe, and it was designed to simulate and help with the analysis of the J-V characteristic curve, spectral response of cell, PCE, FF, J_{SC}, V_{OC}, energy bands of all semiconductor material as well as the concentration of different materials for use in attempting to solve every semiconductor vital type of equation, along with the Poisson equation as well as continuity

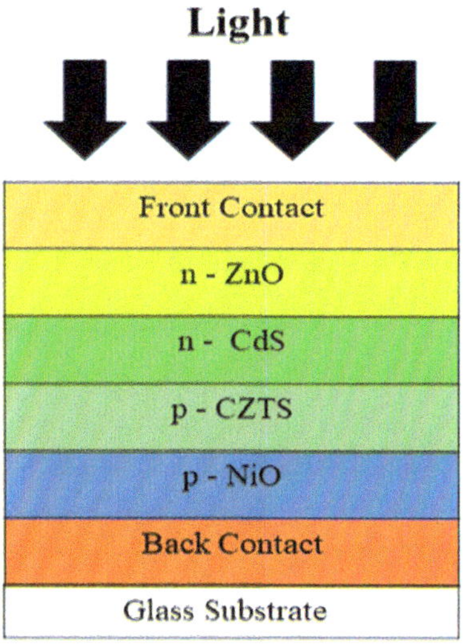

Fig. 1 Schematic diagram of PV cell

equation for both holes and electrons [16, 17]. In this paper, all thin film based solar cell simulations have taken into consideration the PV light spectra AM 1.5 has incident power density of 1000 Wm^{-2}.

As shown in Fig. 2 the NiO HTL valance band energy is a little bit greater than that of the CZTS absorber, allowing for an efficient transfer of the photo created holes from the CZTS layer to the back electrodes initially. Since the NiO HTL conduction band energy is higher than CZTS absorption layer, a barrier at the NiO HTL prevents photo-generated electrons from travelling from of the CZTS to a back electrode. In order to transfer holes, the back electrodes through NiO HTL and prohibit electrons from doing so, NiO and CZTS make a suitable junction. The optical and electrical properties of the various materials have been listed in Table 1. In Table 1 E_g signifies energy gap, χ signifies electron affinity, ε_r signifies permittivity, Nc signifies density of state of conduction band, Nv signifies density of state of valance band, μ_e and μ_p signifies electron and hole mobility, N_D and N_A signifies donor and acceptor concentration.

3 Results and Discussion

3.1 Influence of Series Resistance

Series resistance created because of bulk material junction and junction contacts of the solar cell. Circuit diagram of the proposed PV device is represented in Fig. 3a where R_{sh} is shunt resistance and R_s is series resistance. In this section, series resistance increases from 1 to 6.5 Ω cm^2. Figure 3b represents current density–voltage

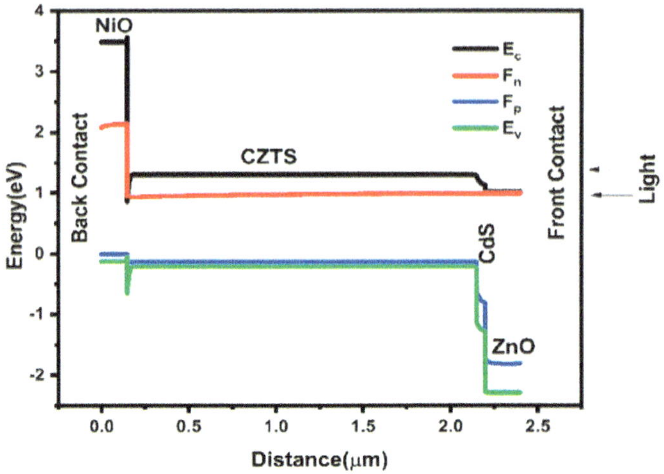

Fig. 2 Energy band diagram of PV Cell

Table 1 Materials that were Simulated for use in modelling the CZTS based solar cell [18, 19]

Parameter	NiO	CZTS	CdS	n—ZnO
Thickness (μm)	0.15	2	0.05	0.2
Eg (eV)	3.6	1.5	2.4	3.3
X (eV)	1.8	4.5	4.5	4.6
ε_r	11.7	10	9	9
N_C (cm^{-3})	2.5×10^{20}	2.2×10^{18}	1.8×10^{19}	2.2×10^{18}
N_V (cm^{-3})	2.5×10^{20}	1.8×10^{19}	2.4×10^{18}	1.8×10^{19}
μ_e ($cm^2 V^{-1} s^{-1}$)	2.8	100	3.5×10^2	100
μ_p ($cm^2 V^{-1} s^{-1}$)	2.8	25	50	25
N_D (cm^{-3})	0	100	1×10^{17}	1×10^{18}
N_A (cm^{-3})	3×10^{18}	2×10^4	1×10^5	1×10^5

curves. As R_s increases V_{oc} increases while J_{sc}, FF and efficiency decrease which are shown in Fig. 4a, b. The decrease in J_{sc} with increase in series resistance (Rs) is due to conduction conductivity of carrier decreases and at 1 Ω cm^2 the value of V_{oc} is 0.9813 V, J_{sc} is 27.854 mA/cm^2, FF is 81.98% and efficiency is 22.41% and for high efficiency of solar cell take low R_s.

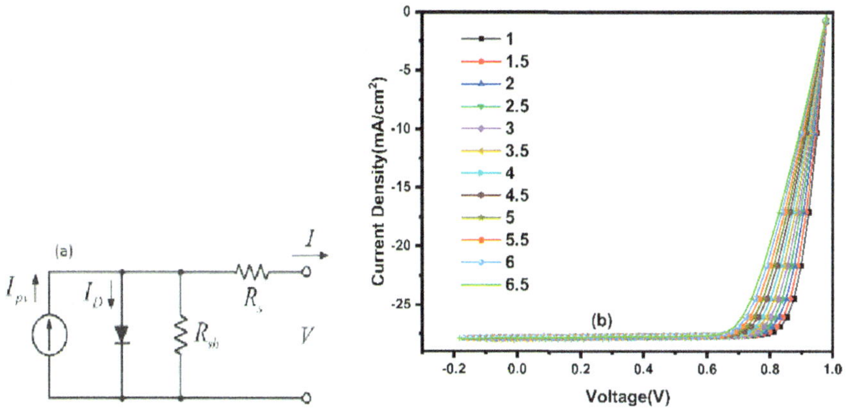

Fig. 3 **a** Circuit diagram of shunt and series resistance **b** Current density (J)–voltage (V) curves to different series resistance

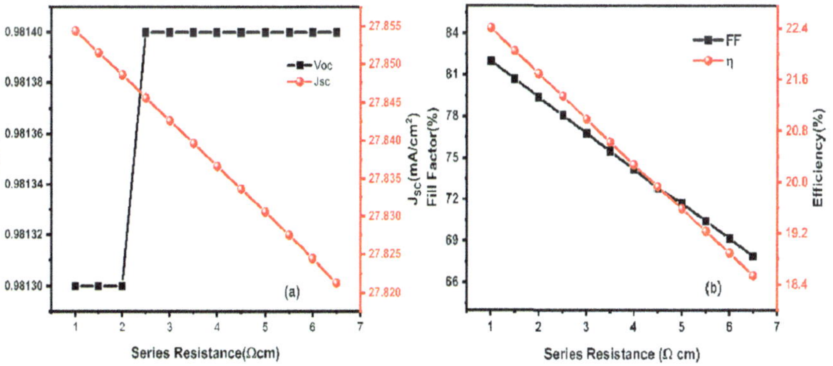

Fig. 4 **a** Series resistance influence on V_{oc} and J_{sc}. **b** Series Resistance influence on FF and PCE

3.2 *Influence of Absorber Layer Thickness*

In this part, the thickness of the CZTS based absorber layer varied between 400 and 2000 nm. Figure 5 shows current density and voltage curves at different thickness of CZTS layer. Figure 6a, b shows thickness influence on V_{oc}, J_{sc}, FF and PCE of PV cell. From Fig. 6 we can observe that the V_{oc}, J_{sc}, FF and PCE improve with increase in the thickness of the CZTS layer. This is due to thicker absorber absorb material large number of photons as compared to thinner absorber layer. Thicker absorber layer having ability to absorb longer wavelengths of photons which contribute to the formation of large number of electron/hole pairs. It was discovered that at 2 μm, thin film solar device shows optimum PCE of 23.13% (Voc = 0.9813 V, Jsc = 27.86 mA/cm^2 and FF = 84.59%).

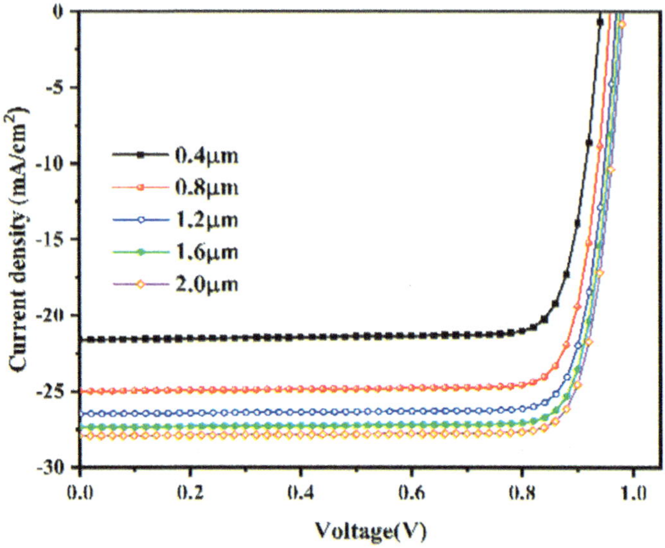

Fig. 5 Current density (J)–voltage (V) curves at different thickness of absorber

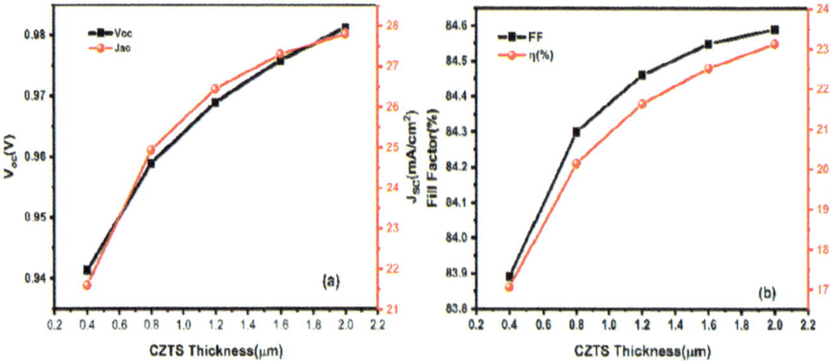

Fig. 6 **a** Thickness influence on solar cell V_{oc} and J_{sc} **b** Thickness influence on solar cell FF and PCE

3.3 Influence of Temperature

The Influence of temperature changes between 300 and 400 K for the CdS/CZTS structure. The absorber layer was maintained at a thickness of 2 µm. V_{oc}, J_{sc}, FF, and quantum efficiency are plotted with respect to temperature which is indicated in Fig. 8. It is evident that when the temperature rises, the V_{oc} declines and the J_{sc} rises. This can be also seen in JV curves in Fig. 7. Due to the development of semiconductor material, the J_{sc} slightly increases because energy band gap reduces with increases in temperature and due to this, a greater number of free electrons reach the conduction band. We can also observe that, rise in temperature causes an increase in reverse saturation current, which is the major factor in the V_{oc} reduction as given in Eq. 1. In this equation J_e signifies current density, J_0 signifies reverse saturation current [20]. This will have an impact on the PV cell PCE and overall FF. It was discovered that at 300 K, thin film PV device shows optimum PCE of 23.13% ($V_{oc} = 0.9813$ V, $J_{sc} = 27.86$ mA/cm^2 and FF = 84.59%).

$$J_e(V) = J_0\left[exp\left(\frac{qV_{oc}}{K_BT}\right) - 1\right] \tag{1}$$

Fig. 7 Voltage and current density curves at different temperature

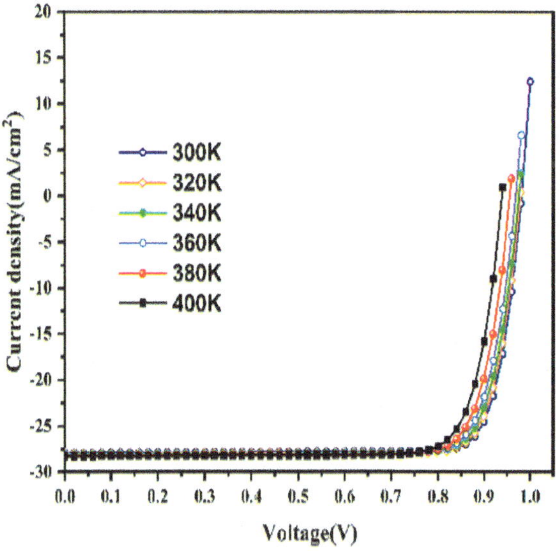

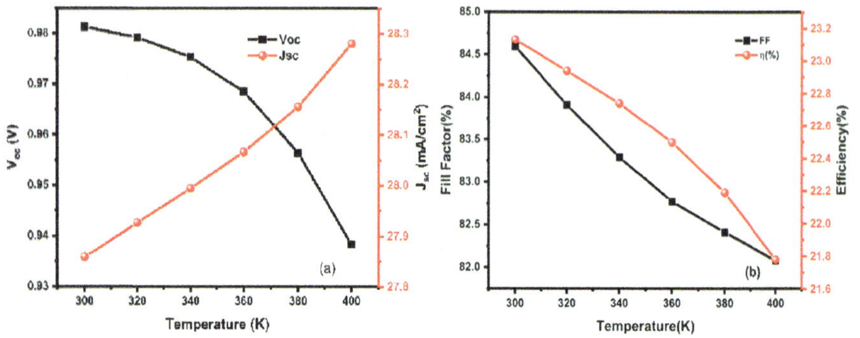

Fig. 8 **a** Temperature influence on solar cell V_{oc} and J_{sc}. **b** Temperature influence on solar cell FF and PCE

4 Conclusion

In this paper, the heterojunction solar cell was simulated based n-ZnO, n-CdS, p-CZTS, and p-NiO on SCAPS software. The output parameters of a hetero structure ZnO/CdS/CZTS/NiO are affected by the thickness of absorber layer, temperature, and series resistance. The optimum thickness of CZTS absorber layer for solar cell performance is 2 μm. By the optimization of device structure and optical properties the efficiency of 23.13% ($V_{oc} = 0.9813$ V, $J_{sc} = 27.8$ mA/cm^2, FF = 84.59%) has been achieved. This proposed simulation provides a path for fabrication of CZTS based thin film solar cell.

Acknowledgements The authors warmly credit Prof. Marc Burgelman of the University of Gent in Belgium for sharing SCAPS-1D simulation software.

References

1. Rahman MA (2021) Design and simulation of a high-performance Cd-free Cu_2SnSe_3 solar cells with SnS electron-blocking hole transport layer and TiO_2 electron transport layer by SCAPS-1D. SN Appl Sci 3
2. Patel AK, Rao PK, Mishra R, Soni SK (2021) Numerical study of a high-performance thin film CIGS solar cell with a-Si and MoTe2 hole transport layer. Optik (Stuttg) 243
3. Green MA, Emery K, Hishikawa Y, Warta W, Dunlop ED (2015) Solar cell efficiency tables (version 45). Prog Photovoltaics Res Appl 23:1–9
4. Jimbo K et al (2007) Cu_2ZnSnS_4-type thin film solar cells using abundant materials. Thin Solid Films 515:5997–5999
5. Hegedus S (2006) Thin film solar modules: The low cost, high throughput and versatile alternative to Si wafers. Progress Photovoltaics Res Appl 14:393–411. https://doi.org/10.1002/pip.704
6. Katagiri H et al (1997) Preparation films by and evaluation of Cu_2ZnSnS_4 thin sulfurization of E-B evaporated precursors. Solar Energy Mater Solar Cells 49

7. Katagiri H et al (2009) Development of CZTS-based thin film solar cells. Thin Solid Films 517:2455–2460
8. Shin B et al (2016) Thin film solar cell with 8.4% power conversion efficiency using an earth-abundant Cu_2ZnSnS_4 absorber. Prog Photovoltaics Res Appl 21:72–76
9. Wang W et al (2014) Device characteristics of CZTSSe thin-film solar cells with 12.6% efficiency. Adv Energy Mater 4
10. Xu Y, Gong T, Munday JN (2015) The generalized Shockley-Queisser limit for nanostructured solar cells. Sci Rep 5
11. Amin N, Hossain MI, Chelvanathan PM, Uzzaman ASM, Sopian K (2010) Prospects of Cu_2ZnSnS_4 (CZTS) solar cells from numerical analysis. Int Conf Electr Comput Eng (ICECE 2010) 4:730–733
12. Wanda MD, Ouédraogo S, Tchoffo F, Zougmoré F, Ndjaka JMB (2016) Numerical investigations and analysis of Cu_2ZnSnS_4 based solar cells by SCAPS-1D. Int J Photoenergy 2016
13. Adewoyin AD, Olopade MA, Chendo M (2017) Enhancement of the conversion efficiency of Cu_2ZnSnS_4 thin film solar cell through the optimization of some device parameters. Optik (Stuttg) 133:122–131
14. Mandadapu U, Vedanayakam SV, Thyagarajan K, Babu BJ (2018) Optimisation of high efficiency tin halide perovskite solar cells using SCAPS-1D. Int J Simul Process Modell 13
15. Patel AK, Mishra R, Soni SK (2022) Performance enhancement of CIGS solar cell with two dimensional MoS_2 hole transport layer. Micro Nanostruct 165
16. Simya OK, Mahaboobbatcha A, Balachander K (2015) A comparative study on the performance of Kesterite based thin film solar cells using SCAPS simulation program. Superlattices Microstruct 82:248–261
17. Decock K, Zabierowski P, Burgelman M (2012) Modeling metastabilities in chalcopyrite-based thin film solar cells. J Appl Phys 111
18. Khan R, Ahmed SF, Khalid M, Joshi B (2021) Investigating effect of CdS buffer layer on the performance of Cu_2ZnSnS_4 based solar cells using SCAPS-1D. Trans Electr Electron Mater 22:177–184
19. Islam MS et al (2021) Defect study and modelling of SnX3-based perovskite solar cells with SCAPS-1D. Nanomaterials 11
20. Sha WEI, Ren X, Chen L, Choy WCH (2015) The efficiency limit of $CH_3NH_3PbI_3$ perovskite solar cells. Appl Phys Lett 106

Perovskite $CH_3NH_3SnI_3$ Solar Cell Performance Investigation with SnS_2 Buffer Layer

Abhimanyu Chaudhari, Rajan Mishra, and Alok Kunar Patel

Abstract A Solar cell, also identified as a photovoltaic cell, is a digital device that converts daylight directly into electric electricity through the photovoltaic effect. The research community is paying close attention to perovskite solar cells (PSCs) because to the sudden increase in their efficiency (PCE) to 25.6% over a short period of time. However, a significant barrier to their commercialization has been the poisonousness of the more prevalent lead created perovskites. Due to their high efficiency in converting power and cheap manufacturing costs, perovskite materials have attracted a lot of attention recently. Perovskite cells are retaining a very high desire within the solar energy world, with a year-over-year increase in efficiency. In this study, the TCO/SnS_2/$CH_3NH_3SnI_3$/p-$CH_3NH_3SnI_3$ solar cell was simulated using SCAPS-1D. For this, each of these nonidealities was carefully examined. The synthetic solar cell is enhanced by changing the buffer layer and expanding the absorber layer. Based on the suggested device architecture, the effect of back-contact metal work function (BMWF), operating temperature, SnS_2 ETL, p-$CH_3NH_3SnI_3$ HTL, absorber layer carrier concentration is investigated to enhance cell performance. The impact of each of these layers was gradually examined. The solar cell's efficiency was measured at 600 nm thickness of absorber layer then Power conversion efficiency is 30.22%, fill factor of 83.72%, V_{OC} of 1.04 V, and J_{SC} of 34.50 mA/cm^2.

Keywords Solar cells · SCAPS · Buffer layer · $CH_3NH_3SnI_3$

1 Introduction

Fossil fuels are nowadays the main source of energy that is consumed all over the world [1]. The need for energy is in rise worldwide, making energy production one of the world's challenges. Solar energy is used as a substitute for fossil fuels. Solar panels are a long-term, cost-effective choice because they require little maintenance

A. Chaudhari (✉) · R. Mishra · A. K. Patel
Department of Electronics and Communication Engineering, Madan Mohan Malviya University of Technology, Gorakhpur, India
e-mail: abhimanyu274207@gmail.com

S. Kumar et al. (eds.), *Control Applications in Modern Power Systems*, Lecture Notes in Electrical Engineering 1128, https://doi.org/10.1007/978-981-99-9054-2_11

and have a long lifespan [2]. We can save time and money by using crucial techniques like numerical modelling or simulation. By addressing a number of issues, we will review experimental research that has already been published and improve the performance of the solar cell. Perovskite solar cells are a type of photovoltaic cell that use a special class of materials known as perovskites to absorb and convert sunlight into electricity. Perovskites are materials that have a unique crystal structure that enables them to efficiently capture and transport electrons. Comparing perovskite solar cells to conventional silicon-based solar cells, there are a number of benefits. They are cheaper to produce, more efficient at converting sunlight into electricity, and can be made flexible, transparent, and in a wide range of colors. Solar cells are important because they can convert a considerable quantity of sunlight into electrical energy. Currently, lead-based materials are mentioned in several papers. It is anticipated that by creating different multilayer thin film architectures, device efficiency based on Pb-free perovskites will be further increased by applying this research and the straightforward vapour processing approach [3]. However, the European Union and other nations have imposed strict restrictions on lead-based electronic products. These materials demonstrated significant power conversion efficiencies. Contrarily, lead is unsustainable and harmful to both people and the environment [4]. Lead-free perovskite CH3NH3SnI3 has been investigated as a photovoltaic material by numerous scientists and researchers to solve this problem and because of its better optoelectronic capabilities [5]. Looking for new alternative lead-free perovskite materials, to achieve higher efficiency and cheaper cost. Some characteristic, such as absorber layer thickness, materials for hole transport (HTL), and electron transport (ETM) layers, have been researched in demand to increase the efficiency and stability of the perovskite-based cells. As demonstrated, it has been discovered that using SnS_2 as a buffer layer between these materials yields the maximum efficiency of 30.22%.

2 Methodology

One-dimensional solar cell capacitance simulator was the simulation programme employed in this research. This programme functions as a general-purpose tool for simulating solar cells and may be used to run numerical calculations, simulate solar cells, and track the behaviour of photovoltaic systems. The programme solves continuity and semi-conductor equations as its main means of characterising the astronomy of the photovoltaic assembly. A variety of physical input parameters that SCAPS employs can be designed under different physical input conditions (Fig. 1).

Fig. 1 Device structure

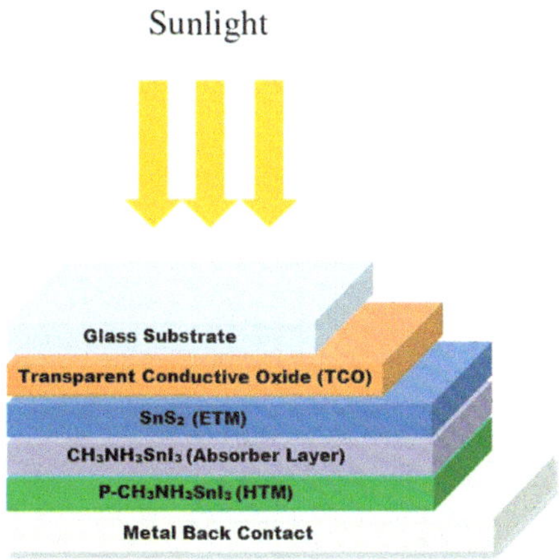

3 Device Schematic and Simulation Data

4 Results and Discussion

Thickness Variations of $CH_3NH_3SnI_3$ Absorber Layer

The effectiveness of the solar cell is greatly influenced by the absorber layer [6]. The efficacy of the solar cell is significantly influenced by the thickness and makeup of the absorber layer. Incoming sunshine must be effectively absorbed by the absorber layer, while still allowing for the efficient transport of electrons to the electrical contacts. The material used in the absorber layer should have a bandgap that matches the energy level of the photons in sunlight, so that the maximum amount of energy can be extracted from the incoming photons. The previously released paper demonstrates how the absorber layer thickness affects photovoltaic metrics as JSC, VOC, FF, and PCE [7]. A layer's absorber's thickness, which played a part in the device simulation, was altered from 100 to 1000 nm while the other variables listed in Table 1 remained the same. When the absorber layer is 600 nm thick, the solar cell operates at its best. The creation of new charge carriers is represented by the efficiency increasing as the thickness increases in Fig. 2a, b. An open circuit's voltage the reason for the smooth VOC drop may be due to the better free charge carrier recombination in the thicker absorber [7]. The high absorption coefficient of the perovskite is primarily responsible for the significant value of the short-circuit current JSC (34.50 mA/ cm^2) with a thickness of 600 nm. The higher series resistance may be the cause of the decreased value of FF with regard to absorber thickness [7]. efficiency of power conversion PCE initially rises, peaks at 600 nm (30.22%), then starts to fall as absorber thickness is increased. That will depend on the absorber layer's breadth.

Table 1 Simulated material parameters [8]

Parameters	TCO	SnS_2	$CH_3NH_3SnI_3$	p-$CH_3NH_3SnI_3$
W (nm)	300	100	800	300
Eg (eV)	3.5	1.85	1.3	1.3
χ(eV)	4	4.26	4.17	3.9
ε_r	9	17.70	6.5	6.5
μp (cm^2/Vs)	10	25	1.6	1.6
μn (cm^2/Vs)	20	50	1.6	1.6
N_c ($1/cm^3$)	2.2×10^{18}	7.3×10^{18}	1×10^{18}	1×10^{18}
N_v ($1/cm^3$)	1.8×10^{19}	1×10^{18}	1×10^{19}	1×10^{19}
N_d ($1/cm^3$)	2×10^{19}	3.2×10^{18}	0	0
N_a ($1/cm^3$)	0	0	1×10^{18}	3.2×10^{18}

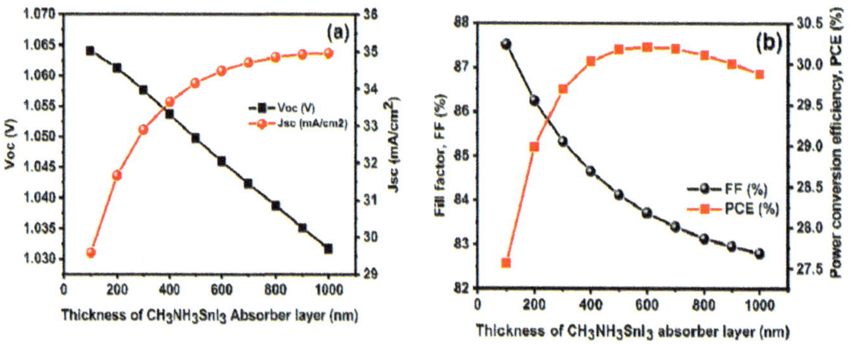

Fig. 2 Variations in thickness of absorber layer

Impact of Buffer Layer SnS_2

An electron transport layer (ETM) is an important component in perovskite cells, as it facilitates the movement of electrons from the light-absorbing layer of the cell to the electrode, where they can be collected and used to generate an electric current. In a perovskite cells, ETL is typically made of a perovskite material that can efficiently absorb sunlight and generate electron–hole pairs. The ETM material must be carefully chosen to have the right properties to enable efficient electron transport. The use of an electron transport layer can help to progress the efficiency of perovskite cells. This learning examines buffer layer SnS_2 for device optimization. We utilised a simulation with an absorbent layer thickness of 600 nm and a buffer layer thickness of 100 nm. When SnS_2 is utilised as a buffer layer, the greatest efficiency of 30.22% is discovered. The mobility of the carriers at the interface materials and associated conductance in band offset (CBO) have the most effects on carrier dynamics [9]. As a result of SnS_2's low CBO, great efficiency was delivered because the produced carriers could pass freely from the absorber to the buffer and subsequent contacts

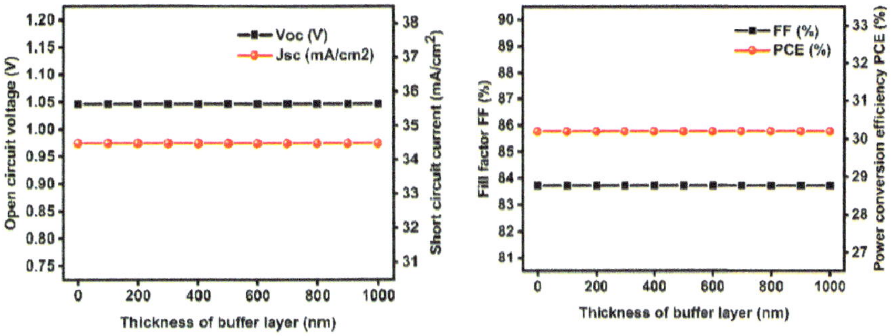

Fig. 3 Variations in thickness of buffer layer

demonstrates how the SnS_2 solar cell's simulated parameters change as the buffer layer goes from 100 to 1000 nm in thickness. Voc and Jsc continue at the same value and there is no change in efficiency or fill factor of fluctuations with the buffer layer (Fig. 3).

Impact of (Na) in Absorber Layer

N-I-P planar structures require a clear layer for electron transport because of the perovskite's ability to absorb light. The carrier attention of the absorbent layer is a measure of the number of free charge carriers (electrons or holes) present in the layer. This layer determines the number of permitted charge carriers available for the conversion of sunlight into electricity. This, in turn, moves the efficiency of the cells. A higher carrier absorption typically leads to higher solar cell efficiency, as more photoexcited carriers can contribute to the generation of an electric current. The carrier concentration of the absorber layer can be controlled through various means, including doping the semiconductor material with impurities or adjusting the growth conditions during fabrication. By optimizing the carrier concentration, it is possible to improve the act of the solar cell and increase its efficiency. Additionally, it is preferable to have lower values for its thickness for quicker light absorption [10]. High ETM thickness results in increased recombination, decreased electron transit to the front contact and increased series cell resistance. This part considered the ideal absorber thickness and investigated the effect of ETM thickness in the 100–1000 nm range. Then while open circuit voltage is steadily rising, short circuit current is falling to a certain degree of doping. Efficiency and fill factor both rise linearly. The findings are shown in Fig. 4c, d as efficiency and recombination plots. Efficiency normally increases as the cell's ETM thickness rises. We may infer that the whole cell's photovoltaic array characteristics decrease in value as a result of increased recombination at larger ETM thicknesses and the absorber layer's ability to absorb late light.

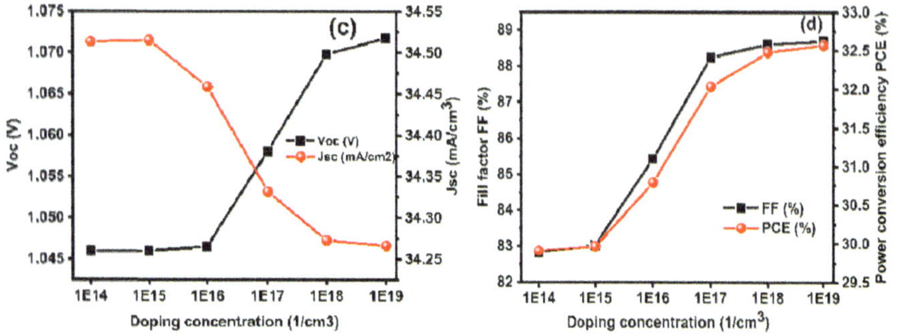

Fig. 4 Variations in the doping concentration (Na)

Influence the Defect Density of Absorber Layer

A critical initial step in improving the presentation of the perovskite device is to alter the defect of the perovskite material to the right value. Plots of recombination and power conversion efficiency would be affected by doping density, just as the absorber layer's thickness is being investigated. The absorbent layer doping density was examined. The performance plots were produced for this purpose in the 1013–1017 cm^{-3} range with the optimised thicknesses [11]. The lifetime will decrease as the defect density rises, which will also raise the recombination rate and increase the recombination rate, which will lower the carrier diffusion length and decrease the PCE [4]. The best solar cell presentation is thought to be at this worth with PCE of 30.22%, FF of 83.72%, J_{SC} of 34.50 mA/cm^2, and V_{OC} of 1.04 V. Therefore, superior perovskite photovoltaic cells with defect densities smaller than 1×10^{14} cm^3 are tough to obtain in experimental effort (Fig. 5).

Paramters	1×10^{13}	1×10^{14}	1×10^{15}	1×10^{16}	1×10^{17}
Voc (V)	1.046	0.95	0.87	0.74	0.58

(continued)

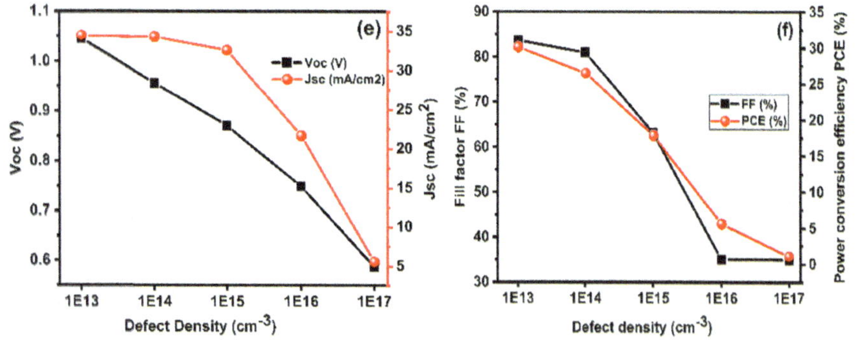

Fig. 5 Impact of defect density (Nt)

(continued)

Paramters	1 × 10^{13}	1 × 10^{14}	1 × 10^{15}	1 × 10^{16}	1 × 10^{17}
Jsc (mA/cm^2)	34.50	34.32	32.62	21.68	5.55
FF (%)	83.72	81.01	63.09	35.05	35.47
PCE (%)	30.22	26.58	17.93	5.70	1.16

Influence of Temperature

The temperature of a solar cell can have a significant influence on its performance and efficiency. Generally, growing the temperature of a cells primes to a decrease in efficiency, although the exact effect depends on the specific type of solar cell and its operating conditions. One reason for this effect is that increasing temperature increases the thermal energy of the semiconductor material in the solar cell, which can cause more electron–hole pairs to be generated through thermal excitation. While this may sound like a good thing, it can also lead to increased recombination of these carriers, resulting in a lower overall efficiency. Another reason is that the Voc of a cell decreases with increasing temperature. This is because the voltage drops across the solar cell's p–n junction, which is responsible for generating the electric field that separates the photoexcited carriers, decreases with increasing temperature. This effect can also prime to a decrease in PCE. Moreover, the fill factor of the solar cell, which is a measure of how effectively the solar cell can convert light into electricity, can also decrease with increasing temperature. This is because the struggle of the solar cell's metal contacts and the resistance of the semiconductor material itself both increase with temperature, leading to a higher series resistance and lower fill factor. Overall, the impact of temperature on a cell performance is a complex interplay between various factors. While it is possible to design solar cells that are optimized for higher temperature operation, in general, solar cells perform best at lower temperatures [12]. It is, therefore, important to consider the operating temperature of a solar cell when designing and installing solar energy systems to maximize their efficiency and output. Variations in the various properties of the investigated structure in relation to operation temperature [13]. Figure 6 illustrates the temperature fluctuation range, which is set between 300 and 480 K. There is no doubt that temperature variations have a negative effect on the performance of solar cells. The PCE dramatically decreases with temperature, going from 30.22% at 300 K to 20.64% at 480 K. A shorter charge carrier diffusion length is related to the efficiency drop that occurs as temperature rises. The bulk of the usual act is finished at a 300 K working temperature [14–16].

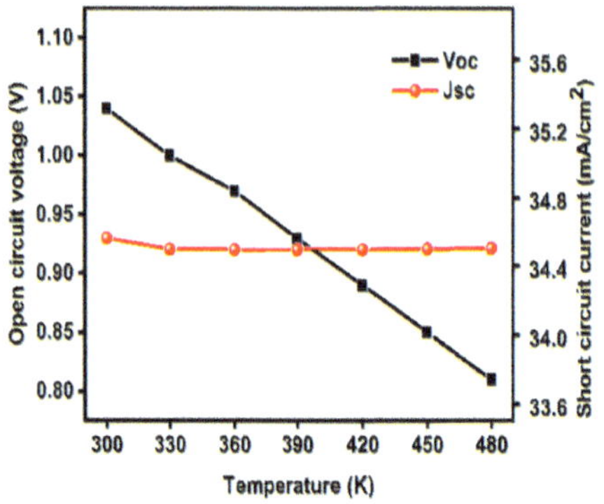

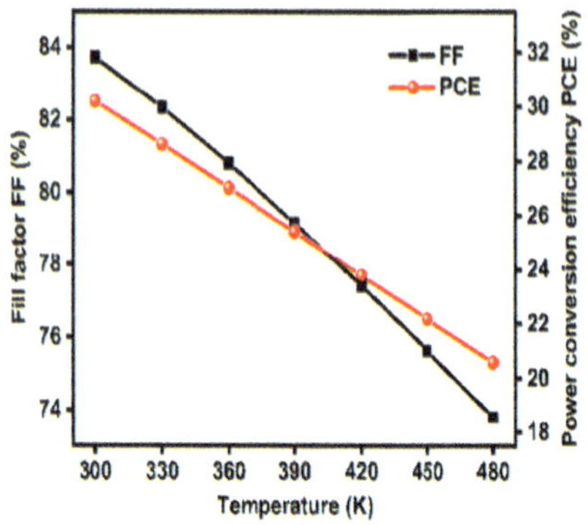

Fig. 6 Variation in temperature (K)

5 Conclusion

The efficiency of a solar cell depends on various factors, including the absorber material, the cell design, and the operating conditions. By optimizing these factors, it is possible to increase the efficiency of a solar cell and make it a viable source of renewable energy. Solar cells have many advantages, including their ability to generate electricity without producing greenhouse gases or other harmful pollutants. They have a wide range of uses, from supplying energy to entire communities to powering solitary devies. Despite their many benefits, solar cells also have some limitations, including their relatively low efficiency compared to other forms of electricity generation, and their dependence on sunlight. However, ongoing research and development are addressing these challenges and driving the development of more efficient and cost-effective solar cells. Overall, solar cells have the potential to play a significant role in the transition to a more sustainable and environmentally friendly energy system. The goal of the current study is to provide a fresh viewpoint on the SCAPS-1D simulation tool's modelling of perovskite solar cells [17]. The results show that the device's PCE may be greatly changed by the absorber layer's thickness, and the optimal absorber layer thickness (600 nm) is established. Furthermore, efficiency has been proven to suffer with an increase in operating temperature. A solar cell performs best at a temperature of 300 K. The solar cells greatest performance was attained at the absorber layer thickness of 600 nm and defect density of 3.2×10^{13} at PCE of 30.22%, J_{SC} of 34.50 mA/cm^2, V of 1.046 V, and FF of 83.72% because of the optimization of the device parameters.

Acknowledgements The SCAPS programme used in this simulation was made available to the writers by Professor Marc Burgelman of gent Unversity.

References

1. Ghanem HT, Saleh AN (2022) Simulation and improvement of the efficiency of the CFTS solar cell. Karbala Int J Modern Sci 8:63–69
2. Tala-Ighil Zaïr R, Oudjehani C, Tighilt K (2021) SCAPS Simulation for perovskite. Solar Cell
3. Yasin S, Al Zoubi T, Moustafa M (2021) Design and simulation of high efficiency lead-free heterostructure perovskite solar cell. Optik
4. Srivastava V, Chauhan RK, Lohia P (2022) Highly efficient cesium-based halide perovskite solar cell using SCAPS-1D software: theoretical study. J Opt
5. Alok Patel AK, Rao PK, Mishra R, Soni SK (2021) Numerical study of a high-performance thin film CIGS solar cell with a-Si and MoTe2 hole transport layer. Optik
6. Chaurasiya R, Gupta GK, Dixit A (2019) Ultrathin Janus WSSe buffer layer for W(S/Se)2 absorber based solar cells. Elsevier
7. Lin L, Jiang L, Li P, Fan B, Qiu Y (2019) A modelled perovskite solar cell structure with a Cu_2O hole-transporting layer enabling over 20% efficiency by low-cost low-temperature processing. J Phys Chem Solids 14:205–211
8. Lazemi M, Asgharizadeh S, Bellucci S (2018) A computational approach to interface engineering of lead-free $CH_3NH_3SnI_3$ highly-efficient perovskite solar cells. Phys Chem Chem Phys 00:1–10
9. Jeyakumar R, Bag A, Nekovei R, Radhakrishnan R (2016) Influence of electron transport layer (TiO_2) thickness and its doping density on the performance of $CH_3NH_3PbI_3$-based planar perovskite solar cells. J Electron Mater 49(6):3533–3539
10. Ayad M, Fathi M, Mellit A (2021) Study and performance analysis of Perovskite solar cell structure based on organic and inorganic thin films. Optik
11. Alok Patel AK, Mishra R, Soni SK (2022) Performance analysis of a Cu(In1−xGax)Se_2 solar cell with nontoxic WS2 and WSSe buffer layers. J Electron Mater 51(11):6168–6179
12. Alok patel AK, Mishra R, Soni SK (2022) Performance enhancement of CIGS solar cell with two dimensional MoS_2 hole transport layer. Micro Nanostruct 165
13. Pandey K, Patel AK, Mishra R (2022) Numerical study on performance enhancement of CZTSSe solar cells with Cu_2O and $MoTe_2$. J Comput Electron
14. Pansuriya T, Malani R, Kheraj V (2022) Investigations on the effect of buffer layer on CMTS based thin film solar cell using SCAPS 1-D. Opt Mater
15. Hosseini S, Bahramgour M (2022) Investigating the effect of non-ideal conditions on the performance of a planar $CH_3NH_3PbI_3$-based perovskite solar cell through SCAPS-1D simulation. Heliyon
16. Srivastava V, Lohia P, Chauhan RK (2022) Theoretical study of a lead-free perovskite solar cell using ZnSe as ETL and PTAA as HTL. Emerg Mater Res 59:1–10
17. Gautam S, Patel AK, Mishra R (2022) Performance analysis of WSe_2 solar cell with Cu_2O hole transport layer by optimization of electrical and optical properties

Field Oriented Control of PV-BESS Fed Traction PMSM Drive for EV Application

Mayank Kumar, R. N. Mahanty, and Pankaj Deosarkar

Abstract This paper presents a study to design a novel electric power-drivetrain for Solar PV-BESS powered EV's driven by traction permanent magnet synchronous motor (PMSM) for an automobile application. The proposed field-oriented control (FOC) technique for controlling the traction motor aims to improve dynamic and steady state performance of drive system. For maximum power extraction from PV array IN-C MPPT algorithm is used in the model. The key features like higher efficiency of drive system, longer battery life, faster dynamic response, long range are considered in this paper. The traction system is supported with the BESS connected in shunt to maintain constant DC link bus voltage. The drive system is subjected to operate in three driving modes i.e., acceleration, cruising (constant speed) and deceleration mode along with variable solar insolation condition to test the robustness behavior of system. The simulation results of model are verified in MATLAB-Simulink environment.

Keywords MPPT · Solar PV · BESS · INC MPPT · FOC · MATLAB Simulink

1 Introduction

Electric vehicles (EVs) have become increasingly popular as a sustainable transportation solution, offering a cleaner and more efficient alternative to conventional fossil fuel-based vehicles. One of the key components of an EV is its motor drive system, which plays a crucial role in determining the vehicle's overall performance

M. Kumar (✉) · R. N. Mahanty
Department of Electrical Engineering, National Institute of Technology Jamshedpur, Jamshedpur, India
e-mail: mayankumar3299@gmail.com

R. N. Mahanty
e-mail: rnmahanty.ee@nitjsr.ac.in

P. Deosarkar
TATA Motors Limited, Pune, India
e-mail: pankaj.deosarkar@tatamotors.com

S. Kumar et al. (eds.), *Control Applications in Modern Power Systems*, Lecture Notes in Electrical Engineering 1128, https://doi.org/10.1007/978-981-99-9054-2_12

and efficiency [1]. In recent years, significant research has been conducted to improve the performance of EV motor drive systems by integrating renewable energy sources, such as photovoltaic (PV) arrays and battery energy storage systems (BESSs).

The use of PV arrays and BESSs can help reduce the reliance of EVs on the power grid and minimize their carbon footprint. However, integrating these components into an EV motor drive system presents several technical challenges, including effective power management and control. Field-oriented control (FOC) is a widely used technique for controlling motor drives and has been shown to be effective in improving the dynamic and steady-state performance of EVs [4].

This paper proposes a novel FOC scheme for a PV-BESS fed traction permanent magnet synchronous motor (PMSM) drive system in EV applications. The proposed control strategy is designed to regulate the torque and speed of the PMSM while ensuring optimal power transfer between the PV-BESS and the motor drive. The effectiveness of the proposed FOC scheme is evaluated through simulation studies, which demonstrate improved performance in terms of efficiency, dynamic response, and stable operation under varying operating conditions.

To improve the driving range of vehicle the motor drive system of EV is connected with the PV-battery combination. This combination will ensure that the vehicle can be operated in all driving condition irrespective of the environmental factors. For PV system solar insolation plays key role for the output efficiency. Thus, to ensure maximum output IN-C MPPT algorithm is employed to track maximum power for supply. The Incremental conductance (IN-C) MPPT is an efficient method which operates under the range of zero voltage difference at MPP [5]. Also, the BESS is used in the drive system to maintain DC-bus voltage constant throughout the operation. In EV powertrain for propulsion of vehicle earlier DC motors are now replaced with PMSM motor due to high power density, torque to weight ratio, excellent flux distribution. On construction basis it is divided into two categories i.e. I-PMSM and S-PMSM [10]. In PMSM motor the major drawback of DC motor i.e., commutation problem due to brushes-collectors is eliminated and in place position sensors are used.

The paper is structured as follows. Section 2 provides a detailed description of the proposed PV-BESS fed traction PMSM drive system and the. Section 3 presents system design and modelling of PMSM motor, single-diode model of the PV cell and space vector modulation and its implementation in the given system. Section 4 discusses the control methodology of PV array and vector control of PMSM motor. Section 5 discusses the simulation results and their performance analysis. And finally, Section 6 presents the conclusions and suggestions for future research.

2 Configuration of Proposed System

The EV traction PMSM drive system is designed which is fed with combination of PV-battery system and subjected with constant and dynamic loading and simulated in MATLAB-Simulink. The EV drive system consists of PV arrays, connected to a DC link bus via boost converter and controlled with IN-C MPPT controller [11]. With the same DC link bus, a battery energy storage system (BESS) is connected in shunt with a bi-directional DC-DC converter, which enables proper controlling for battery charging as well as regulation of DC link voltage which is maintained constant for traction inverter supplying AC to the motor of vehicle [6]. The control of overall motor drive system is done by the field-oriented control (vector control) technique. And the regulated output of VSI's powers the PMSM drive of vehicle. And the gating pulses for switching the power devices of inverter is done with the space vector modulation technique.

The operating stages of proposed PV-BESS fed PMSM traction motor drive controlled with vector control (FOC) technique is illustrated in Fig. 1. i.e., PMSM traction motor, battery pack, 3-phase VSI are selected to get the desired result. In below sections the detail mathematical modelling of PMSM motor, PV array modelling is discussed. Along with this the mathematical analysis of space vector pulse width modulation technique is also done in this section.

2.1 PMSM Motor Mathematical Modelling

To examine the EV motor drive system performance, the mathematical modelling of motor plays the crucial role. The motor mathematical model is generally obtained by the space phasor theory. It is done on the basis of following assumptions to have good control on motor for operation [13]:

(1) Space Harmonics in the air gaps between rotor and stator are neglected
(2) Hysteresis losses and eddy currents are neglected.
(3) The air-gap reluctance is considered to be constant and saturation is neglected

The dynamic equation of the 3-phase PMSM motor in the synchronous reference frame (d-q) are as follows [7]:

$$V_{sd'} = R_{s'} \cdot i_{sd'} + L_{d'} \frac{di_{sd'}}{dt} - \omega_{r'} L_{q'} i_{sq'} \tag{1}$$

$$V_{sq'} = R_{s'} \cdot i_{sq'} + L_{q'} \frac{di_{sq'}}{dt} - \omega_{r'} L_{d'} i_{sd'} \tag{2}$$

$$T_{e'} = \frac{3}{2} \cdot \frac{P}{2} \cdot \left(\psi_{sd'} i_{sq'} - \psi_{sq'} i_{sd'} \right) \tag{3}$$

$$\begin{aligned} \psi_{sd'} &= L_{d'} i_{sd'} + \lambda_{m'} \\ \psi_{sq'} &= L_{q'} i_{sq'} \end{aligned} \tag{4}$$

The flux linked with the permanent magnet of motor i.e., the rotor flux linkage vector given by $\psi_{r'}$. The d-axis current, flux linked with motor and stator voltage is

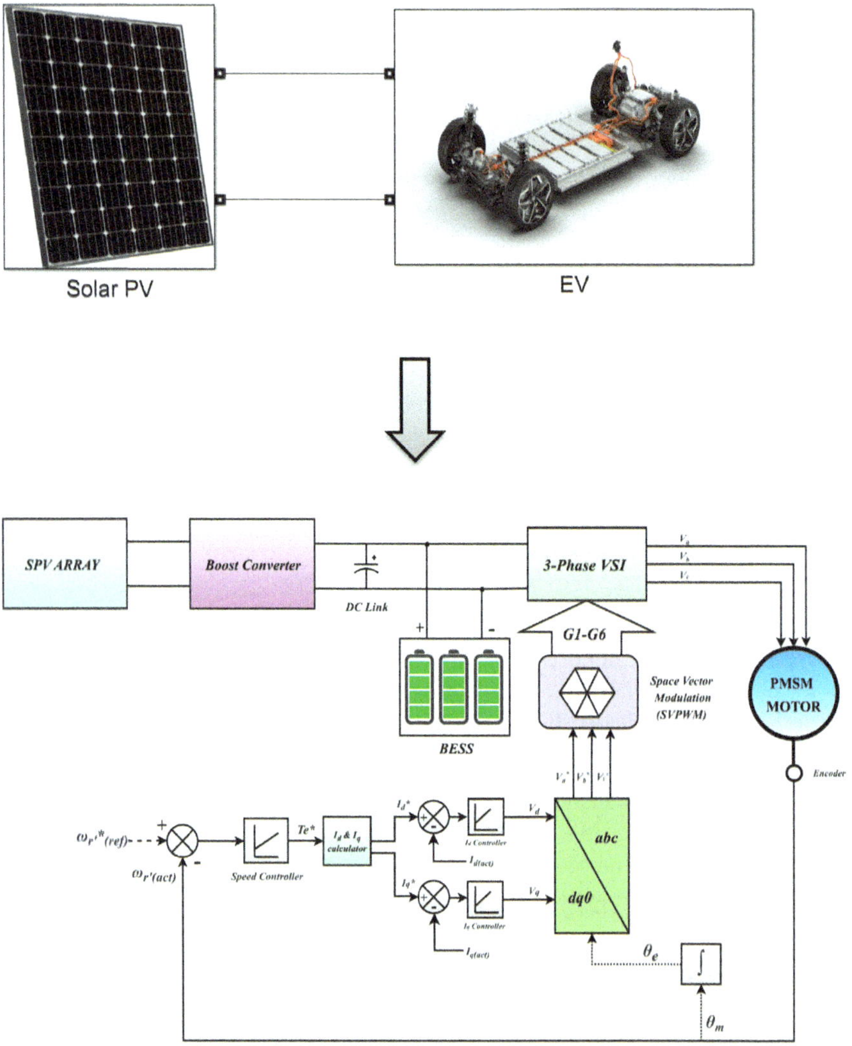

Fig. 1 Control Scheme of PV-BESS coupled PMSM motor drive system

given by $i_{sd'}$, $\psi_{sd'}$, $V_{sd'}$. Similarly on the q-axis the voltage, motor flux and current associated in d-q reference frame is given by $V_{sq'}$, $i_{sq'}$, $\psi_{sq'}$. The stator winding resistance of motor, stator flux and rotor angular speed is denoted by $R_{s'}$, $\omega_{r'}$ and $\psi_{s'}$. On substituting all values, the PMSM motor electromagnetic torque equation $T_{e'}$ is given by:

$$T_{e'} = \left(\frac{3}{2}\right)\left(\frac{P}{2}\right)\left[\lambda_{m'} i_{sq'} - \left(L_{q'} - L_{d'}\right) i_{sd'} i_{s'}\right] \tag{5}$$

PM-synchronous motor mechanical equation is given by:

$$J_m \frac{d\omega_{m'}}{dt} = T_{e'} - T_{ld'} - B_{m'}\omega_{m'} \tag{6}$$

$$\omega_{r(rotor)'} = \left(\frac{P}{2}\right)\omega_{m'} \tag{7}$$

When shaft of motor is subjected with loading the load torque is given by $T_{load'}$. The moment of inertia of motor is given by J_m, and flux linked with PM-motor produced by permanent magnets mounted on rotor denoted by $\lambda_{m'}$. The poles of motor by P, while q & d axis stator inductances are given by $L_{q'}$ and $L_{d'}$.

2.2 Mathematical Modelling of PV Arrays

Several individual PV panels are connected in a series–parallel arrangement to form PV arrays, while each PV panel is composed of PV cells connected in a series–parallel configuration. PV cells are made up of a current source that is connected anti-parallel to a diode and an internal resistance represented by a low-resistance series resistance, $R_{s'}$, and a high-resistance shunt resistance, $R_{sh'}$, which represents the cell's leakage resistance [15]. Figure 2 depicts the equivalent circuit diagram of a typical single diode model of a PV cell. When a load is connected across the cell's terminals, a potential V is developed, and the current source supplies current, I_l. As the diode is forward-biased, current, $I_{diode'}$, flows through it [3]. The values of the cell's leakage and internal resistance are represented by resistors $R_{sh'}$ and $R_{s'}$, respectively. The governing mathematical equations for PV cell are as follows:

$$I = I_{ph'} - I_s\left[\left\{e^{\left(\frac{e(V+IR_{s'})}{m\cdot K\cdot T}\right)}\right\} - 1\right] - \left(\frac{(V + IR_{s'})}{R_{sh'}}\right) \tag{8}$$

And,

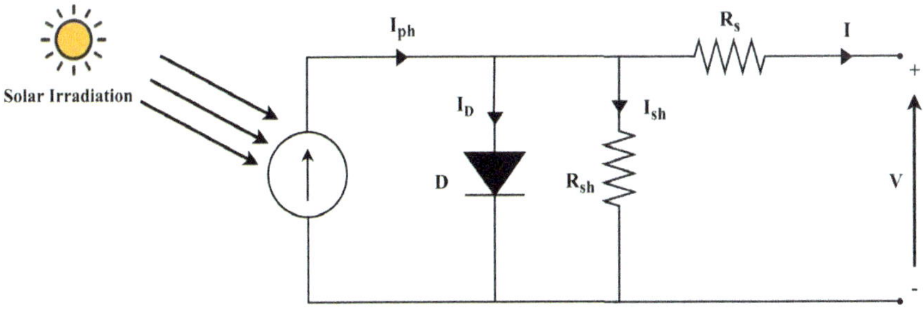

Fig. 2 A PV cell model with single diode configuration

$$I_{diode\prime} = I_o\left[e^{\left(\frac{e(V+I.R_{s\prime})}{mKT}\right)} - 1\right] \tag{9}$$

where I_0 is reverse saturation current, e is charge of electron, n is ideality factor, I_{ph} is light generated current, K is Boltzmann constant and T is temperature.

The maximum power that can be obtained in a given solar irradiance and temperature is,

$$P_m = \left[\frac{\left(eV_m^2/KT\right)}{1+\left(eV_m/KT\right)}\right]\cdot(I_{sc}+I_o) \tag{10}$$

The current generated from the photons depends on the solar insolation and temperature on which PV cell is working. The governing equation is given by:

$$I_{ph\prime} = \left[I_{sc} + k_i\left(T - T_{ref\prime}\right)\right]\cdot\left(\frac{G}{1000}\right) \tag{11}$$

The PV cell saturation current (I_s) is characterized by a cubic relationship with the PV cell temperature (T), which can be expressed as:

$$I_s = I_{rs\prime}^3\left(\frac{T}{T_{ref\prime}}\right)^3\left[e^{\left(\frac{e}{k}\right)\cdot\left(\frac{E_{g\prime}}{m}\right)\cdot\left(\frac{T-T_{ref\prime}}{T_{ref\prime}\cdot T}\right)}\right] \tag{12}$$

And the reverse saturation current of PV cell $I_{rs\prime}$ is given by equation (Fig. 3):

$$I_{rs\prime} = \frac{I_{sc}}{\left[\left\{e^{\left(\frac{eV_{OC}}{Nsr\cdot k\cdot m\cdot T}\right)}\right\} - 1\right]} \tag{13}$$

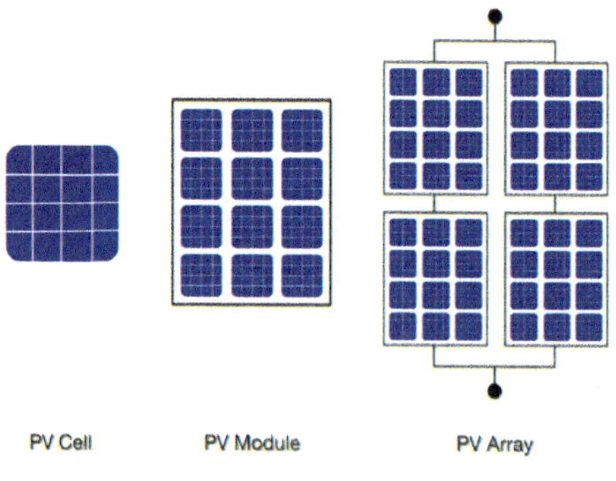

Fig. 3 PV cells, modules and arrays

2.3 *Space Vector Modulation*

In contrast to PWM techniques, Space Vector Modulation (SVM) operates differently. PWMs involve using three individual push–pull driver stages in the inverter to generate waveforms for each phase independently. On the other hand, SVM considers the inverter as a unified entity, where it can be stimulated to attain eight distinct states. The inverter's state is switched to achieve modulation, and this process is carried out using digital control strategies. One such technique is SVM, which aims to generate load line voltages that match a given or reference voltage on average [2]. This is achieved by selecting the appropriate switching states for the inverter and calculating the corresponding time periods during each sampling period (Figs. 4 and 5).

The resultant space vector for load phase voltage is given by:

$$V_{R'}(t) = \frac{2}{3}\left[V_{an'}(t) + V_{bn'}(t)e^{j^{2\pi}/3} + V_{cn'}(t)e^{j^{4\pi}/3}\right] \tag{14}$$

The Space vectors $V_{R'}$(t) have both magnitude and angle. The individual voltages can be balanced or unbalanced and need not to be sinusoidal quantity. The space vectors for all combination given in Table 1:

3 Control Methodology

The methodology of control algorithm for the proposed PV-BESS fed EV is divided into two sections first is the controlling of PV array with INC MPPT technique and second part is controlling speed of PMSM motor drive using field-oriented control (vector control) technique. For switching the power devices of the inverter, space vector modulation (SVPWM) technique is used [12]. In the vector control

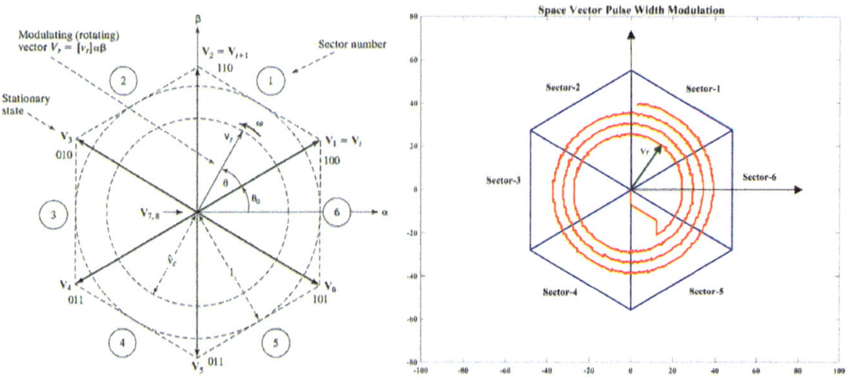

Fig. 4 Illustration of sectors of space vector [2]

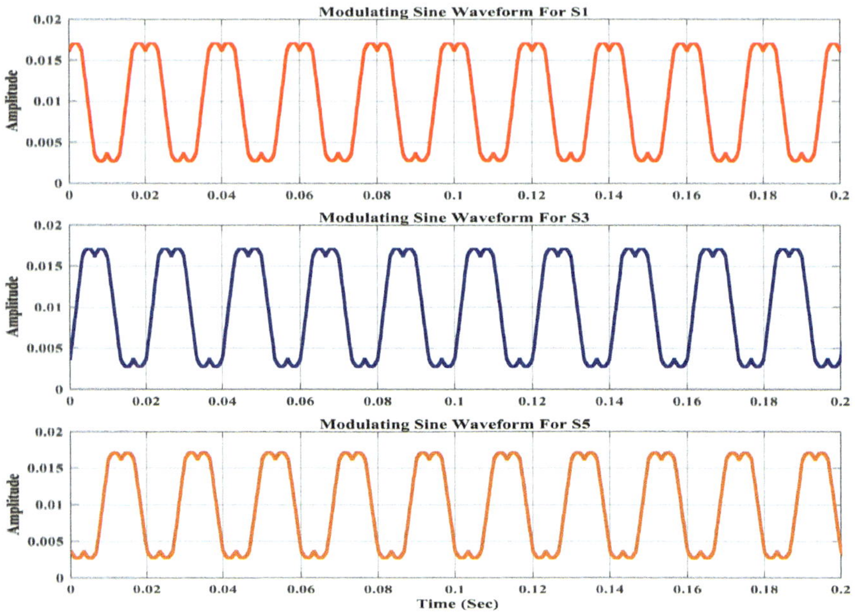

Fig. 5 Space vector modulated sine waveform

Table 1 Space vector for all switching combination

Space vectors	Switching states	Resultant space vector $V_{R'}(t)$	Vector states
V_0	000	$\vec{V_0} = 0$	Zero vector
V_1	100	$\vec{V_1} = \frac{2}{3} V_{D'} e^{j0}$	Active vectors
V_2	110	$\vec{V_2} = \frac{2}{3} V_{D'} e^{j\pi/3}$	
V_3	010	$\vec{V_3} = \frac{2}{3} V_{D'} e^{j2\pi/3}$	
V_4	011	$\vec{V_4} = \frac{2}{3} V_{D'} e^{j3\pi/3}$	
V_5	001	$\vec{V_5} = \frac{2}{3} V_{D'} e^{j4\pi/3}$	
V_6	101	$\vec{V_6} = \frac{2}{3} V_{D'} e^{j5\pi/3}$	
V_7	111	$\vec{V_7} = 0$	Zero vector

methodology initially the speed of the motor which is sensed by the rotor position sensor $\omega_{r'(act)}$ is fed to the speed controller as input where it is compared with the reference speed of motor $\omega_{r'(ref)}$. The output of the speed controller generates the reference torque T_e*. The reference currents, $i_{sd'}$ and $i_{sq'}$ are generated from the torque T_e* which is output of speed controller which is fed into two inner current loops. And the output of these loops is then fed into the SVM block to generate switching pulses to turn on the devices of the inverter.

3.1 IN-C MPPT Control of PV Array

The Incremental Conductance (INC) Maximum Power Point Tracking (MPPT) algorithm is a widely used technique for controlling the PV array. If there is a variation in irradiance and temperature, it will impact the I-V curve, P–V curve, and maximum power point. Hence, to obtain maximum power, adjustments to the operating voltage of the PV array are necessary. This means that the converter's impedance needs to be altered in line with that of the PV array. To achieve this, changes to the duty cycle of the converter are required [8, 9].

This algorithm continuously adjusts the operating voltage of the PV array to maintain maximum power output. It uses the rate of change of power with respect to voltage $^{dP}/_{dV}$ to determine the direction in which the operating voltage should be adjusted. When $^{dP}/_{dV}$ is zero, the operating voltage is at the maximum power point. If $^{dP}/_{dV}$ is positive, the algorithm increases the voltage, and if it is negative, the voltage is decreased [14, 15]. The INC MPPT algorithm is known for its fast-tracking speed and accuracy, making it suitable for applications where there are frequent changes in irradiance and temperature. Overall, the INC MPPT algorithm is an efficient way to extract the maximum power from a PV array, providing a stable and reliable source of renewable energy.

Under Normal condition the equation governing IN-C method are as follows:

$$\frac{\Delta I_{spv(s)}}{\Delta V_{spv(s)}} = -\frac{I_{spv(s)}}{V_{spv(s)}}, \text{ at MPP in } P_{spv} \text{ versus } V_{spv} \text{ curve}$$
$$\frac{\Delta I_{spv(s)}}{\Delta V_{spv(s)}} < -\frac{I_{spv(s)}}{V_{spv(s)}}, \text{ right of MPP in } P_{spv} \text{ versus } V_{spv} \text{ curve}$$
$$\frac{\Delta I_{spv(s)}}{\Delta V_{spv(s)}} > -\frac{I_{spv(s)}}{V_{spv(s)}}, \text{ left of MPP in } P_{spv} \text{ versus } V_{spv} \text{ curve}$$

Where ΔI_{spv} and ΔV_{spv} are deviations in PV array voltage and current respectively (Fig. 6).

3.2 Field Oriented Control of PMSM Motor

The vector control (FOC) algorithm is utilized to regulate the stator current of a PMSM. The PMs mounted on the rotor generate a constant flux field. As the motor should not be operated above the base speed, there is no need for field weakening operation. Consequently, the d-axis current $I_{sd'}*$ is kept zero [6].

From Eq. (15) below the flux linked with the direct axis is equivalent to the flux generated by the excitation of PMs which is given by:

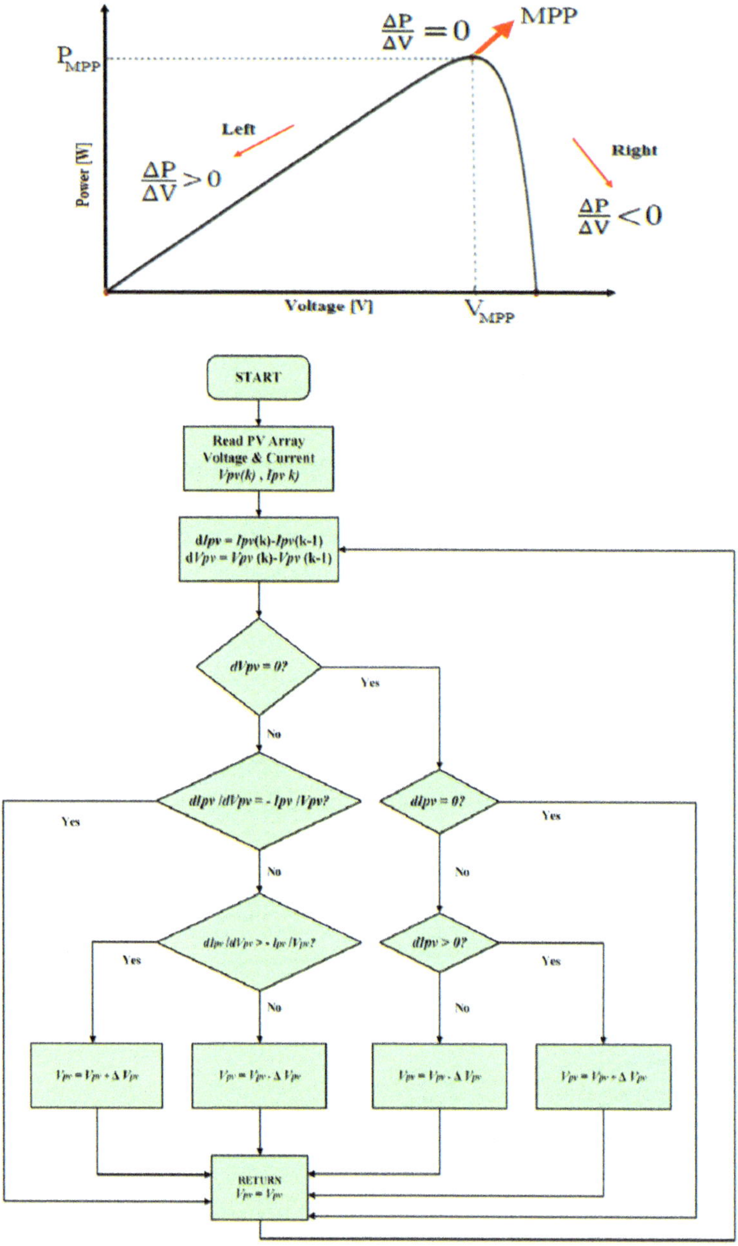

Fig. 6 I-NC MPPT algorithm flowchart and PV curve diagram

$$\lambda_{sd'} = \lambda_{Psm'} \quad (15)$$

In the d-q synchronous reference frame the electromagnetic torque equation of PMSM motor is given by:

$$T_{e\prime} = \left(\frac{3}{2}\right) P\prime \lambda_{Psm\prime} i_{sq\prime} \quad (16)$$

P' is the pole pair of motor, and the torque reference denoted by T_e* is expressed as:

$$T_e* = K_{e\prime} i_{sq\prime} \quad (17)$$

The calculation of the current in the quadrature axis (q-axis), denoted as $i_{sq\prime}*$, is given by:

$$i_{sq'} = \frac{T_e*}{K_{e'}} \quad (18)$$

The speed controller produces T_e* as output, and to obtain the reference current in the quadrature axis, $i_{sq'}$, T_e* it is divided by the torque constant, K_e.

$$T*_{e(i)\prime} = T*_{e(i-1)\prime} + k'_{p\omega}\left\{\omega_{error(i)\prime} - \omega_{error(i-1)\prime}\right\} + k'_{i\omega}\omega_{error(i)\prime} \quad (19)$$

where,

$$\omega'_{error} = \omega * - \omega_{act\prime}$$

Here $k'_{p\omega}$ and $k'_{i\omega}$ are proportional and the integral gain of PI speed controller.

Using dq0 to abc (Inverse Clarke) transformation, the reference voltage signals are generated and fed to the space vector modulation block, the three phase reference signals of PMSM is transformed into α-β component. Then resultant voltage vector V_r is calculated along with sector selection and modulation index and then output is fed to switching sequence generator to generate the gating pulses.

4 Simulation Results and Discussions

The proposed PV-BESS motor drive system is subjected to different driving and weather condition and simulated in MATLAB-Simulink. The performance analysis is shown from Figs. 7, 8, 9, 10, 11 which includes PMSM output, solar PV and BESS characteristics in steady state and dynamic condition.

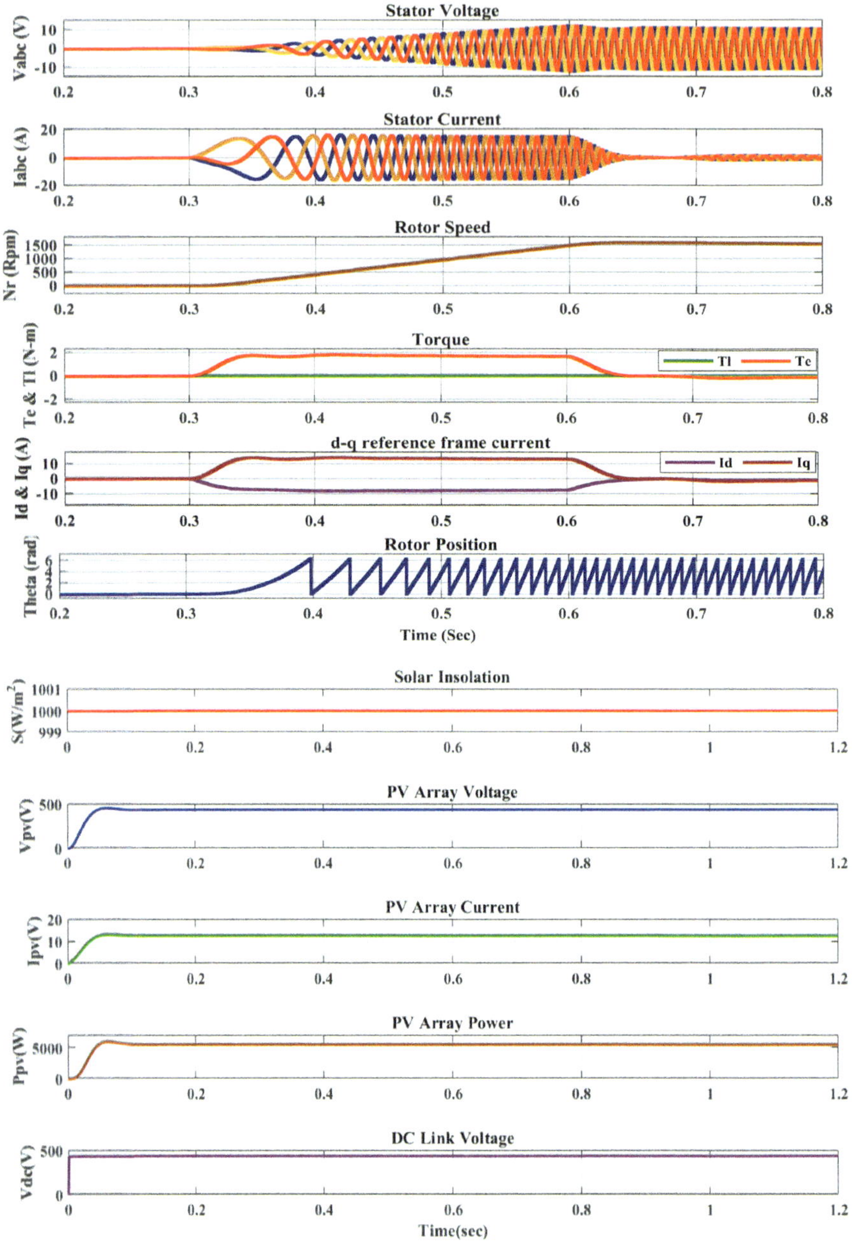

Fig. 7 Output characteristics of PMSM, PV and BESS at 1000 W/m^2 in acceleration mode

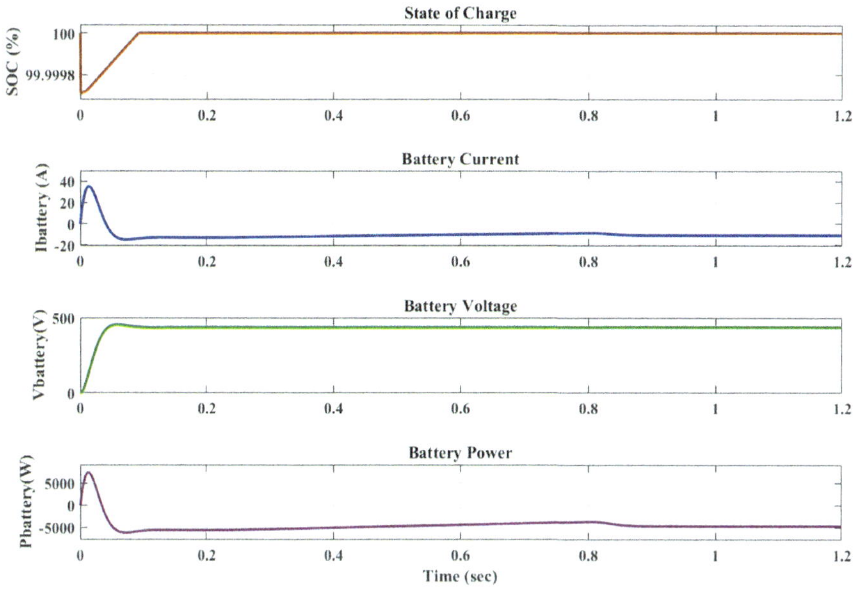

Fig. 7 (continued)

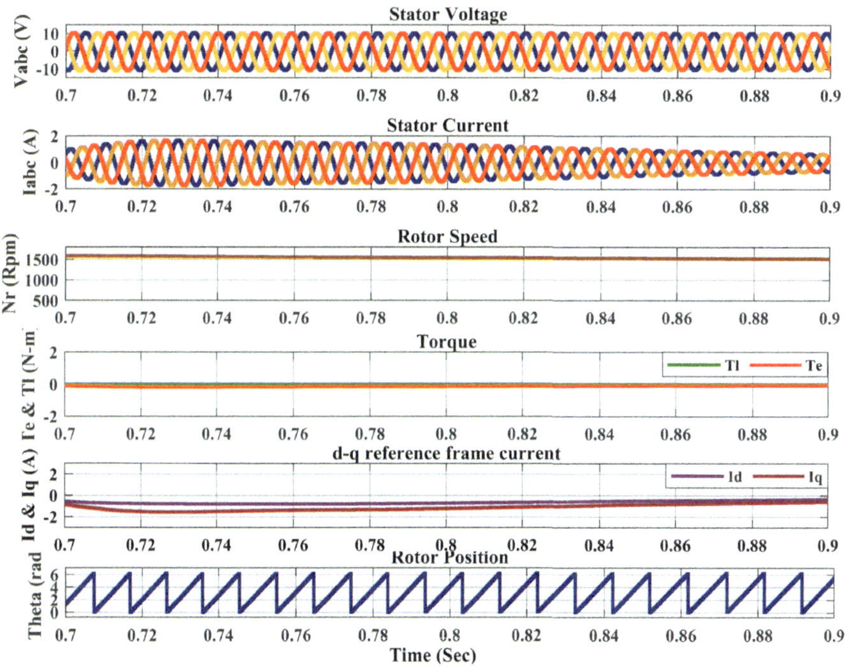

Fig. 8 Output characteristics of PMSM, PV and BESS at 1000 W/m^2 solar radiation in cruising mode

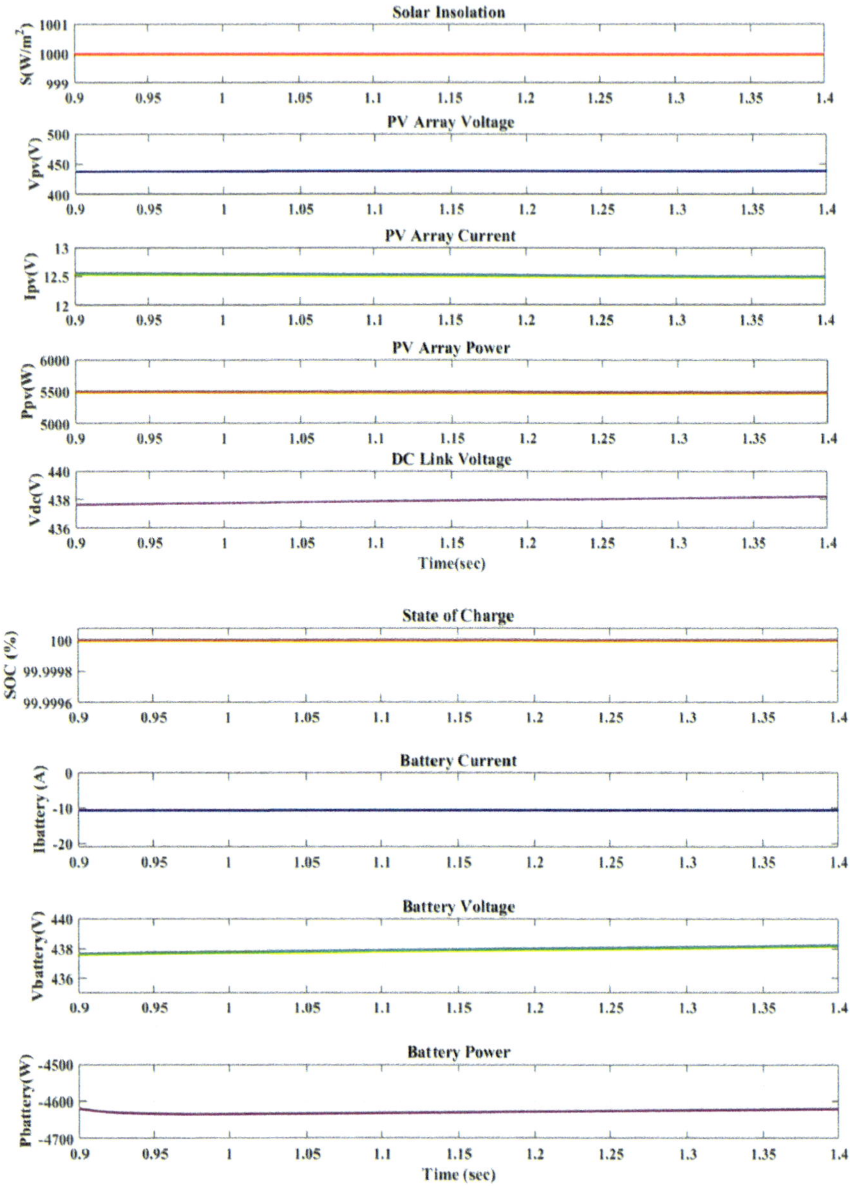

Fig. 8 (continued)

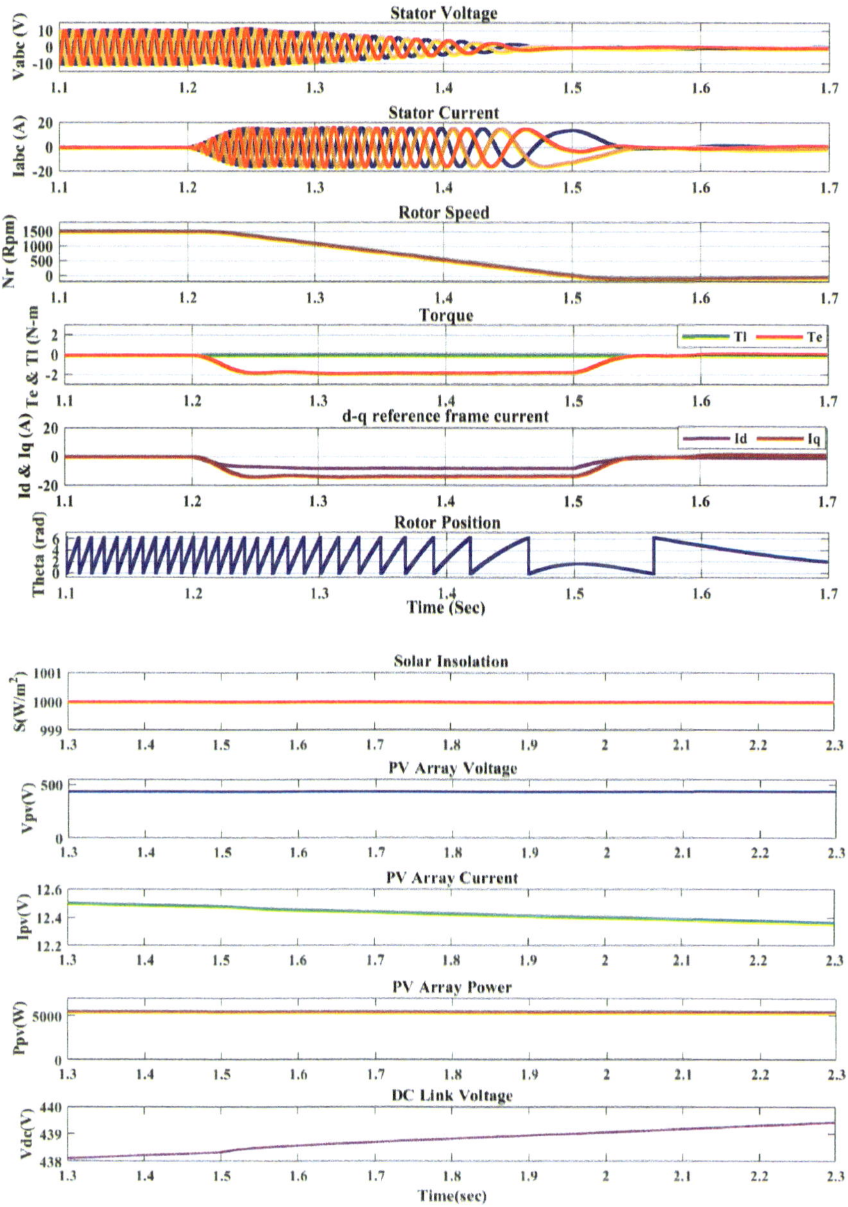

Fig. 9 Output characteristics of PMSM, PV and BESS at 1000 W/m^2 solar radiation in deceleration mode

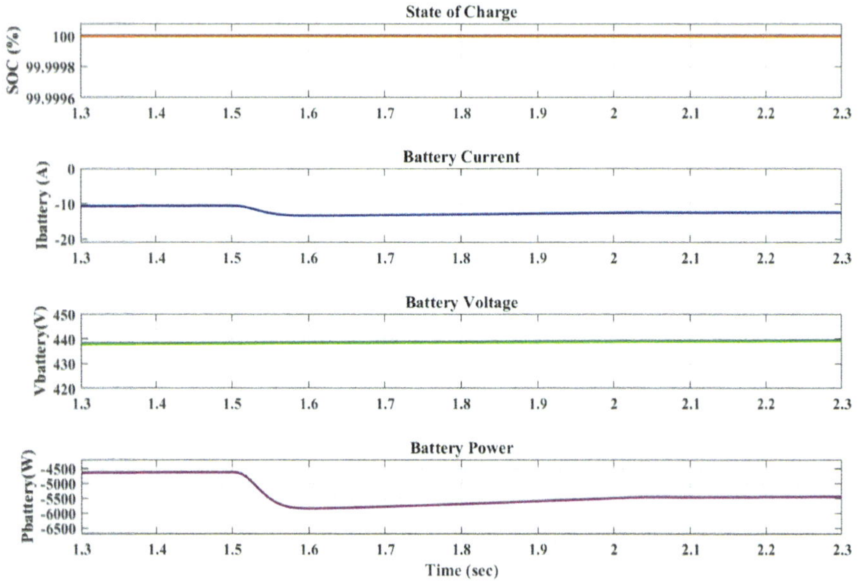

Fig. 9 (continued)

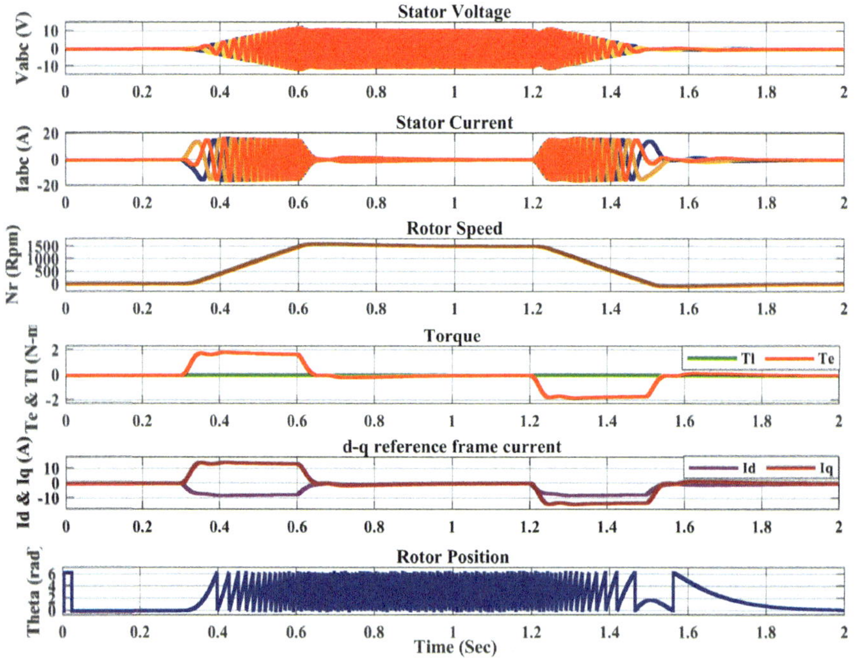

Fig. 10 Dynamic output characteristics of PMSM, PV and BESS at (450-1000-450) W/m^2 solar radiation

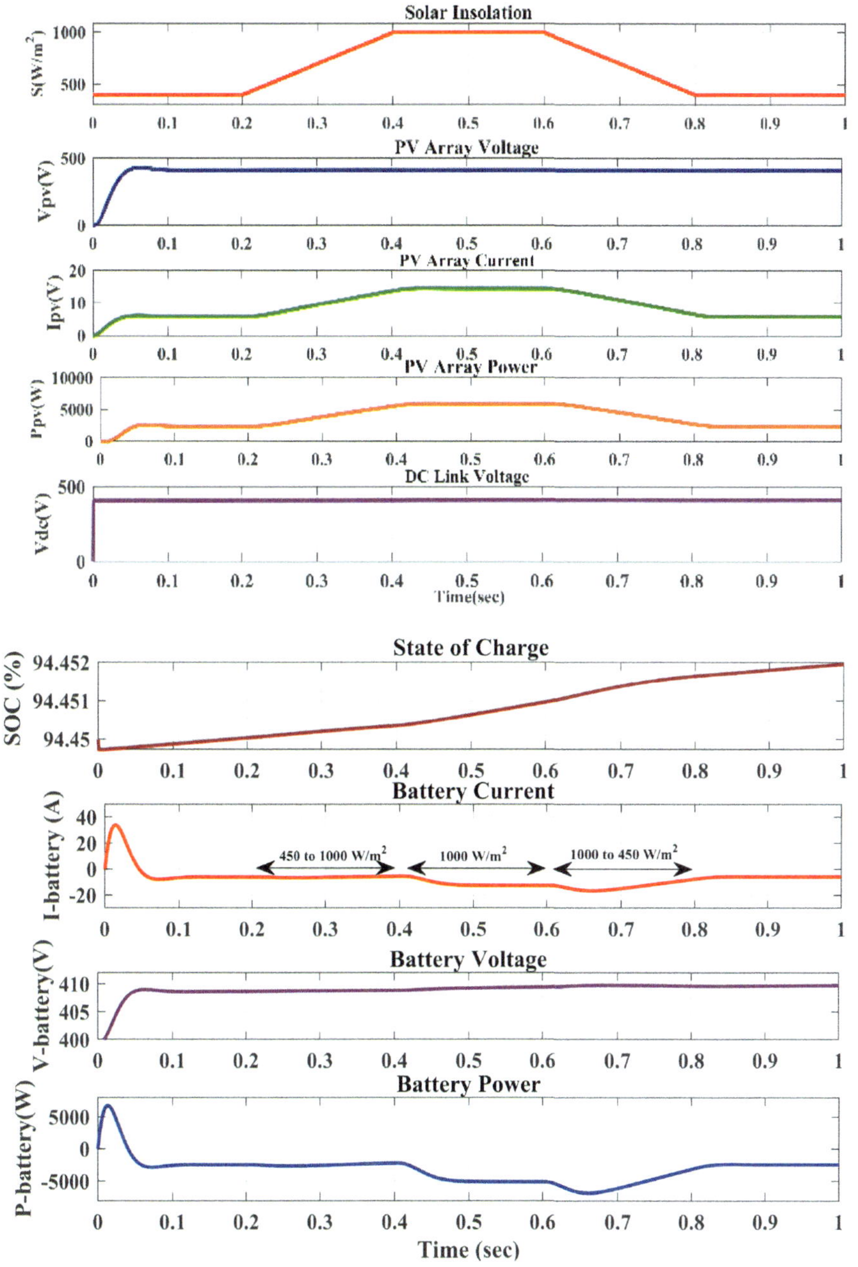

Fig. 10 (continued)

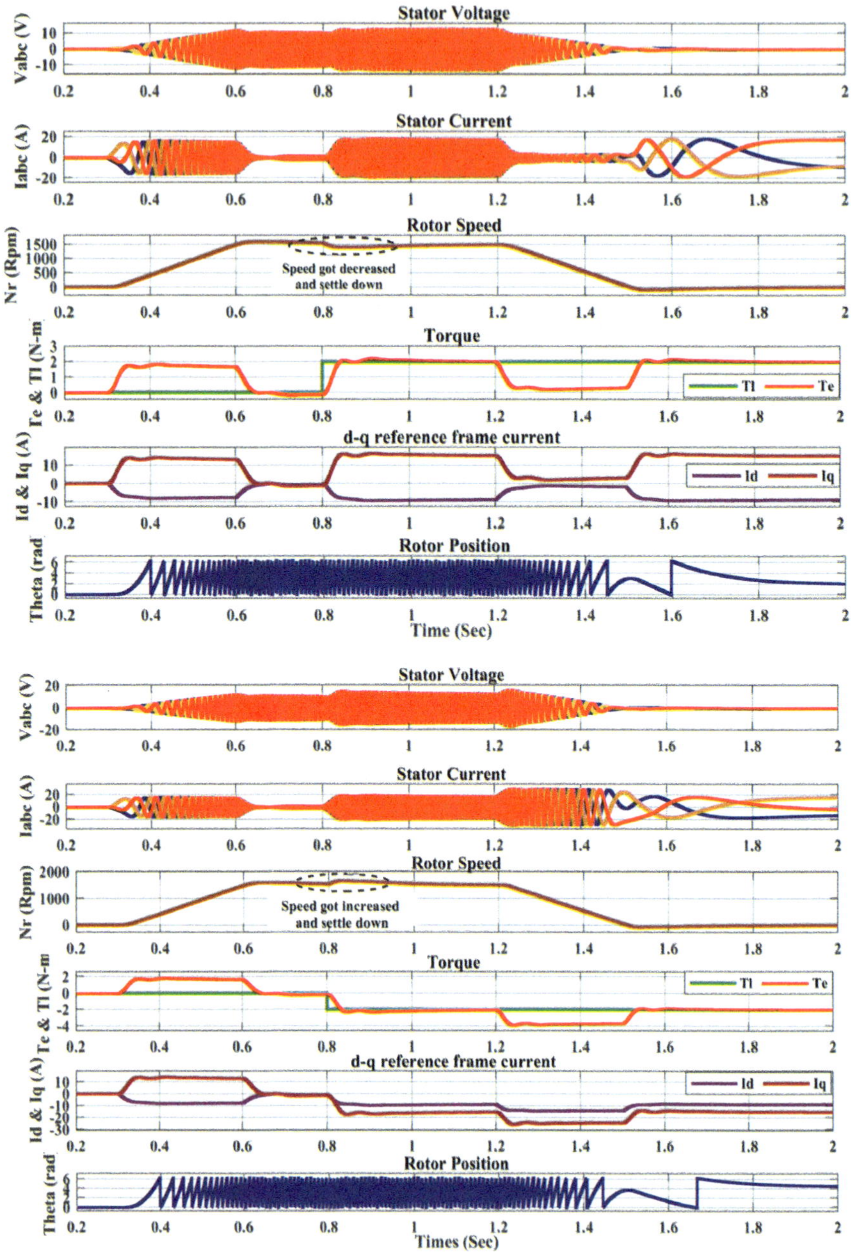

Fig. 11 Dynamic output characteristics of PMSM motor with varying load torque

4.1 EV PMSM Drive Operating in Acceleration Mode at 1000 W/m²

See Fig. 7.

4.2 EV PMSM Drive Operating in Cruising Mode at 1000 W/m²

See Fig. 8.

4.3 EV PMSM Drive Operating in Deceleration Mode at 1000 W/m²

See Fig. 9.

4.4 EV PMSM Drive Operating in Dynamic Mode at (450-1000-450) W/m²

See Fig. 10.

4.5 EV PMSM Drive Operating in Varying Load Torque Condition

In acceleration mode the traction PMSM motor drive of vehicle is operating under constant solar sunshine radiation of 1000 W/m^2. In Fig. 7 the rotor speed of motor is increased in ramp from time period of 0.3 to 0.6 s during this period the stator current buildup and varies from +20 to −20-amp peak to peak. As vehicle is under no load condition the electromagnetic torque (T_e) initially increased to 2 N-m and at 0.6 s instant becomes equal to load torque (T_L) when motor reaches to its rated speed of 1500 rpm. During acceleration mode the PV output power is 6 kW and DC link bus voltage is maintained constant. And battery gets charged and SOC is constant throughout the operation.

In cruising mode from time instant of 0.7 to 0.9 s as shown in Fig. 8 the stator current of motor gets reduced to +2 to −2 amps peak to peak and running at constant speed of 1500 rpm. During entire operation the load torque (T_L) and electromagnetic torque remains (T_e) equal. The battery is continuously charged by PV system in this entire mode as solar radiation remained constant.

From Fig. 9 between 1.2 and 1.5 s time instant, the vehicle enters the decelerating mode of operation. During this mode, when braking action is done the vehicle gets retarded which decreases the vehicle speed from 1500 to 0 rpm. But the frequency gets changed as speed decreases while stator current of motor remains sinusoidal in nature. Also, the torque of the motor (T_e) becomes negative during this phase.

When the traction system is subjected into dynamic mode operation when solar radiation varied from (450-1000-450) W/m^2 and speed of motor also increased in ramp.

During the drop in solar radiation condition from 1000 to 450, the solar pv output is not sufficient to maintain the DC link bus voltage for supplying the 3-phase inverter then BESS supply the difference of voltage and maintains the DC voltage constant. In Fig. 10 the battery current increased during time instant of 0.6 to 0.8. And SOC of battery increased from 94.45 and it is in continuous charging mode.

In loaded condition the load torque across motor is varied under positive torque and negative torque condition shown in Fig. 11. When positive torque is applied at instant of 0.8 s the speed of motor gets reduced and with comparison between the reference the actual speed the error gets settles around 1 s. The direct axis current remains in negative region and quadrature axis current is +10 amp and increases slightly at loaded condition. Thus, this verifies the two inner current loops working properly in FOC mode. Similarly, when negative load torque is applied the speed gets increased and error gets settles down.

5 Conclusion

In this work, a comprehensive study on the field-oriented control (FOC) of a photovoltaic-battery energy storage system (PV-BESS) fed traction permanent magnet synchronous motor (PMSM) drive for electric vehicle (EV) applications has been presented. With proper selection of rating of Li-ion battery pack for BESS and PV array system, ensures optimal energy utilization and extended battery life by using a I-NC MPPT algorithm for the PV array and a state-of-charge (SOC) control algorithm for the battery. The FOC of the PMSM provides precise torque and speed control, resulting in efficient and smooth operation of the EV. Furthermore, the performance of the proposed PV-BESS fed traction PMSM drive has been evaluated under varying weather and dynamic loading conditions. This proposed system can reduce the burden on the power grid due to EV charging and can mitigate the power quality issues due to BEVs. It can provide sustainable standalone solution to make vehicle self-independent to charge in driving mode as well as in standalone mode.

Appendix

Parameters of Solar PV Array: Parallel String = 2, Series String = 14, Open circuit voltage (V_{oc}) = 36.4, Maximum power point voltage (V_{mpp}) = 29 V, Maximum power point current (I_{mpp}) = 7.36 A, Short circuit current (I_{sc}) = 7.85, Diode ideality factor (n) = 0.98.

Parameters of PMSM: Stator resistance (R_s) = 1.4 Ω, Direct axis inductance (L_d) = 6.6 mH, Quadrature axis inductance (L_q) = 5.9 mH, Voltage constant (*V_peak L-L/krpm*) = 280.77, No. of phases = 3, pole pair = 3, Moment of Inertia ($J\ Kg.m^2$) = 0.028.

Parameters of BESS: Initial SOC (%) = 92, Battery type = Li-Ion, Rated capacity = 100Ah, Nominal voltage = 400 V.

References

1. Jain D, Gupta S, Jain M, Dewadi SK, Tandekar J (2021)Variable speed PMSM based PV-battery powered electric vehicle. In: 2021 IEEE 2nd international conference on electrical power and energy systems (ICEPES), pp 1–5
2. Mohan N, Undeland TM, Robbins WP (2003) Power electronics: converters, applications, and design. John Wiley & Sons
3. Rai RK, Dubey M, Kirar M, Kumar M (2022) Analytical analysis of solar PV module using MATLAB. In: 2022 IEEE international students' conference on electrical, electronics and computer science (SCEECS), pp 1–6
4. Kumar M, Deosarkar P, Tayade N, Yenare S (2023)Design and control of solar-battery fed PMSM drive for LEVs. In: 2023 IEEE international students' conference on electrical, electronics and computer science (SCEECS), pp 1–9
5. Kumar M, Deosarkar P, Mahanty RN (2023)Design and development of BLDC motor drive for solar-PV irrigation system using MATLAB. In: 2023 IEEE international students' conference on electrical, electronics and computer science (SCEECS), pp 1–4
6. Sreejith R, Singh B (2020)Position sensorless PMSM drive for solar PV-battery light electric vehicle with regenerative braking capability. In: 2020 IEEE energy conversion congress and exposition (ECCE), pp 1447–1452
7. Sharma RK, Sanadhya V, Behera L, Bhattacharya S (2008) Vector control of a permanent magnet synchronous motor. In: 2008 annual IEEE India conference, pp 81–86
8. De Brito MAG et al (2012) Evaluation of the main MPPT techniques for photovoltaic applications. IEEE Trans Indus Electron 60(3):1156–1167
9. Li F, Alshareef M, Lin Z, Jiang W (2016) A modified MPPT algorithm with integrated active power control for PV-battery systems. In: 2016 IEEE international conference on renewable energy research and applications (ICRERA)
10. Murshid S, Singh B (2018)Utility grid interfaced solar water pumping system using PMSM drive. In: 2018 IEEE international conference on power electronics, drives and energy systems (PEDES), pp 1–6
11. Kumar R, Singh B (2018) Brushless DC motor-driven grid-interfaced solar water pumping system. IET Power Electron 11(12):1875–1885
12. Ducar I, Marinescu C (2014)Increasing the efficiency of motor-pump systems using a vector controlled drive for PMSM application. In: 2014 international symposium on fundamentals of electrical engineering (ISFEE), pp 1–5
13. Hitesh S, Sunanda C (2020) Modelling and performance analysis of an electric vehicle with MATLAB/Simulink. Int Res J Eng Technol. e-ISSN: 2395-0056

14. Balasubrahmanyam CS, Gupta OH (2020) Detailed study of solar energy conversion system using boost converter—a new MPPT technique. J Inst Eng India Ser B 101:631–639. https://doi.org/10.1007/s40031-020-00478-1
15. Kumar P, Mahanty RN (2023) An off-MPPT controller for standalone PV systems. In: Gupta OH, Singh SN, Malik OP (eds) Recent advances in power systems. Lecture notes in electrical engineering, vol 960. Springer, Singapore. https://doi.org/10.1007/978-981-19-6605-7_12

Novel GA-Based Complex Fractional Order PID Controller Design for First-Order Integer and Non-integer Systems

Omar Hanif, R. Ranganayakulu, Sonu Singha, Shivani Richhariya, and G. Uday Bhaskar Babu

Abstract In this paper, a Complex Fractional Order Proportional Integral Derivative (CFOPID) controller is designed for a stable first-order system and complex order system. The increased complexity in tuning of the CFOPID controller with seven parameters needs to be addressed compared to PID with three parameters and Fractional Order PID (FOPID) controller with five parameters. The CFOPID tuning becomes tedious with conventional tuning methods. Hence, Genetic Algorithm (GA) is used for tuning the CFOPID controller with cost function as integral absolute error (IAE). The closed-loop response of the chosen systems with the CFOPID controller is compared against PID and FOPID controllers tuned using GA method. The indicators used to assess the performance are settling time and IAE for servo and regulation. A Bode plot is provided to check the stability.

Keywords CFOPID controller · FOPID controller · Complex order system · Genetic algorithm · IAE

O. Hanif
The University of Sheffield, Western Bank, Sheffield, UK
e-mail: ohanif1@sheffield.ac.uk

R. Ranganayakulu · S. Singha · S. Richhariya · G. U. B. Babu (✉)
National Institute of Technology Warangal, Telangana 506004, India
e-mail: udaybhaskar@nitw.ac.in

R. Ranganayakulu
e-mail: rayalla.ranga@student.nitw.ac.in

S. Singha
e-mail: ks22chm2r03@student.nitw.ac.in

S. Richhariya
e-mail: ks22chm2r06@student.nitw.ac.in

S. Kumar et al. (eds.), *Control Applications in Modern Power Systems*, Lecture Notes in Electrical Engineering 1128, https://doi.org/10.1007/978-981-99-9054-2_13

1 Introduction

The initial attempt on the controller's tuning was implemented through charts [1]. The Proportional integral derivative (PID) controller was later taken over by the industry and used even today due to the availability of tuning rules but gives inaccurate results owing to limited degrees of freedom [2]. The advancement of calculus [3, 4] facilitated the control engineers to develop fractional-order PID (FOPID) controllers [5–7].

There are several approaches for designing the fractional controllers [8] using frequency domain specifications [9] and performance contours [10]. However, there is a scope to increase the FOPID controller performance by adding additional degrees of freedom with complex orders to the integrator and differentiator terms. One such complex controller was proposed as proportional complex order integral controller [11].

There are several works on the design of FOPID controller for first-order system. A hybrid tuning procedure of fractional order $PI^{\lambda}D$ controller is proposed for first-order systems utilizing frequency domain and time domain specifications [12]. An FO[PD] controller was designed for a family of fractional order systems using frequency domain specifications [13]. Further, FOPID controller was used to stabilize an inverted pendulum [14]. A complex fractional order PID controller was designed through the use of curve fitting-based algorithm [15]. The complex-order controller was implemented in Simulink for a buck converter [16].

A complex FOPID (CFOPID) controller was designed for a real system modeled as first-order system using specified frequency domain parameters and is compared with FOPID controller [17]. Further, the first-order system was controlled by CFOPI controller [18] tuned based on maximum sensitivity (M_s). Next, a proportional integral complex order derivative controller was designed for a second-order system. The controller is tuned using numerical optimization to maintain robust performance [19]. There were limited methods in the literature for tuning the CFOPID controller designed for stable systems with integer order but not for complex order systems.

Genetic algorithm (GA) is a promising approach for tuning CFOPID controller than classical tuning methods due to the local optimization ability [20]. The CFOPID controller tuning is challenging with more tuning parameters than FOPID controller. The GA-based tuning of PID controller gains was reported considering a generalized feedback loop and proved to give fruitful results compared to other PID tuning methods [21]. Also, the GA technique was applied for tuning FOPID controller [22].

In this article, the CFOPID controller is designed for first-order system and a complex-order system and tuned using GA method. The performance is compared with PID and FOPID controller tuned by GA technique. Integral Absolute Error (IAE) is used as a cost function in the GA technique. The closed-loop performance is compared using settling time (t_s) and IAE.

2 Preliminaries

A complex-order transfer function (COTF): $G(s) = s^{a+ib}$ is decomposed into real and imaginary transfer functions using complex algebra and trigonometry rules.

$$G_R = R(s^{a+ib}) = R[s^a\cos(b\log s) + js^b(b\log s)] = s^a\cos(b\log s) \quad (1)$$

$$G_I = I(s^{a+ib}) = I[s^a\cos(b\log s) + js^b(b\log s)] = s^b(b\log s) \quad (2)$$

The present work uses Levy's method [23] of frequency domain for the identification of COTF as fractional-order transfer function (FOTF) [24] in the form of Eq. (3).

$$G(s) = \frac{b_m s^{m\gamma} + b_{m-1}s^{(m-1)\gamma} + \cdots + b_1 s^{\gamma} + b_0}{a_n s^{n\gamma} + a_{n-1}s^{(n-1)\gamma} + \cdots + a_1 s^{\gamma} + a_0} \quad (3)$$

where 'γ' is the commensurate order in fractional form; 'n' and 'm' are real polynomial orders.

The PID ($C_I(s)$), FOPID ($C_F(s)$), and CFOPID ($C_C(s)$) controllers realized in the current work are presented in (4)–(6):

$$C_I(s): k_p + \frac{k_i}{s} + k_d s \quad (4)$$

$$C_F(s): k_p + \frac{k_i}{s^{\lambda}} + k_d s^{\mu} \quad (5)$$

$$C_C(s): K_p + K_i\left(\text{Real}_{/i}\left[\frac{1}{s^{f+ig}}\right]\right) + K_d\left(\text{Real}_{/i}\left[s^{q+ir}\right]\right) \quad (6)$$

where k_p—proportional gain, k_i—integral gain and k_d—derivative gains; λ—fractional order of integrator, and μ—fractional order of differentiator; f, q and g, r are the real and imaginary parts of the order of integrator and differentiator, respectively; $\text{Real}_{/i}$ denotes the real component of the expression. The system response with the designed controllers is verified using t_s and IAE (Eq. 7).

$$\text{IAE} = \int_0^{\infty} |e(t)|dt \quad (7)$$

3 Proposed Approach

The present method proposes CFOPID controller and their tuning with GA technique. Two more controllers PID and FOPID are also tuned using GA for fair comparison with CFOPID controller. The GA technique uses IAE as the cost function. Table 1 gives the GA properties used and the proposed control algorithm is depicted in Fig. 1. The ranges of real and imaginary parameters of the controllers used for tuning through GA method are also listed in Table 1.

4 Results and Discussion

The performance of the proposed approach is explained with two examples: one representing the first-order system and the other is a complex-order system identified as non-integer system. The tracking response and regulation are observed for a unit step change in variables by using the measures: IAE and t_s along with stability plot.

4.1 *Example 1*

The first-order system used for simulation is given in Eq. (8). The ranges of controller gains K_p, K_i, and K_d used for tuning the three controllers, respectively, are (1–20), (1–20), and (1–40). The ranges of real and imaginary parameters for $G_1(s)$ are given in Table 1. GA-tuned parameters of the three controllers are presented in Table 2. It is interesting to note that the K_i and K_d are almost close to one another for the three controllers while the varying parameter is K_p.

$$G_1(s)=\frac{1}{s+10} \tag{8}$$

Table 1 GA properties and controller tuning parameter ranges used for tuning

Properties		Ranges of controller tuning parameters					
Parameter	Types/value	Controller	Parameter	Re	Im	Re	Im
Maximum generation	100	$C_I(s)$	Minimum	–	–	–	–
Population size	100		Maximum	–	–	–	–
Encoding	Binary	$C_F(s)$	Minimum	0.1	–	0.1	–
Selection	Uniform		Maximum	2	–	2	–
Crossover fraction	0.65	$C_C(s)$	Minimum	0.1	0.1	0.1	0.1
Mutation fraction	0.35		Maximum	2	2	2	2

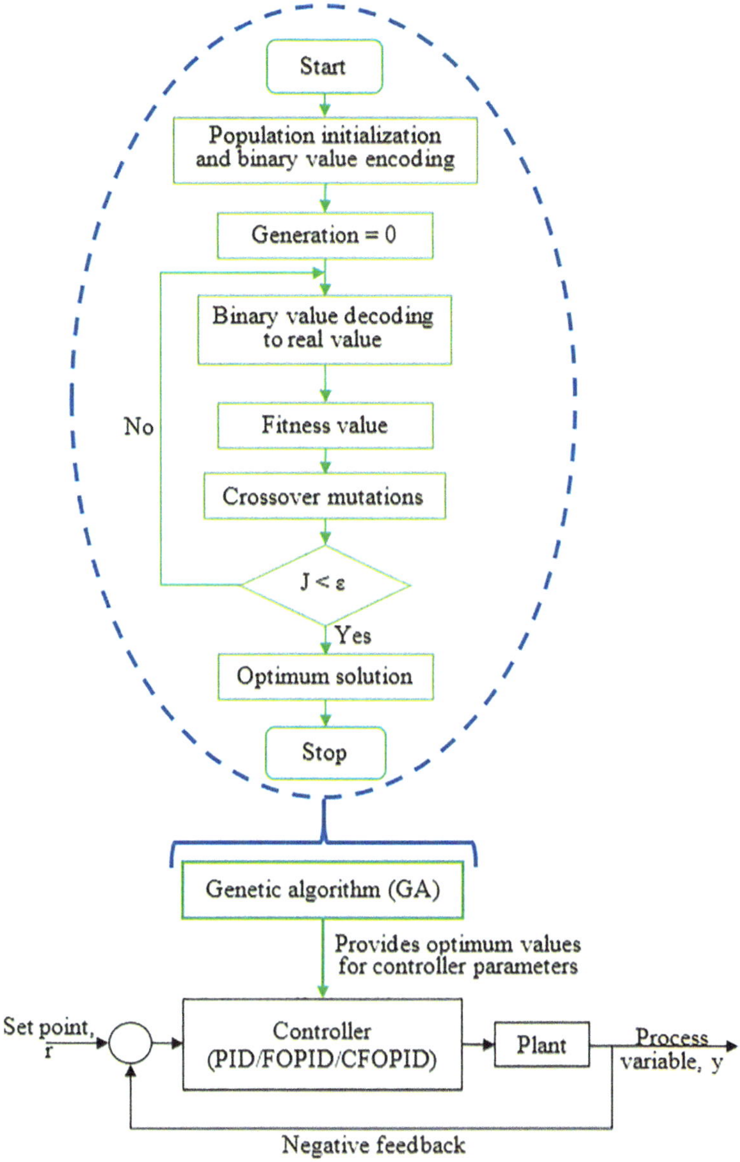

Fig. 1 Feedback loop with GA-tuned controller

The servo response of the system for step input is illustrated in Fig. 2 and the performance indicators are listed in Table 3. It is observed that the plant with CFOPID controller shows better tracking than with PID and FOPID controllers witnessing the smaller IAE. The FOPID-controlled plant is settling (t_s) slightly faster than CFOPID-controlled system. The system response for a disturbance given at $t = 2$ s is shown

Table 2 GA tuned controller parameters for Example 1

Controller	K_p	K_i	K_d	Re	Im	Re	Im
$C_I(s)$	6.3001	19.3371	1.0051	–	–	–	–
$C_F(s)$	1.0001	19.9783	1.0005	0.9987	–	0.1001	–
$C_C(s)$	1.0062	19.9958	1.0077	1.0204	0.1052	0.1029	0.4862

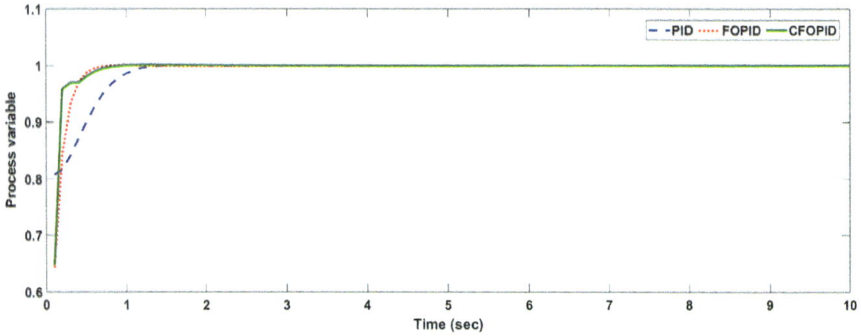

Fig. 2 Servo response for Example 1

Table 3 Comparison of performance indicators for Example 1

Controller	Servo response		Regulatory response	
	t_s	IAE	t_s	IAE
$C_I(s)$	1.1359	0.9158	3.1937	0.0526
$C_F(s)$	0.5379	0.4561	2.7198	0.0557
$C_C(s)$	0.5529	0.3292	2.6552	0.0565

in Fig. 3. It is evident that CFOPID-controlled plant response is faster than the other two controllers. This is proven with low value of t_s, while the IAE specification is slightly lower for PID-controlled plant. The CFOPID controller is relatively more stable which is observed from the Bode plot in Fig. 4.

4.2 *Example 2*

A novel complex-order transfer function (COTF) [19] is considered (Eq. 9) from which the following FOTF is identified (Eq. 10) using Levy's method. The frequency response (both magnitude and phase) of the identified and the original plant is plotted, and the best-fit transfer function is obtained in Fig. 5. This FOTF has a commensurate value $\gamma = 0.45$.

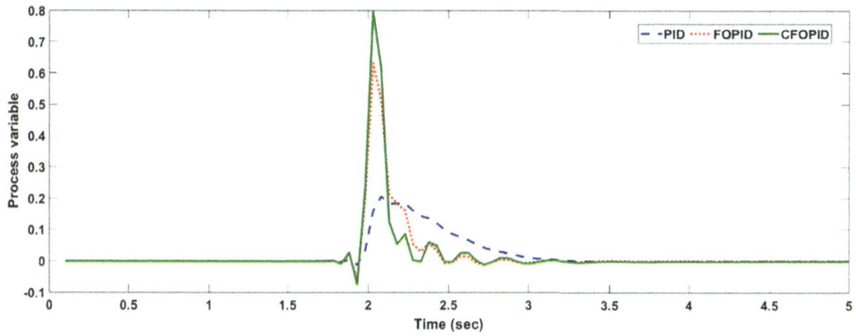

Fig. 3 Regulation for example 1

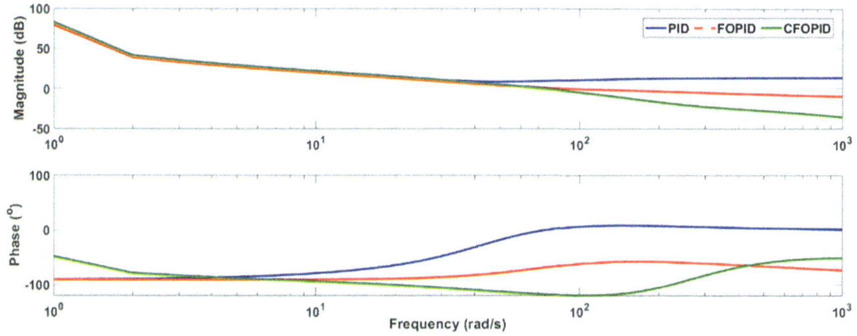

Fig. 4 Example 1: bode plot

$$G(s) = \frac{1}{s^{0.5+i0.5} + 1} \tag{9}$$

$$\begin{aligned}G_{\text{identified}} &= (6.1333 \times 10^{-14}s^{3.15} + 4.0333 \times 10^{-10}s^{2.7} - 2.1947 \times 10^{-7}s^{2.25} \\ &+ 3.6907 \times 10^{-5}s^{1.8} - 0.0024236s^{1.35} + 0.06108s^{0.9} - 0.4724s^{0.45} + 0.98446)/ \\ &(6.241 \times 10^{-16}s^{4.05} + 1.1046 \times 10^{-13}s^{3.6} - 3.655 \times 10^{-11}s^{3.15} \\ &- 4.6362 \times 10^{-8}s^{2.7} + 3.7617 \times 10^{-6}s^{2.25} + 1.674 \times 10^{-4}s^{1.8} \\ &- 0.012625s^{1.35} + 0.08195s^{0.9} - 0.28135s^{0.45} + 1)\end{aligned} \tag{10}$$

The ranges of controller gains K_p, K_i, and K_d used for tuning the three controllers, respectively, are (1–50), (1–50), and (1–70). The optimized controller settings are given in Table 4. An important observation here is that the gains are very close for all the controllers with a variation in their decimal values.

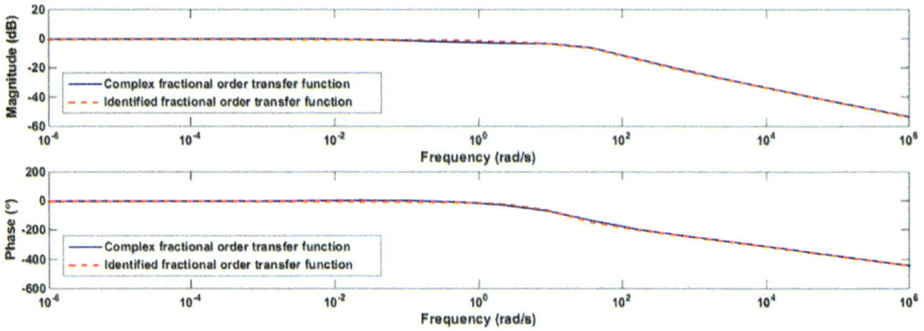

Fig. 5 Frequency domain identification of COTF

Table 4 GA-tuned controller parameters for Example 2

Controller	K_p	K_i	K_d	Re	Im	Re	Im
$C_I(s)$	49.869	49.731	69.741	–	–	–	–
$C_F(s)$	49.991	49.995	69.997	1.714	–	0.1002	–
$C_C(s)$	49.998	49.97	69.981	1.744	0.2088	0.1003	0.1001

The closed-loop response for servo and regulation is presented in Fig. 6 and Fig. 7 and the observed indicators during simulation are in Table 5. The closed loop with CFOPID controllers shows superior performance with minimum t_s and IAE for both the cases compared to other controller variants and it is stable (Bode plot in Fig. 8).

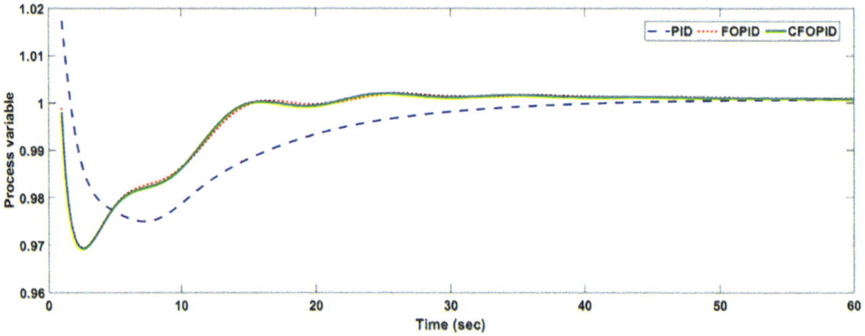

Fig. 6 Servo response for Example 2

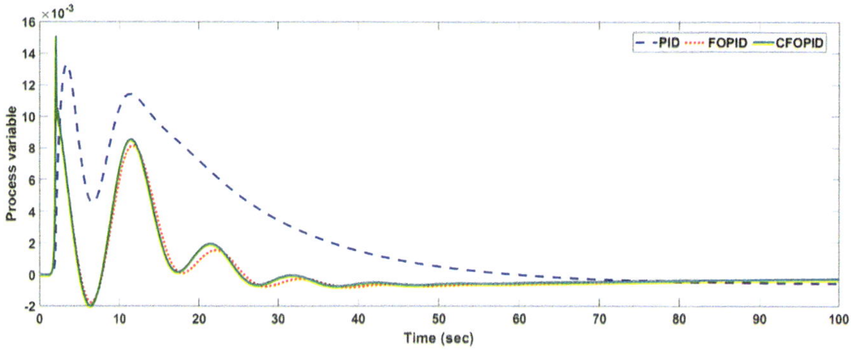

Fig. 7 Regulation for Example 2

Table 5 Comparison of performance indicators for Example 2

Controller	Servo response		Regulatory response	
	t_s	IAE	t_s	IAE
$C_I(s)$	44.982	0.365	69.913	0.132
$C_F(s)$	38.866	0.275	53.508	0.054
$C_C(s)$	35.534	0.254	49.951	0.053

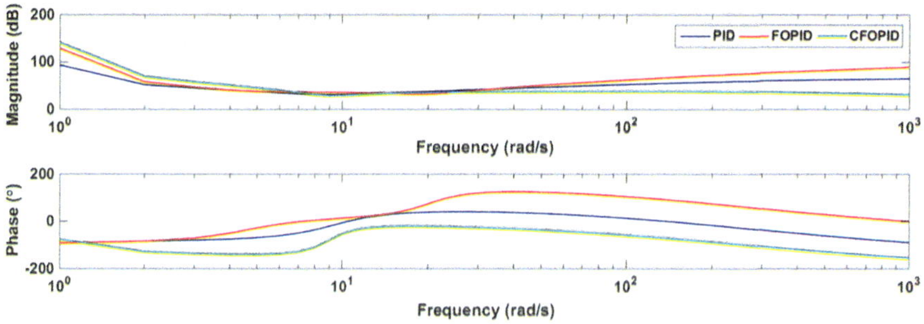

Fig. 8 Example 2: bode plot

5 Conclusion

A CFOPID controller is designed for first-order and complex-order systems approximated as non-integer or fractional system. GA technique is used to tune CFOPID controller and the performance is compared with GA-tuned PID and FOPID controllers. The simulation results for both the examples confirm that the CFOPID controller gives enhanced closed-loop performance with less settling time and IAE compared to the other two controllers. This is verified with step changes in set point and disturbance. The proposed CFOPID controller design approach can be further applied to control non-minimum phase systems and unstable systems.

References

1. Breedveld PC (2004) Port-based modeling of mechatronic systems. Math Comput Simul 66(2–3):99–128
2. Ziegler JG, Nichols NB (1993) Optimum settings for automatic controllers. J Dyn Syst Meas Contr 115(2B):220–222
3. Miller KS, Ross B (1993) An introduction to the fractional calculus and fractional differential equations. John Wiley & sons, New York
4. Das S (2011) Functional fractional calculus. Springer, Berlin Heidelberg
5. Valério D, Da Costa JS (2011) Introduction to single-input, single-output fractional control. IET Control Theory Appl 5(8):1033–1057
6. Podlubny I, Petras I, Vinagre BM, Chen Y, O'Leary P, Dorcak L (2003) Realization of fractional order controllers. Acta Mont Slovaca 8(4):233–235
7. Ranganayakulu R, Babu GUB, Rao AS, Patle DS (2016) A comparative study of fractional order $PI\lambda/PI\lambda D\mu$ tuning rules for stable first order plus time delay processes. Resour-Effic Technol 2:S136–S152
8. Ranganayakulu R, Babu GUB (2021) Analytical design of fractional filter PID controller for improved performance of first order plus time delay plants. ICIC Express Lett 15(2):135–142
9. Ahn HS, Bhambhani V, Chen Y (2008) Fractional-order integral and derivative controller design for temperature profile control. In 2008 Chinese control and decision conference. IEEE, Yantai, China, pp 4766–4771
10. Oustaloup A, Pommier-Budinger V, Lanusse P (2003) Design of a fractional control using performance contours. Application to an electromechanical system. Fract Calc Appl Anal 6(1):1–24
11. Shahiri M, Ranjbar A, Karami MR, Ghaderi R (2016) New tuning design schemes of fractional complex-order PI controller. Nonlinear Dyn 84(3):1813–1835
12. Chen P, Luo Y, Peng Y, Chen Y (2021) Optimal robust fractional order $PI^{\lambda}D$ controller synthesis for first order plus time delay systems. ISA Trans 114:136–149
13. Luo Y, Chen Y (2009) Fractional order [proportional derivative] controller for a class of fractional order systems. Automatica 45(10):2446–2450
14. Mondal R, Chakraborty A, Dey J, Halder S (2020) Optimal fractional order $PI^{\lambda}D^{\mu}$ controller for stabilization of cart-inverted pendulum system: experimental results. Asian J Control 22(3):1345–1359
15. Bingi K, Kulkarni RR, Mantri R (2021) Design and analysis of complex fractional-order PID controllers. In 2021 IEEE madras section conference (MASCON). IEEE, Chennai, India, pp 1–6
16. Warrier P, Shah P (2023) Design of an Optimal Fractional complex order PID controller for buck converter. J Robot Control (JRC) 4(3):243–262
17. Abdulwahhab OW (2020) Design of a complex fractional order PID controller for a first order plus time delay system. ISA Trans 99:154–158
18. Moghadam MG, Padula F, Ntogramatzidis L (2018) Tuning and performance assessment of complex fractional-order PI controllers. IFAC-PapersOnLine 51(4):757–762
19. Guefrachi A, Najar S, Amairi M, Aoun M (2017) Tuning of fractional complex order PID controller. IFAC-PapersOnLine 50(1):14563–14568
20. Katoch S, Chauhan SS, Kumar V (2021) A review on genetic algorithm: past, present, and future. Multimed Tools Appl 80(5):8091–8126
21. Elsisi M (2021) Optimal design of non-fragile PID controller. Asian J Control. 23(2):729–738

22. Jaiswal S, Suresh Kumar C, Seepana MM, Babu GUB (2020) Design of fractional order PID controller using genetic algorithm optimization technique for nonlinear system. Chem. Prod. Process. Model. 15(2):20190072
23. Valério D, Sá da Costa J (2005) Levy's identification method extended to commensurate fractional order transfer functions. In Fifth EUROMECH nonlinear dynamics conference. Eindhoven, Netherlands, pp 1357–1366
24. Tepljakov A, Petlenkov E, Belikov J (2011) FOMCOM: A MATLAB toolbox for fractional order system identification and control. Int J Microelectron Comput Sci 2(2):51–62

DG Placement of an Islanded Microgrid

Mudasir Farooq, Sachin Mishra, and Javed Dhillon

Abstract As renewable energy is progressively incorporated into the power system, microgrids operating in the islanding mode may experience system frequency deviation, voltage profile deviation, and increase in losses from the required level due to the unpredictable and inconsistent production from renewable energy. Distributed generation (DG) services are required in the power system due to environmental concerns and the rising demand for electrical energy. The placement and sizing of DGs have been employed in this study to achieve the aims of lowering losses and enhancing the microgrid's voltage profile. From a variety of artificial intelligence techniques, the proposed Genetic Algorithm (GA) is described as a solution to the issue. On the network of IEEE 33 buses, the technique is put into practice. This study is presented to show the impact of the selection and placement of DGs in order to cut losses, enhance the voltage profile, and improve the frequency profile of island.

Keywords Newton–Raphson method · Islanded microgrid · Genetic algorithm

1 Introduction

Natural disasters including storms, tornadoes, earthquakes, and others have the capacity to seriously harm electricity grid networks. Depending on the extent of the damage, this might cause the network to experience extensive power outages that last for hours, days, or even weeks. Utility-provided power is necessary for the continued functioning of essential services for the general public. By providing emergency power during such grid outages, microgrids, also known as localized low-voltage grids, can assist utilities in enhancing the resilience of crucial public assets (i.e., key public resources). Distributed generation sources, both conventional and non-conventional, power storage devices, loads or supply nodes, an interconnection switch, a controller, and an energy management system are all components of microgrids [1]. The ability of island-based microgrids to function independently

M. Farooq · S. Mishra · J. Dhillon (✉)
School of Electronics and Electrical Engineering, Lovely Professional University, Phagwara, India
e-mail: javeddah@gmail.com

S. Kumar et al. (eds.), *Control Applications in Modern Power Systems*, Lecture Notes in Electrical Engineering 1128, https://doi.org/10.1007/978-981-99-9054-2_14

of the main grid during natural disasters, known as islanded mode operation, makes them important resources for utility corporations.

An islanded microgrid often uses wind or solar/photovoltaic-based renewable DGs. Due to the need for land space to build wind turbines, wind-based DGs are more frequently encountered in rural regions. This contrasts with solar/photovoltaic (PV) DGs, which can be quickly installed on rooftops in both rural and urban settings, but the most important thing is where we place the DG in the microgrid because it is most important to decrease the losses in the system, enhance the voltage profile, and enhance the frequency profile of the islanded microgrid. In our case study, we are trying to find the optimal place for the DG in the microgrid for that we are using optimization technique, i.e., genetic algorithm that helps us in finding optimal place for our DG, modified Newton–Raphson load flow technique is used for calculating the losses in the islanded microgrid and also analyzed the frequency and voltage variation across the system.

2 Literature Review

In [2], the author identifies the location of several distributed generators in a primary distribution network using improved analytical technique to decrease the overall losses across the network but voltage profile improvement was not considered in this work. The positioning and sizing of DGs were also optimized by Gangwar et al. [3] using a multi-objective PSO (particle swarm optimization) to minimize power losses and enhance system reliability.

In [4], the author economically increases the system reliability by assigning dispatchable distributed generating units in distribution systems. The primary goal of this research is to identify the best possible configuration of distributed generation units and to calculate the number of loads to be shed throughout the system to maintain the flat voltage profile. The author used a probabilistic load model to determine the ideal size of allocated DG units across the system. In [5], the author used an evolutionary algorithm for optimization, and load levels such as low load, average load, and high load are taken into consideration during scheduling, allowing the utilization of the outcomes for all network conditions. In the paper, placement and ideal capacity of DGs are chosen using the plant growth simulator algorithm (PGSA) method to lower losses and enhance the voltage profile of the grid.

The latest DG integration designs, have been evaluated by Prakash and Khatod in [6], Yang and Ehsan in [7], and Adetunji et al. [8]. In the given research, authors optimized the location and size of integrated DG (for instance, exact location), with the goals to minimize power losses, operational cost, demand satisfied (system reliability), and system resilience, which is typically attained by improving voltage level across the network. To the best of our knowledge, from the given literature studies, it has been determined that the location of the DG in the islanded microgrid is nonexistent because, in the microgrid, DG is required to regulate both voltage and frequency regulation. so, in our case study, a model has been designed to find the optimal place for DG in the islanded microgrid so that it can minimize the losses and provide supply where it is needed. As a result, the overall objective function accounts for the reliability, resiliency, expenses, and possible power losses of the microgrid in order to reduce the overall cost to a utility. This enables utility providers to assess the investment viability, dependability, and resilience of islanded microgrids when tasked with supplying emergency power to vital public resources in the event that the main grid is compromised.

3 Problem Formulation

In this section, objective function is presented. As already discussed above that the objective of DG placement is required to minimize losses, and improve frequency and voltage in the islanded microgrid. Mathematical formulation for different terms of the objective function is presented as follows:

Objective Function: min (VDI + FDI − ELI).

Voltage deviation index (VDI): As everyone is aware, maintaining bus voltage within limitations is the most crucial factor, particularly in an island microgrid. A small change in voltage has an impact on the entire system and may cause a block-out condition. The VDI indicates the amount of voltage that deviates from the voltage rated.

$$VDI = \sum_{i}^{NL} \frac{|V_{rated} - V_i|}{V_{rated}} \tag{1}$$

where V_{rated} is the nominal voltage of the system.
V_i is the voltage at the ith bus.
N_B is the number of buses.

Frequency deviation index (FDI): Maintaining frequency within the limits is the most important criteria, especially in an islanded microgrid. A slight variation in the frequency affects the whole system and may result in block-out conditions. The amount of frequency that deviates from the rated frequency is given by the FDI:

$$FDI = \frac{|F_{rated} - F_i|}{F_{rated}} \tag{2}$$

where F_{rated} is the nominal frequency of the system.
F_i is the frequency of the ith bus.

Energy loss index: In order to make our system more efficient, we have to minimize the losses as much as possible. When the power losses increases, the system affects the rated voltage, so it affects the performance of the system. In order to use the maximum energy, we have to minimize the losses. The change in energy from the base energy due to losses is given by the energy deviation index.

$$\mathrm{ELI} = \sum_{ij}^{Nl} \frac{\mathrm{Losses}_{\text{in case A}} - \mathrm{losses}_{\text{in case B}}}{\mathrm{Losses}_{\text{in case A}}} \tag{3}$$

In case "A", DG is connected on bus 1, and in case "B", DGs are connected at optimal places in an islanded microgrid.

4 Modified Newton–Rapson Method

The load flow problem for islanded microgrids is solved by the modified Newton–Raphson method (MNR) utilizing a novel strategy that combines the traditional Newton–Raphson (NR) method with the DG's droop control.

4.1 Describing the System's Buses

The first step in resolving any system's power flow is to determine the different types of buses that are there. The pre-specified quantities determine the bus's type. When tackling the power flow challenge using the traditional model, the NR technique employs three well-defined buses, the PV bus, the load bus, and the slack bus. The premise that any DG will operate as a slack bus in a various-source isolated microgrid is invalid since no single DG can maintain a steady local bus voltage and system frequency. The bus kinds for islanded microgrids are categorized in this research as follows:

- The real and imaginary power of PQ bus is known.
- PV bus: The real power and voltage magnitude are known.
- VF-dependent bus: The real and imaginary power of the bus rely on the system frequency as well as the bus voltage. For such voltage ranges within system, a swing bus is chosen to act as a reference. A generator bus, load bus, or voltage frequency-dependent bus can serve as a swing bus.

4.2 Modified Newton–Raphson (MNR)

The real and imaginary power of all the droop buses must be determined to be able to determine angle of the voltage and intensity of a droop bus at the (I + 1) iteration (to be added to the mismatch matrix). The real and imaginary powers of the droop bus k can be calculated under the assumption that the output impedance of distributed generator is inductive and by setting the references Po and Qo to zero.

$$Q_{GK^{i+1}} = \frac{1}{n_{pk}}\left(|V_0| - \left|V_K^i\right|\right) \tag{4}$$

$$P_{GK^{i+1}} = \frac{1}{m_{pk}}\left(\omega_0 - \omega^i\right) \tag{5}$$

In the event that the droop bus's active and reactive powers go beyond the limit, the relevant power is adjusted to its restricted value. Like what is shown in (1), the frequency of the system affects the active power of the droop bus. All of the droop control buses communicate using the system frequency to be able to distribute the system's loads. In other words, the system angular frequency, which is shared by all droop buses in the microgrid, will be used to deliver active power. The total active powers for d droop buses in the system are expressed as follows:

$$P_{\text{sys}} = \sum_{k=1}^{d} P_{GK} = \sum_{k=1}^{d} \frac{1}{m_{pk}}(\omega_0 - \omega) \tag{6}$$

$$Q_{\text{sys}} = \sum_{k=1}^{d} Q_{GK} = \sum_{k=1}^{d}(|V_0| - |V_K|) \tag{7}$$

In the case of the modified Newton–Raphson load flow approach, starting system frequency of one p.u. is anticipated, as are voltages of the buses at the start of the load flow. Bus 1 voltage magnitude ($|V_1|$) is one more parameter in the updated Newton–Raphson approach (because of the omission of slack bus). The formula for the differential vector (X′) is

$$X' = [|V|^T \omega |V_1|]^T \tag{8}$$

$$\Delta' = \left[P^T - P_c^T Q^T - Q_c^T P_{\text{tot}} - P_{\text{sys}} Q_{\text{tot}} - Q_{\text{sys}}\right]^T \tag{9}$$

The updated mismatch matrix now contains four new terms. P_{tot} is the result of adding the cumulative real power and P_{loss}. The overall real power requirement in the (P_{load}) system and the real power losses of the system can be used to replace the total real powers of the distribution generators (P_{tot}). The system's cumulative reactive power (Q_{tot}) is equal to the product of its reactive power loss (Q_{loss}) and total reactive load (Q_{load}).

$$P_{\text{tot}} = P_{\text{load}} + P_{\text{loss}} \tag{10}$$

$$Q_{\text{tot}} = Q_{\text{load}} + Q_{\text{loss}} \tag{11}$$

Losses must therefore be determined in each cycle and are determined using the below equation:

$$P_{\text{loss}} = \frac{1}{2}\sum_{k}^{N}\sum_{n}^{N}\{Y_{kn}(V_k * V_n + V_n * V_k)\} \tag{12}$$

$$Q_{\text{Loss}} = \frac{1}{2}\sum_{k}^{N}\sum_{n}^{N}\{Y_{kn}(V_k * V_n + V_n * V_k)\} \tag{13}$$

Next, it is estimated for each bus what their organized real power (P), anticipated imaginary power (Q), computed real power (P_c), and computed imaginary power (Q_C) are. Making the mismatch matrix (Δ') equal to zero is the goal. The revised Jacobian matrix is acquired by differentiating Eqs. (19) and (20) with regard to, |V|, δ, ω, and $|V_1|$ to identify the whole modified Jacobian matrix. The updated Jacobian matrix has been provided below

$$\begin{bmatrix} J_{11} & J_{12} & J_{13} & J_{14} \\ J_{21} & J_{22} & J_{23} & J_{24} \\ J_{31} & J_{32} & J_{33} & J_{34} \\ J_{41} & J_{42} & J_{43} & J_{44} \end{bmatrix}\begin{bmatrix} \Delta\delta \\ \Delta V \\ \Delta\omega \\ \Delta V_1 \end{bmatrix} = \begin{bmatrix} \Delta P \\ \Delta Q \\ \Delta P_{\text{sys}} \\ \Delta Q_{\text{sys}} \end{bmatrix} \tag{14}$$

The updated Jacobian's first sub-matrix can be determined using the standard Newton–Raphson approach because it is the same as the Jacobian of that method. The final twelve sub-matrices, which are updated Jacobian sub-matrices for the updated Newton–Raphson approach, are computed utilizing

$$J_{41} = \left[\frac{\partial Q_{\text{sys}}}{\partial \delta_2} \cdots \frac{\partial Q_{\text{sys}}}{\partial \delta_N}\right] J_{42} = \left[\frac{\partial Q_{\text{sys}}}{\partial |V_2|} \cdots \frac{\partial Q_{\text{sys}}}{\partial |V_N|}\right] \tag{15}$$

$$J_{31} = \left[\frac{\partial P_{\text{sys}}}{\partial \delta_2} \cdots \frac{\partial P_{\text{sys}}}{\partial \delta_N}\right] J_{32} = \left[\frac{\partial P_{\text{sys}}}{\partial |V_2|} \cdots \frac{\partial P_{\text{sys}}}{\partial |V_N|}\right] \tag{16}$$

$$J_{43} = \left[\frac{\partial Q_{\text{sys}}}{\partial \omega}\right] \quad J_{23} = \left[\frac{\partial Q_{c2}}{\partial \omega} \cdots \frac{\partial Q_{cN}}{\partial \omega}\right]^T \tag{17}$$

$$J_{44} = \left[\frac{\partial Q_{\text{sys}}}{\partial |V_1|}\right] J_{24} = \left[\frac{\partial Q_{c2}}{\partial |V_1|} \cdots \frac{\partial Q_{cN}}{\partial |V_1|}\right]^T \tag{18}$$

$$J_{33} = \left[\frac{\partial P_{\text{sys}}}{\partial \omega}\right] J_{13} = \left[\frac{\partial P_{c2}}{\partial \omega} \cdots \frac{\partial P_{cN}}{\partial \omega}\right]^T \tag{19}$$

$$J_{34} = \left[\frac{\partial P_{\text{sys}}}{\partial |V_1|}\right] J_{14} = \left[\frac{\partial P_{c2}}{\partial |V_1|} \cdots \frac{\partial P_{cN}}{\partial |V_1|}\right]^T \tag{20}$$

5 Genetic Algorithm

The Darwinian idea of "survival of the fittest" serves as the foundation for the optimization technique known as the genetic algorithm. It is a search method that makes advantage of crossover and mutation, two natural genetic processes. It uses artificial intelligence (AI) to find the best site for each DG in an island-based microgrid. Most practical issues can be solved using a basic genetic algorithm that uses the three operators of reproduction, crossover, and mutation.

The following details/facts show how genetic algorithm method is apart from all other optimization methods:

- Rather than coding the parameters directly, GA employs the coding of a group of parameters.
- GA encourages the use of objective function values rather than relying on additional information.
- GA never looks for a single point; it always searches the population.
- GA never employs deterministic rules; instead, it employs probabilistic rules.

For each DG source, the GA takes into account the production of a random population of a location. Following the load flow program's execution, it determines the losses in the anticipated states for the placement of the DG per bus and then distinguishes between optimal states and non-optimal modes. And, the parent uses these ideal states to produce the population in the following generation. This procedure is repeated until the algorithm's stopping requirements are met.

The algorithm for optimal DG placement in an islanded microgrid is given below (Fig. 1).

6 Results and Discussions

For solving the optimal DG placement problem for an islanded microgrid optimization technique is required with a load flow algorithm to calculate the voltage, frequency, and losses. It also helps to find the optimal placement of DGs. For calculating voltage frequency and losses, modified Newton–Raphson method is used and for optimal location of DG's genetic algorithm is used. The results obtained are from 10 bus systems and 38 bus systems.

Table 1 shows the voltages of all the buses for the 10 bus systems when the DGs are placed at optimal locations. It is also seen the frequency is also improved when DGs are at optimal locations. Figure 2 shows the comparison of voltages when all the DGs are placed at optimal location with one of the DGs is randomly placed at Bus 1.

From the above graph, it can be seen that voltage profile of all the 10 buses is improved when DGs are connected at the best location in the islanded microgrid, i.e., the deviation in voltage of all the buses is minimum.

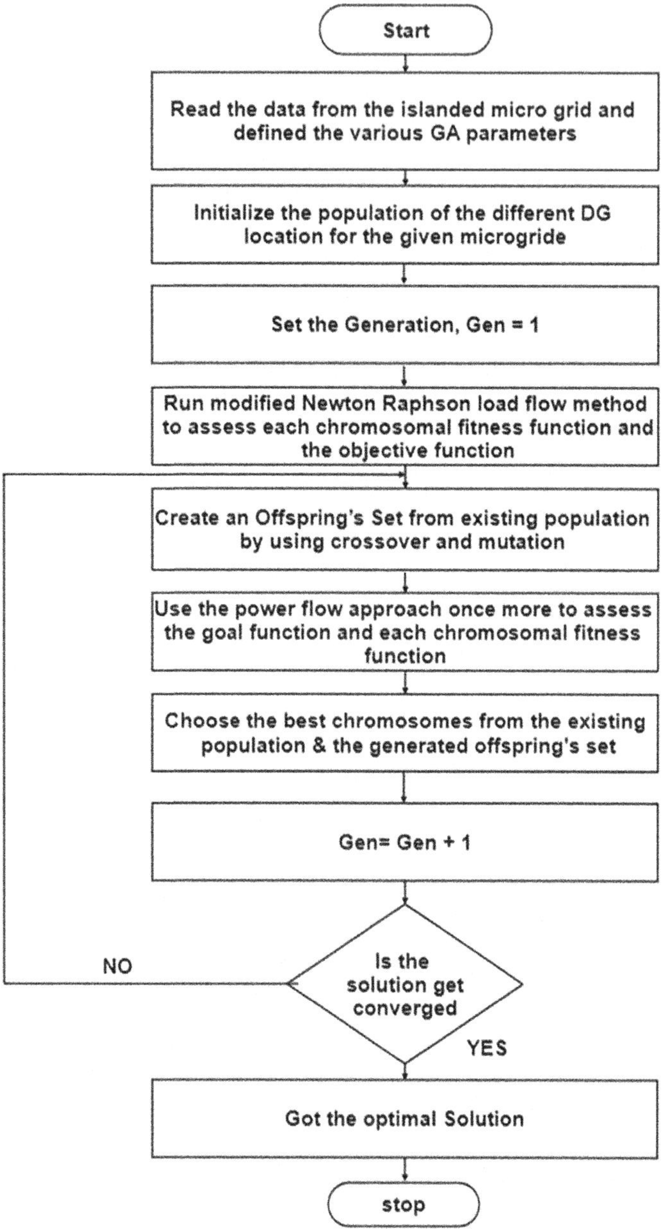

Fig. 1 Genetic algorithm flowchart for DG placement

Table 1 Voltages for 10 bus islanded microgrids

Bus	Voltage	Voltage angle
1	0.999059	0
2	0.999659	0.000155
3	1.001188	0.003364
4	1.003009	0.005802
5	1.004742	0.007848
6	1.001023	0.006794
7	0.998863	0.007517
8	0.999741	0.010302
9	1.001106	0.016416
10	1.00063	0.015726

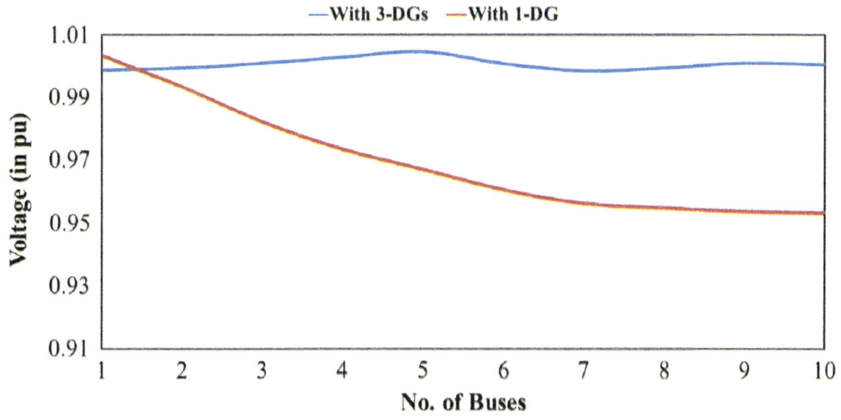

Fig. 2 Voltage profile for 10 bus islanded microgrids

The results obtained when applying genetic algorithm approach in the 10 bus islanded microgrids are given in Table 2. From the below table, it is observed that bus voltages, frequency, and losses are minimized by placing DGs at the best locations in the islanded microgrid, and the system becomes stable and works effectively.

Table 2 Optimal DG placement for 10 bus microgrids

DG parameters	With 1 DG connected on bus 1	With 3 DGs connected at the best locations
Minimum bus voltage	0.9535	0.998863
Frequency (pu)	0.94821	0.999567
Total active power loss (pu)	0.0136	0.000678
Total reactive power loss (pu)	0.0177	0.001731

Table 3 Voltages for 38 bus islanded microgrids

Bus	Voltage	Voltage angle	Bus	Voltage	Voltage angle
1	0.990737	0	20	0.986726	−0.00148
2	0.990737	0	21	0.98604	−0.00185
3	0.99248	0.00007	22	0.98542	−0.00223
4	0.990007	−0.00183	23	0.988668	−0.00063
5	0.987809	−0.00388	24	0.981517	−0.0026
6	0.984007	−0.01025	25	0.977552	−0.00374
7	0.988482	−0.01178	26	0.982303	−0.00974
8	0.991092	−0.01743	27	0.980044	−0.009
9	0.991135	−0.0223	28	0.970128	−0.00829
10	0.992787	−0.02652	29	0.963018	−0.00749
11	0.991585	−0.02653	30	0.959906	−0.00604
12	0.989422	−0.02657	31	0.95626	−0.00763
13	0.98213	−0.02943	32	0.955463	−0.00806
14	0.979402	−0.03155	33	0.955221	−0.00819
15	0.977399	−0.03277	34	0.98813	−0.0201
16	0.975256	−0.03377	35	0.988782	−0.02384
17	0.971759	−0.03702	36	0.987207	−0.02821
18	0.97031	−0.03782	37	0.969672	−0.03836
19	0.99022	−0.0002	38	0.977066	−0.00402

Table 4 Optimal DG placement for 38 bus microgrids

DG parameters	With 1 DG connected on bus 1	With 3 DGs connected at the best locations
Minimum bus voltage	0.7331 V	0.955221 V
Frequency (pu)	0.9829	0.995871
Total active power loss (pu)	0.1725	0.053470
Total reactive power loss (pu)	0.1061VAR	0.042933 VAR

The genetic algorithm optimization technique is tested on the 38 bus systems also. Table 3 shows the voltages of all the buses when DGs are placed at optimal location. Table 4 show the comparison when three DGs are placed at optimal location with a single DG placed at random location on Bus 1. Figure 3 shows the voltage profile of all the buses of 38 bus systems in an islanded microgrid.

From the above graph, it can be seen that the deviation in the bus voltages is very less when DGs are placed at the best locations in the islanded microgrid.

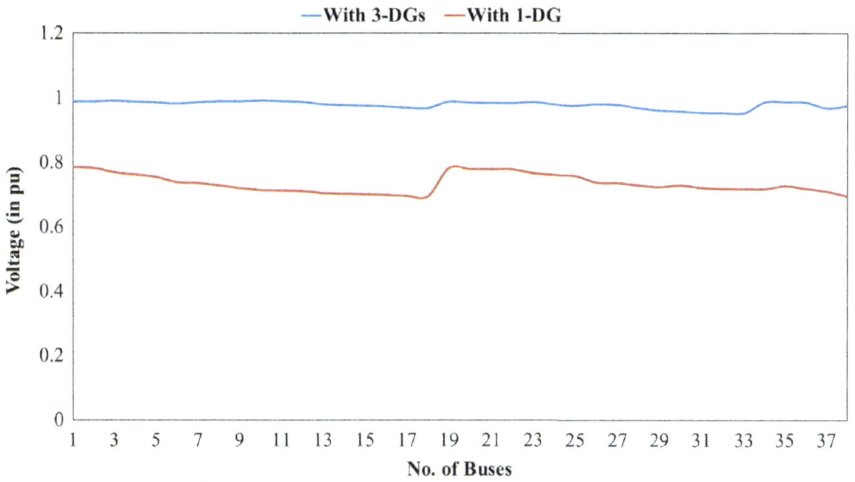

Fig. 3 Voltage profile for 38 bus islanded microgrids

7 Conclusion

In this research, the genetic algorithm technique is used for optimal placement of DGs in an islanded microgrid to minimize losses, enhance voltage profile and frequency. The research shows that by placing the DGs at optimal locations the voltage profile is enhanced, frequency improved, and losses are reduced. Before simulating the load under actual network conditions, decisions cannot be made with certainty. After examining the data, it has been determined that the best DG placement is advantageous since it reduces power loss and significantly raises the voltage profile and maintains the frequency at its rated value. All industrial, residential, and commercial subscribers should be taken into account to provide more precise results. In other words, it is impossible to generalize to all levels of load from the results obtained for the single-level load under nominal conditions.

References

1. Haidar N, Attia M, Senouci SM, Aglzim EH, Kribeche A, Asus ZB (2018) New consumer-dependent energy management system to reduce cost and carbon impact in smart buildings. Sustain Cities Soc
2. Hung DQ, Mithulananthan N (2013) Multiple distributed generator placement in primary distribution networks for loss reduction. IEEE Trans Ind Electron 60(4):1700–1708
3. Gangwar P, Singh S, Chakrabarti S (2019) Multi-objective planning model for multi-phase distribution system under uncertainty considering reconfiguration. IET Renew Power Gener 13:2070–2083

4. Awad A, Salama MMA, Awad ASA, El-Fouly THM (2014) Optimal distributed generation allocation and load shedding for improving distribution. Electr Power Compon Syst 42(6):576–584
5. Singh D, Singh D, Verma KS (2009) Multiobjective optimization for DG planning with load models. IEEE Trans Power Syst 24(1):427–436
6. Prakash P, Khatod DK (2016) Optimal sizing and siting techniques for distributed generation in distribution systems: a review. Renew Sustain Energy Rev 57:111–130
7. Ehsan A, Yang Q (2018) Optimal integration and planning of renewable distributed generation in the power distribution networks: a review of analytical techniques. Appl Energy 210(October 2017):44–59
8. Adetunji KE, Hofsajer IW, Abu-Mahfouz AM, Cheng L (2021) A review of metaheuristic techniques for optimal integration of electrical units in distribution networks. IEEE Access 9:5046–5068

TIDF Controller for Frequency Regulation of Two Area Power System Network with the Supplementary Control of SMES and IPFC

Ch. Naga Sai Kalyan, B. Sekhar, Ramakrishna Adabala, and Chintalapudi V. Suresh

Abstract This work initiates the implementation of a tilt-integral-derivative plus filter controller (TIDF) fine-tuned with a pathfinder algorithm (PFA) for a two-area interconnected power system network (TAIPSN). However, the efficacy of TIDF controller performance is tested with other widely accepted controllers. The considered TAIPSN is investigated subjected to the perturbations in step load (SLP) of 10% in area-1 and further to conduct the most appropriate realistic analysis, the TAIPSN is deliberated with the communication time delays (CTDs). Furthermore, the supplementary control mechanism of the interline power flow controller (IPFC) and superconducting magnetic energy storage (SMES) is implemented with the TAIPSN to enhance dynamical behavior.

Keywords Path finder algorithm · TIDF controller · SMES-IPFC strategy · Communication time delays · 10%SLP

1 Introduction

The electrical system is a complex and highly connected network that integrates various generation sources. Moreover, the highly connected system is always subjugated to some adverse effects even for the light deviations in the electric demand. The frequency of the IPSN must be regulated in variance with the electric demand for obtaining optimal performance. The acceptable deviations in IPSN frequency increase with the mismatch in real power (RPM) [1], that is the gap in generation and demand. Hence, the deployment of the automatic controller in addressing the regulation of RPM for IPSN stability is inevitable.

Ch. N. S. Kalyan (✉) · B. Sekhar · C. V. Suresh
EEE Department, Vasireddy Venkatadri Institute of Technology, Guntur 522508, India
e-mail: kalyanchallapalli@vvit.net

R. Adabala
EEE Department, BVC Engineering College, Odalarevu 533210, India

S. Kumar et al. (eds.), *Control Applications in Modern Power Systems*, Lecture Notes in Electrical Engineering 1128, https://doi.org/10.1007/978-981-99-9054-2_15

Recently, researchers have tried to put forward automatic regulatory control mechanisms for IPSN stability. Most of the control methods in the reported literature are closely associated with the conventional controllers of I and PID [2]. Moreover, some progressive modifications are suggested to the conventional PID in order to extend the range of its performance efficacy. However, the conventional and modified conventional controller's performance predominantly lies on the tunable parameters. Several research works addressed various population and stochastic-based soft computing techniques for optimal tuning of the designed controllers. Algorithms like water cycle algorithm, frog leap algorithm, rabbit search algorithm, grey wolf optimization [3], donkey-smuggler algorithm, class topper algorithm, gravitational search algorithm [4], differential evolution, differential search algorithm [5], fish schooling technique, Egyptian eagle optimization, backtracking search algorithm [6], African vulture optimization, beehives algorithm, culinary chef algorithm, blue whale algorithm, biogeography algorithm, symbiotic organisms search algorithm [7], butterfly optimization algorithm, wild horse algorithm, cryogenic optimization algorithm, atom search algorithm [8], lion optimization technique, gradient descent algorithm, elephant herd optimization [9], levy flight algorithm, hunger game search algorithm [10], moth flame algorithm, Harris hawks algorithm, group search technique, spotted hyena technique, Runge–Kutta algorithm, fractal search algorithm, mayfly algorithm, local-unimodal search technique, equilibrium optimization [11], ant lion algorithm, frog leaping algorithm, and fruit fly algorithm etc. are reported.

Instead of the regulators from the family of classical controllers, the intelligent type fuzzy logic controllers (FLC) are adopted extensively for the IPSN with non-linear realistic constraints. For the power systems with non-linearity features, the classical regulators couldn't sustain the network stability. The ambiguity in the selection of membership functions, the approximation in FLC rules design, and the complexity in implementing the realistic environment the ANN and FLC regulators do not prefer in the modern environment. In substitute to the FLC and ANN, the fractional order (FO) aided controllers are becoming very popular in recent times. Additionally, degree-of-freedom (DOF) controllers are also playing a vital role in establishing the IPSN stability with the benefits of having the individual closed control loops. The DOF and the FO-based controllers also require soft computing algorithms in locating the parameters optimally. In the LFC study, the water cycle algorithm [12], JAYA algorithm, dragonfly algorithm, sine–cosine algorithm [13], artificial flora algorithm, teaching–learning optimization, bat-inspired algorithm, bull lion algorithm, ant colony optimization, volleyball premier league optimization [14], krill herd optimization, etc. are reported for optimizing the FO and DOF-based regulators implemented in various diversified IPS networks. But, the so many tunable parameters in the FO controllers of FOPI/FOPD/FOPID/FOPI-FOPD, etc., and the DOF-aided [15, 16] controllers of 2DOFPID/3DOFPID laid a substantial burden on the computational algorithms. This burden might delay locating the parametric gains and thereby affect the regulator's optimal performance for the automatic actions. Furthermore, other alternative controllers under the FO type is the TIDF controller, whose efficacy over the other FO, conventional PI/PID, Integral-double-derivative with filter (IDDF), PIDD, and FLC-aided controllers are validated in [17, 18]. Though

TIDF is under the family of FO, it is possessed with few parametric optimizations and further its design is as simple as the traditional controllers which made their implementation easy in the practical environment. Hence, the simulation study in the domain of LFC carried out in this paper considered the TIDF controller optimized with one of the recent and efficient optimization algorithms called the PFA.

This work makes the following contributions:

(a) Implemented the PFA-optimized TIDF controller for IPS stability is a primary attempt.
(b) The effectiveness of the PFA-tuned TIDF is validated using FPID and PID controllers.
(c) The TAIPSN is conceived with the CTDs.
(d) The CTDs impact on TAIPSN operation is highlighted.
(e) SMES-IPFC-based supplementary control strategy is implemented and attained considerable performance improvement.

2 Power System Model

The TAIPSN [19] has a thermal unit of capacity 1250 MW in area-1 and the hydro unit with 750 MW generation capacity is placed in area-2 as shown in Fig. 1. Moreover, the reheat type turbine is enacted with the thermal plant and for the hydro unit the electrical type governing system is conceived. Furthermore, the TAIPSN is deliberated with the practical non-linear features of CTDs to bring the investigation closer to reality. The IPSN is of a wide extent and often the measured data is transmitted and received between different apparatus located at numerous locations. This data transmission and reception is carried through the communication peripherals which are having the inherited property of delay. These peripherals send the data with some delay but not instantly. The signal received from the control center by the secondary controller in order to perform the alteration in the IPSN operating point is also with the delay. Hence, with this delay, there might be an effect on the IPSN performance, and this work is addressed to reveal its impact. The modeling of CTDs is given in Eq. (1). The practical delay [20, 21] of 0.25 s is considered for investigation analysis in this paper.

$$e^{-s\tau_d} = \frac{1 - \frac{\tau_d}{2}s}{1 + \frac{\tau_d}{2}s} \tag{1}$$

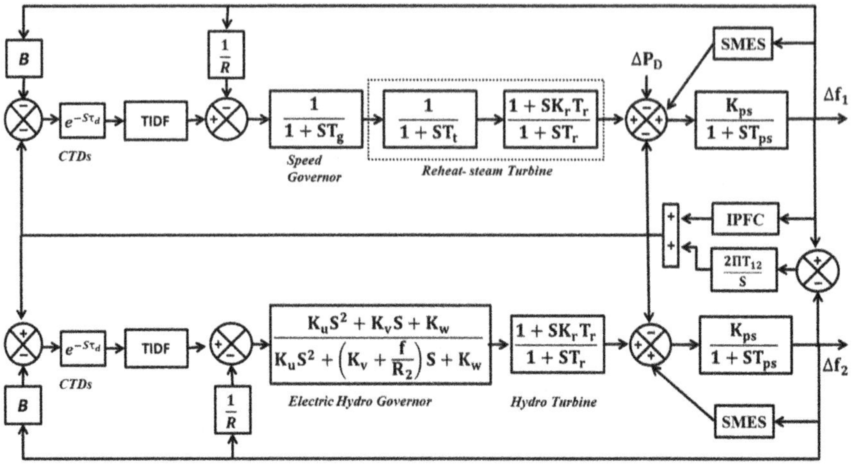

Fig. 1 Model of TAIPSN with SMES-IPFC strategy

3 Controller and Objective Function

One of the most noticeable drawbacks of the traditional first-order regulators is their efficacy subjected to uncertainty in parameters and loading. This drawback can be overcome by employing the FOC, the TIDF is one such regulator whose design is simple and can be enacted with the complex IPS easily. The structure of TIDF is shown in Fig. 2 and the parameters K_T, K_I, K_D, n, and N are the parameters to be optimized. However, the optimization can be done with respect to any of the time domain indices. The integral time square error (ITSE) [22] in Eq. (2) is chosen in this work and subjected to this index the TIDF parameters are found optimally.

$$J_{ITSE} = \int_{0}^{T_{Sim}} \left(\Delta f_1^2 + \Delta P_{tie12}^2 + \Delta f_2^2\right) * T\,dt \tag{2}$$

4 Path Finder Algorithm

PFA [23] is the newest heuristic optimization algorithm that imitates the attacking strategy of the animal swarm. This strategy is population-based and the decision-making can be performed without and with concern from the leader, the leader in the PFA is not a permanent one as the other animals might be aware of the abundant food location. The researchers tested the PFA efficacy with other approaches like the galaxy search technique, tree seed algorithm, artificial bee colony, enhanced bat

Fig. 2 TIDF controller

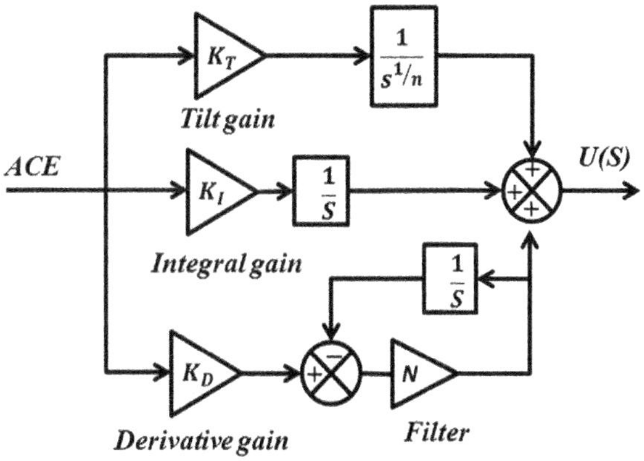

approach, and teaching–learning algorithm. The pathfinder's position at the initial stage while identifying the prey is modeled as in Eq. (3) and the update in the pathfinder position is presented in Eq. (4).

$$x(k + \Delta k) = x^0(k) \cdot n + f_i + f_p + \varepsilon \quad (3)$$

$$x_p(k + \Delta k) = x_p(k) + \Delta x + A \quad (4)$$

where 'x_p' represents the pathfinder position, 'k' time, and the 'f_p' and 'f_i' are the force parameters of global optimum and pathfinder interaction respectively. The 'ε' is a vibration vector. The update in the pathfinder position as given in Eq. (3) cannot be applied to realistic optimization problems. Thus, the modification of Eq. (3) is presented in Eqs. (5) and (6).

$$x_i^{f+1} = x_i^f + rand_1 \cdot (x_j^f - x_i^f) + rand_2 \cdot (x_p^f - x_i^f) + \varepsilon, \quad i \geq 2 \quad (5)$$

$$x_p^{f+1} = x_p^f + 2\, rand_3 \cdot (x_p^f - x_p^{f-1}) \quad (6)$$

where 'f' represents the iteration and the $rand_1$ and $rand_2$ are chosen from 1 to 2. While the vibration vector is evaluated as

$$\varepsilon = \left(1 - \frac{f}{f_{max}}\right) \cdot u_1 \cdot D_{ij}, \; D_{ij} = \left\| x_i - x_j \right\| \quad (7)$$

If the 'f' is less than 'f_{max}' the pathfinder's position is updated with Eq. (6) and the new pathfinder's fitness value is compared to that of the old pathfinder and then the positions get updated. Suppose, the fitness of the previous pathfinder is more effective than that of the new pathfinder then the generation of the vibration vector is updated with Eq. (7). In this work, the PFA is adopted in optimizing the TIDF for maintaining the IPS network stability and as a result, when the technique reaches its maximum quantity of iterations, the most effective settings are presented.

5 SMES-IPFC Control Strategy

SMEs are the most useful and the most convenient ESDs that best function as the spinning reserve. It works on the principle of storing the energy by charging up the semiconductor coil on the magnet and the whole unit is maintained at cryogenic temperature. By doing so, the superconductivity is acquired by the coil and the energy storage is in the DC form. Apart from that, the SMES have the control and the power conditioning unit for delivering energy to the grid. They are effective in delivering energy to the grid and can charge up during the availability of surplus generation. The modeling of SMES [24] is given in Eq. (8).

$$G_{SMES} = \frac{K_{SMES}}{1 + sT_{SMES}} \tag{8}$$

The effective flexible AC transmission device among the family of combined controllers is the IPFC. It comprises multiple series voltage regulators and is joined with a DC link in common. In this work, the IPFC is placed with the tie-line so that the line reactance might be controlled and here by the improvement in line power flow. Moreover, the IPFC has the capacity in restoring the tie-line stability by dampening the power oscillations. The damping control structure of IPFC is shown in Fig. 3.

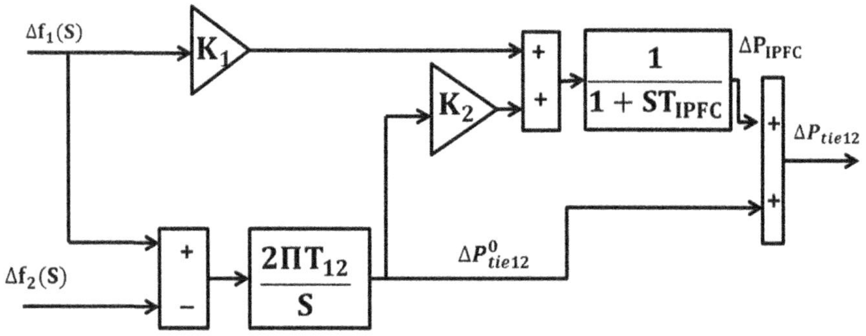

Fig. 3 IPFC as damping controller

6 Simulation Results

6.1 Case-I: Analysis of TAIPSN Under Different Controllers

The TAIPSN is enacted with PID, FPID, and TIDF individuals in the two areas for the assessment of dynamic behavior. The analysis is initiated by injecting a 10% SLP in area-1. To obtain the performance assessment comparatively, all the regulators are optimized with PFA one after the other for the same parametric initializations. The performance of the PID, FPID, and TIDF in terms of TAIPSN system dynamical behavior is compared in Fig. 4. Moreover, the settling time of the TAIPSN responses under the PFA-tuned PID, FPID, and TIDF controllers are noted in Table 1. Observing the Fig. 4 and noticing the settling times in Table 1 it is clear that the TIDF dominates the other two in restricting the deviations in the TAIPSN responses and also concedes more time in reaching the stable position. Further, the ITSE is effectively handled by the TIDF and its performance is improved by 72.09% and 62.16% with PID and FPID. The gains of PID, FPID, and TIDF found with PFA are noted in Table 2.

6.2 Case-II: Analysis of TAIPSN Performance with and Without Considering the CTDs

The TAIPSN is used to consider the non-linear practical nature of CTDs for assessing the dynamical behavior exactly with realistic practice. Hence, it is requisite to showcase its impact on frequency fluctuations and system performance. The TAIPSN dynamic analysis in the above subsection reveals that the PFA-tuned TIDF is the most suitable regulator that exhibits sovereignty over the other PID and FPID regulators. Therefore, under the regulation of the PFA-based TIDF controller, the performance of TAIPSN is analyzed in this case by considering the CTDs and by omitting the CTDs. The TAIPSN responses for 10%SLP are shown in Fig. 5 for this case and are visible that the responses for the context of considering the CTDs are more fluctuated than that of omitting. While considering, because of the delay in getting the signal for control from the control center, there will probably be some lag in changing the operating point. Though more deviations in TAIPSN with deliberating the CTDs are reported it is endorsed to deliberate while assessing the regulator performance. Since the developed regulator for the context of omitting the delays might not be robust enough for the regulation of IPSN subjugated with the unpredictable delays.

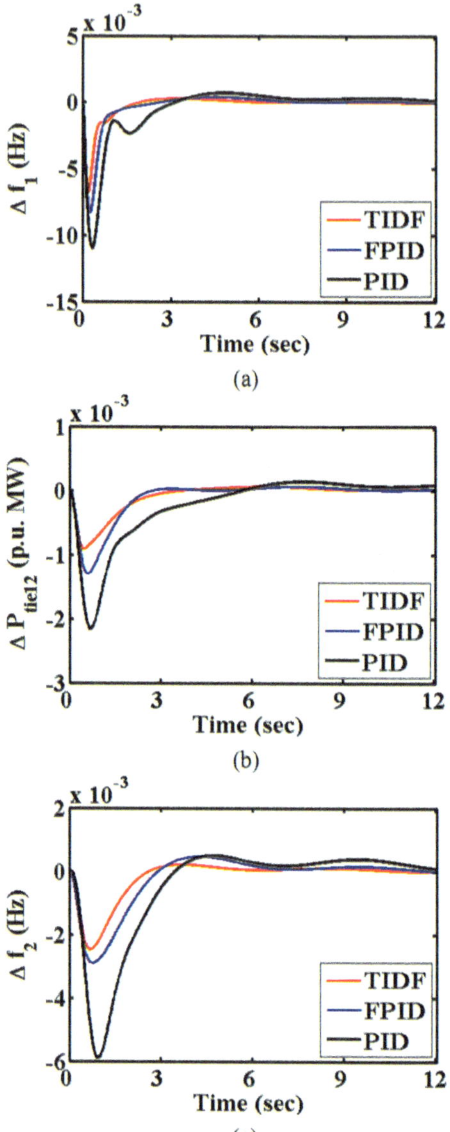

Fig. 4 Case-I responses. **a** Δf_1, **b** ΔP_{tie12}, **c** Δf_2

6.3 Case-III: Analysis of TAIPSN with SMES-IPFC Control Strategy

The additional fluctuation that emerged with the dynamical behavior of the TAIPSN for deliberating the CTDs is to be quenched by adopting territorial control techniques. However, the developed regulator can't balance the mismatch in real power

Table 1 Responses settling time

Settling time (s)	Δf1	ΔPtie12	Δf2	ITSE*10^{-3}
PID	11.45	13.52	11.26	73.29
FPID	7.78	11.05	9.35	54.05
TIDF	6.39	9.57	8.65	20.45
SMES	5.98	8.36	6.43	–
SMES-IPFC	4.65	7.15	5.85	–

Table 2 Controller optimal gains

Parameters	Area-1			Area-2		
	PID	FPID	TIDF	PID	FPID	TIDF
KP/KT	0.688	0.273	0.365	0.778	0.316	0.542
KI	0.019	0.168	0.190	0.165	0.206	0.215
KD	0.217	0.717	0.331	0.405	0.819	0.255
n	–	–	0.063	–	–	0.179
N	–	–	221.61	–	–	218.15

by varying the set point valve alone during large perturbations. Thus, adopting supplementary control techniques for the stability of IPSN is always desired. In this paper, the SMES are integrated in two areas under the supervision of PFA-tuned TIDF, and the responses are shown in Fig. 6. Later, an IPFC is connected to the tie-line, and the SMES insertion is continued; the resulting responses are shown in Fig. 6. From Fig. 6 it is finalized that an improvement in TAIPSN performance is noticed with SMES integration and significant enhancement is reported with SMES-IPFC strategy. For easy identification the TAIPSN responses settling time for various cases are indicated in Fig. 7. The SMES is the most helpful storage device and a better spinning reserve which balances the RPM by delivering the energy to the grid instantly and on the other hand the IPFC effectively regulates the power flow. Hence the SMES-IPFC is recommended as the effective supplementary control approach for IPSN stability.

Fig. 5 Case-II responses. **a** Δf_1, **b** ΔP_{tie12}, **c** Δf_2

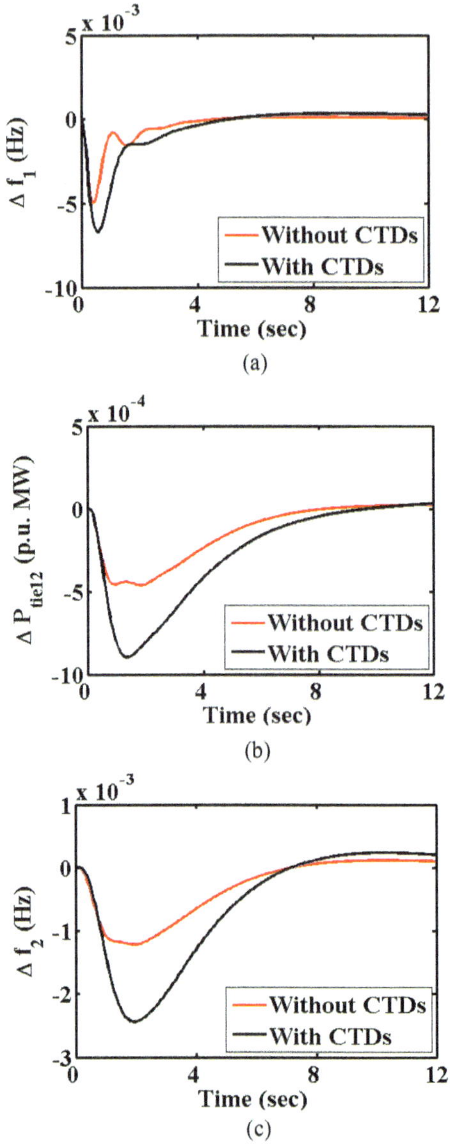

Fig. 6 Case-III responses. **a** Δf_1, **b** ΔP_{tie12}, **c** Δf_2

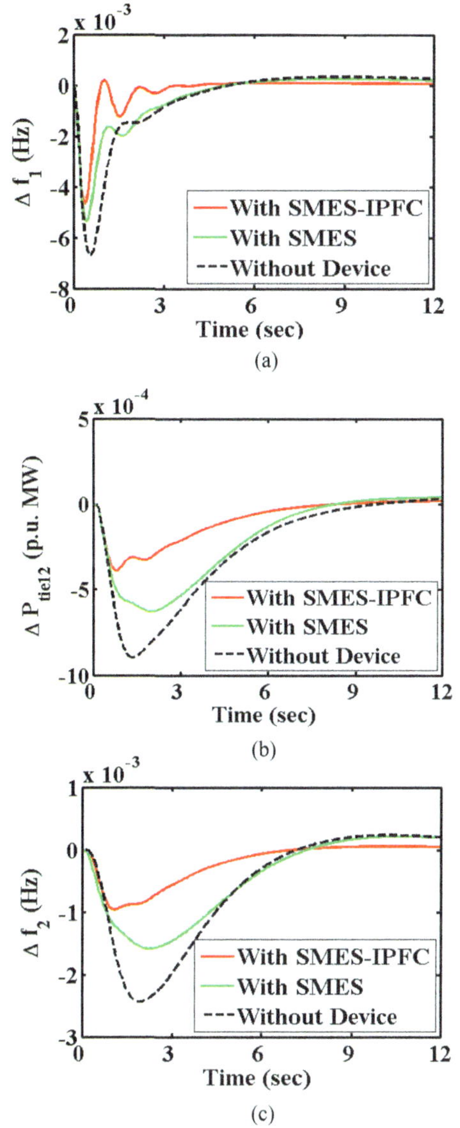

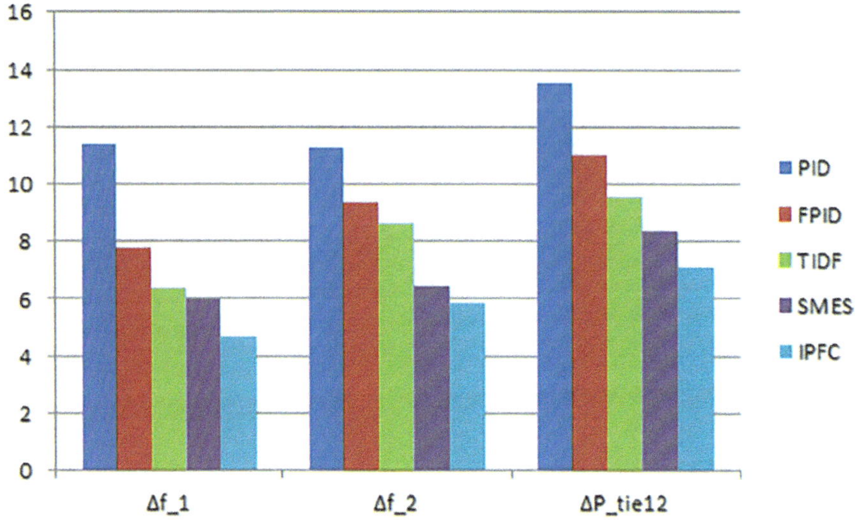

Fig. 7 Settling time (Y axis) comparison for TAIPSN responses for cases I and III

7 Conclusion

This paper puts forward the possibility of implementing the PFA-based TIDF controller to the IPS network for controlling the frequency under load perturbations and is a primary attempt. The efficacious of the suggested controller is validated with the other accepted controllers like FPID and traditional PID. Further, the investigative model of TAIPSN is deliberated with the non-linearity nature of CTDs in order to obtain the dynamical analysis near to reality. The impact of the CTDs on TAIPSN is showcased clearly and the requirement of deliberating the CTDs for assessing the IPS network behavior is justified. Moreover, the TAIPSN is incorporated with the mechanism of SMES-IPFC, and the improvement in system performance is noticed. In the end, it is determined that the PFA-tuned TIDF is recommended as the frequency regulator to the IPS network, and the SMES-IPFC is suggested as the territory controller. The analysis in this work can be extended to the restructured environment and incorporation of distributed energy sources.

References

1. Chang S, Qi S (2022) Design and analysis of a load frequency control system based on improved artificial intelligence control algorithm. Alex Eng J 61:11779–11786
2. Kalyan CNS, Rao GS (2020) Frequency and voltage stabilization in combined load frequency control and automatic voltage regulation of multiarea system with hybrid generation utilities by AC/DC links. Int J Sustain Energ 39(10):1009–1029

3. Paliwal N, Srivastava L, Pandit M (2022) Application of grey wolf optimization algorithm for load frequency control in multi-source single area power system. Evol Intel 15:563–584
4. Sahu RK, Panda S, Padhan S (2015) A novel hybrid gravitational search and pattern search algorithm for load frequency control of nonlinear power system. Appl Soft Comput 29:310–327
5. Guha D, Roy PK, Banerjee S (2016) Quasi-oppositional differential search algorithm applied to load frequency control. Eng Sci Technol 19(4):1635–1654
6. Guha D, Roy PK, Banerjee S (2018) Application of backtracking search algorithm in load frequency control of multi-area interconnected power system. Ain Shams Eng J 9(2):257–276
7. Guha D, Roy PK, Banerjee S (2018) Symbiotic organism search algorithm is applied to load frequency control of multi-area power system. Energy Syst 9:439–468
8. Irudayaraj AXR, Wahab Z, Kumar M, Radzi M, Sulaiman N, Veerasamy V, Farade R, Islam M (2022) Renewable sources based automatic load frequency control of interconnected systems using chaotic atom search optimization. Appl Soft Comput 119:108574
9. Dewangan S, Prakash T, Singh VP (2020) Design and performance analysis of elephant herding optimization based controller for load frequency control in thermal interconnected power system. Optim Control Appl Methods 42(1):144–159
10. Fathy A, Yousri D, Rezk H, Babu S, Hasanien HM (2022) A robust fractional order PID controller based load frequency control using modified hunger games search optimizer. Energies 15(1):361
11. Gulicer AK, Ayas MS (2022) Employing equilibrium optimizer in two area load frequency control for a TIDDF controller design. In: International congress on human-computer interaction, optimization and robotic applications, pp 1–5, Turkey, 9–11 June 2022
12. Kalyan CNS (2021) Determination of appropriate GRC modeling for optimal LFC of multi area thermal system. In: IEEE international power and renewable energy conference (IPRECON), pp 1–6, Kollam, 24–26 September 2021
13. Rajbongshi R, Saikia LC (2017) Performance of coordinated FACTS and energy storage devices in combined multi area ALFC and AVR system. J Renew Sustain Energy 9(6):064101
14. Sariki M, Shankar R (2022) Optimal CC-2DOF(PI)-PDF controller for LFC of restructured multi-area power system with IES-based modified HVDC tie-line and electric vehicles. Eng Sci Technol 32:101058
15. Kalyan CNS, Suresh CV (2022) Higher order degree of freedom controller for load frequency control of multi area interconnected power system with time delays. Global Trans Proc 3(1):332–337
16. Kalyan CNS, Goud BS, Reddy CR, Bajaj M, Rao GS (2023) SMES and TCSC coordinated strategy for multi-area multi-source system with water cycle algorithm based 3DOF-PID controller. Smart Sci 11(1):1–15
17. Ahmed EM, Mohammed EA, Selim A, Aly M, Alsadi A, Alhosaini M, Alnuman H, Ramadan HA (2023) Improving load frequency control performance in interconnected power systems with a new optimal high degree of freedom cascaded FOTPID-TIDF controller. Ain Shams Eng J. https://doi.org/10.1016/j.asej.2023.102207
18. Ahmed A, Magdy G, Khamies M, Kamel S (2022) Modified TID controller for load frequency control of a two-area interconnected diverse-unit power system. Electr Power Energy Syst 135:107528
19. Chaine S, Tripathy M (2019) Performance of CA optimized controllers of DFIGs and AGC to improve frequency regulation of a wind integrated hydrothermal power system. Alex Eng J 58:579–590
20. Kalyan CNS, Rao GS (2021) Impact of communication time delays on combined LFC and AVR of a multi-area hybrid system with IPFC-RFBs coordinated control strategy. Protect Control Mod Power Syst 6:1–20
21. Kalyan CNS, Rao GS (2022) Coordinated control strategy for simultaneous frequency and voltage stabilization of the multi-area interconnected system considering communication time delays. Int J Ambient Energy 43(1):5512–5524
22. Kurakula N, Nithya M, Kalyan CNS (2022) Enhancement in AGC performance of multi area interconnected power system with practical constraints using WCA based PIDD controller. In:

2nd international conference on power electronics & IoT applications in renewable energy and its control, pp. 1–6, Mathura 21–22 January 2022.

23. Yapici H, Cetinkaya N (2019) A new meta-heuristic optimizer: pathfinder algorithm. Appl Soft Comput 78:545–568
24. Kalyan CNS, Rao GS (2022) Combined frequency and voltage stabilization of multi-area multi source system by DE-AEFA optimized PID controller with coordinated performance of IPFC and RFBs. Int J Ambient Energy 43(1):3815–3831

Robust Controller Design based on IMC Scheme for Motion Control of DC Servo Systems

S. Sai Lakshmi, R. Jeyasenthil, and Uday Bhaskar Babu

Abstract In this article, a robust controller is designed based on the internal mode control (IMC) scheme for the motion control of the DC servo system. The challenging part in servo systems is the robust tracking in the presence of plant uncertainties and load variations. The proposed controller design has an interesting structure, i.e., integer filter term cascaded with PD controller. This work identified the integer IMC filter whose single optimal tuning parameter is selected using a simple iterative approach based on the minimization of integral absolute error (IAE) and total variation (TV). Control effort minimization is also equally significant because of the actuator saturation constraints which eventually cause the wear and tear and can make the system unstable at the worst case. The tuning parameter is the filter time constant, which trades off the performance, robustness, and control effort. As a result, the proposed controller handles the parametric uncertainty and load disturbances effectively not at the cost of higher control effort. This paper compares the tracking performance and control effort with different filter structures which cascaded with the PD controller, and it concludes with a novel controller structure with the introduction of 2-degree of freedom (DOF) pre-filter along with the controller which provides a good compromise between faster tracking performance and control effort minimization and it is compared with the literature regarding performance indices, i.e., IAE, TV, and Ms. This controller prevents the actuator to saturate when there is abrupt change in the reference tracking command. The simulation results show that the proposed control enhances performance (IAE), stability (M_s) and minimizes the control effort (TV). Also, robustness analysis has been carried out with perturbations in process model parameter.

S. Sai Lakshmi (✉)
Department of Electrical Engineering, New Delhi, India
e-mail: sl22eer2r06@student.nitw.ac.in

R. Jeyasenthil
Department of Chemical Engineering, New Delhi, India
e-mail: jey@nitw.ac.in

U. B. Babu
National Institute of Technology Warangal, Telangana 506004, India
e-mail: udaybhaskar@nitw.ac.in

S. Kumar et al. (eds.), *Control Applications in Modern Power Systems*, Lecture Notes in Electrical Engineering 1128, https://doi.org/10.1007/978-981-99-9054-2_16

Keywords Saturation · Robustness · Maximum sensitivity · Gain margin · Phase margin · Small gain theorem · Multiplicative uncertainty

1 Introduction

The proportional-integral-derivative (PID) controller is a widely used control strategy in process industries over the last 2 decades because of its simple tuning structure and coherent performance. The first most popular PID tuning method was proposed by Ziegler and Nichols [1], but it is not widely used because of its unreliability performance. There are numerous guidelines for selecting this tuning parameter as reported in the literature from classical like Cohen-Coon tuning, direct synthesis to modern adaptive, and auto tuning method. The most widely accepted model-based PID tuning strategy is IMC which provides a generic, simple transparent framework for controller synthesis. The distinctive feature of IMC is that it has only one tuning parameter, dual stability, perfect control, and perfect set-point tracking performance. The relationship between Zeigler-Nichols, Cohen-Coon, and IMC-PID tuning rules is discussed in [2].

The challenging issue in IMC design is the guidelines for the selection of the tuning parameter. The tuning parameter should be chosen in such a way that it is lesser than the closed loop bandwidth for which the system is valid as suggested in [3]. According to [4], they endorsed that the filter parameter should be chosen such that the low frequency gain should be 20 times more than high frequency gain. In work [5] recommended that filter parameter should be less than the largest time constant. The tuning rules are formulated in [6] based on the gain and phase margin specifications. For stable first and second order systems with time delay, the empirical relationship between the IMC filter time constant and the percentage overshoot is analysed in [7]. The well-accepted simple PI/PID controller tuning method for delay integrating processes is proposed by [8]. The modified IMC structure is proposed by introducing an additional controller with basic IMC PID controller which enhances closed loop bandwidth thereby having faster tracking response for the second order delayed process in [9]. The tuning parameters are selected based on stability margins. In order to achieve a suitable compromise between performance and robustness for both servo and regulatory concerns for the first order plus time delay system, the IMC filter is constructed as an inappropriate transfer function in [10]. An IMC controller is recommended for integrating systems with time delays, such as pure integrators with dead time, double integrating systems with time delays, and stable and unstable first order plus time delay (FOPTD) systems with integrators in [11]. A new design methodology for designing IMC-based FOPID controller is proposed in [12] where the design of an external IMC filter is eliminated by modelling the fractional order plant using binomial expansion. For speed control and disturbance rejection of the heavy-duty vehicle (HDV), an efficient internal model control (IMC) incorporating robust filter, adaptive, and H-∞ control is proposed [13]. To determine the optimal IMC-PID controller settings for wing vibration systems, the innovative method

GNPSO—a hybrid of the modified Grey-Particle swarm optimization (GPSO) and novel Luus-Jaakola approach (NLJ)—was addressed in [14]. The IMC controller with the fractional first order and higher order filter is proposed in [15] for a series cascade control system for different time delay process models in chemical industries. For the stable/unstable first and second processes with time delay, respectively, the IMC-PID controller with imaginary filter or compensator based on pole zero transformation is explained in [16].

The motivation of this article is to develop an integer filter IMC controller whose tuning parameter is selected which overcomes the trade-off between tracking performance and control effort minimization for DC servo systems. The paper is organized as follows: Sect. 2 describes the system identification of DC servo motor using the first principle and bump test. IMC with two types of integer filter structure is proposed in Sect. 3. Performance and robustness analysis of the IMC controller are discussed in Sect. 4. Section 5 compares the performance of different proposed IMC design and existing PV with simulation results.

2 Mathematical Modelling of DC Servo Motor

For the experimental setup, Quanser QUBE-Servo 2 motor is used. The motor armature circuit is shown in Fig. 1. The motor shaft has a moment of inertia J_m which is connected to load hub with moment of inertia J_h. The load hub is a metal disc used to mount the disc with moment of inertia J_d.

By Kirchhoff's Voltage law, the dynamics of the DC motor is given as follows:

$$v_m = R_m i_m + L_m \frac{di_m}{dt} + e_b \tag{1}$$

Since motor inductance L_m effect is very much less, it can be ignored.

Where e_b is the back-emf that depends on the speed of the motor shaft ω_m with back-emf constant of k_m, and it is expressed as $e_b = k_m \omega_m$.

The motor shaft equation is given as $J_{eq}\dot{\omega}_m = \tau_m$,

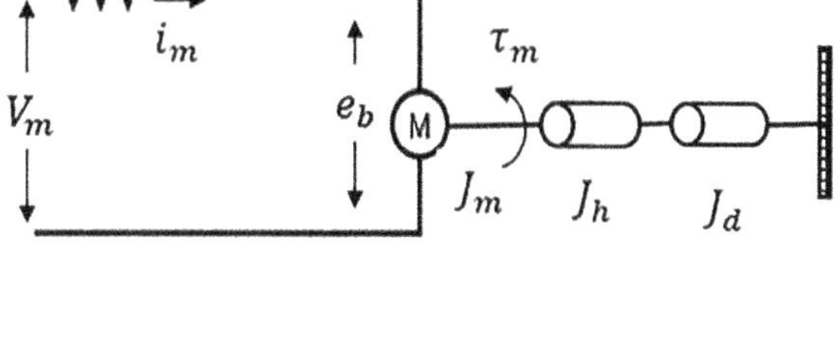

Fig. 1 DC motor armature circuit

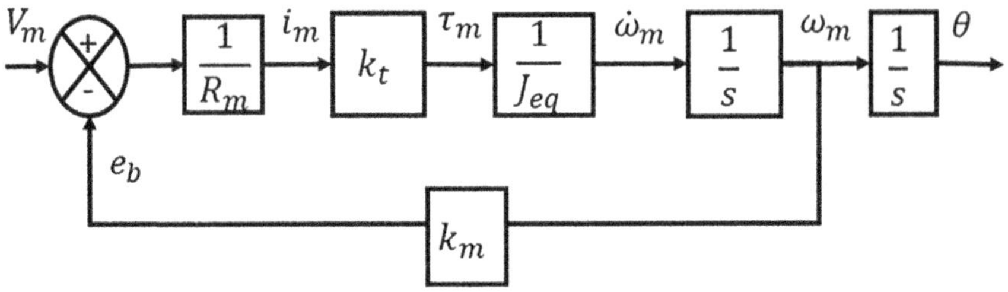

Fig. 2 Mathematical modelling of DC Servo motor

where $J_{eq} = J_m + J_h + J_d$ is the total moment of inertia acting on the shaft and $\tau_m = k_t i_m$ is the motor torque according to the current applied, where k_t is the torque constant.

The block diagram of DC servo motor is shown in Fig. 2.

The transfer function from input voltage to motor shaft speed ω_m

$$\frac{\omega_m(s)}{V_m(s)} = \frac{1/k_m}{s(\frac{J_{eq}R_m}{k_t k_m}) + 1} \tag{2}$$

The transfer function from input voltage to motor shaft position θ_m

$$\frac{\theta_m(s)}{V_m(s)} = \frac{1/k_m}{s(s(\frac{J_{eq}R_m}{k_t k_m}) + 1)} \tag{3}$$

The above transfer function is of the form

$$G_p(s) = \frac{\theta_m(s)}{V_m(s)} = \frac{K}{s(s\tau + 1)} \tag{4}$$

The modelling of DC (QUBE) servo motor is carried out by means of system identification. The Bump test is performed to identify the speed transfer function of the DC (QUBE) servo motor.

The identified speed transfer function of DC (QUBE) servo motor is given as

$$G_p(s) = \frac{\theta_m(s)}{V_m(s)} = \frac{22.83}{s(0.1286s + 1)} \tag{5}$$

3 Proposed Integer Filter IMC-PID Controller Design

The DC servo motor Eq. (5) is in the form of integrating first order plus time delay (IFPTD) where the parameters are the gain (K) and motor time constant (τ). The proposed IMC structure is shown in Figs. 4a and 3b.

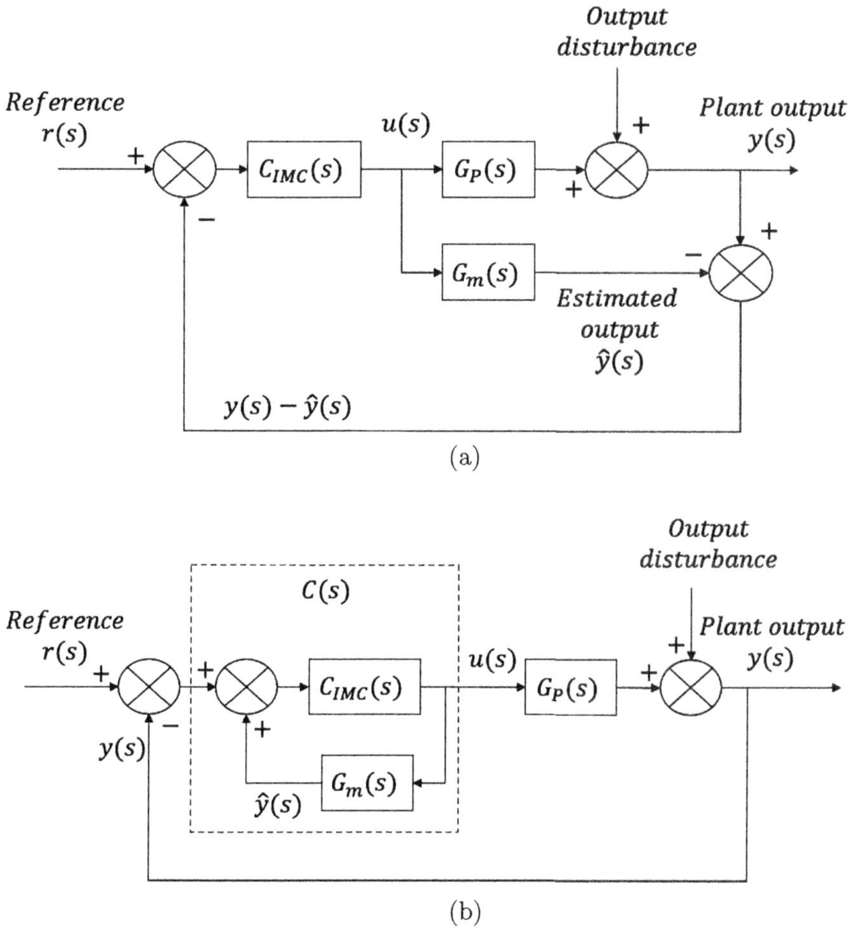

Fig. 3 **a** IMC control structure; **b** equivalent IMC structure

The IMC equivalent controller can be written as a PID structure with integer filter (IF) as given in Eq. (6):

$$C(s) = IF(s)K_p[1 + \frac{1}{T_i s} + T_d s] \tag{6}$$

where K_p, T_i, and T_d are the proportional gain, integral time, and derivative time. According to the IMC method, the feedback controller is designed as follows:

$$C(s) = \frac{C_{IMC}}{1 - C_{IMC}G_m(s)} \tag{7}$$

where C_{IMC} is the IMC controller and $G_m(s)$ is the DC motor model.

Here, the controller C_{IMC} is designed as follows:

$$C_{IMC}(s) = \frac{1}{G_m^{-1}(s)} F(s) \tag{8}$$

where $G_m^{-1}(s)$ is the invertible part of $G_m(s)$ which has only minimum phase poles and zeros and $F(s)$ is the integer order filter whose structure can be any one of the following:

- **Type 1**:

$$F(s) = \frac{1}{(\lambda s + 1)^n}; n = 1, 2 \tag{9}$$

- **Type 2**:

$$F(s) = \frac{\beta s + 1}{(\lambda s + 1)^{n+1}}; n = 1 \tag{10}$$

Here, $F(s)$ acts as an error low pass filter (LPF) whose parameter λ is the cut-off frequency of LPF and β is the additional tuning parameter which adds more flexibility in achieving the desired faster transient response.

The DC motor model $G_m(s)$ (Eq. 5) has only an invertible part, i.e., minimum phase poles and no zeros:

$$G_m^{-1}(s) = \frac{K}{s(\tau s + 1)} \tag{11}$$

Case 1: Using Type 1 Filter Structure

$$F(s) = \frac{1}{(\lambda s + 1)^n}; n = 1, 2 \tag{12}$$

$$C_{IMC}(s) = \frac{1}{G_m^{-1}(s)} F(s) \tag{13}$$

On substituting Eqs. (11) and (12) in Eq. (13), we get

$$C_{IMC}(s) = \left(\frac{s(\tau s + 1}{K}\right)\left(\frac{1}{(\lambda s + 1)^n}\right) \tag{14}$$

The final controller is designed according to Eq. (7) as follows:

$$C(s) = \frac{s(\tau s + 1)}{K[(\lambda s + 1)^n - 1]} \tag{15}$$

Case 2: Using Type 2 Filter Structure

$$F(s) = \frac{\beta s + 1}{(\lambda s + 1)^{n+1}}; n = 1 \tag{16}$$

$$C_{IMC}(s) = \frac{1}{G_m^{-1}(s)} F(s) \tag{17}$$

On substituting Eqs. (11) and (16) in Eq. (17), we get

$$C_{IMC}(s) = \left(\frac{s(\tau s + 1)}{K}\right)\left(\frac{\beta s + 1}{(\lambda s + 1)^{n+1}}\right) \tag{18}$$

The final controller is designed according to Eq. (7) as follows:

$$C(s) = \frac{s(\tau s + 1)(\beta s + 1)}{K[(\lambda s + 1)^n - (\beta s + 1)]} \tag{19}$$

4 Controller Validation

4.1 *Performance Analysis*

The tracking performance is assessed using IAE, ISE as follows:

$$IAE = \int_0^\infty |e(t)|\, dt \tag{20}$$

$$ISE = \int_0^\infty e^2(t)\, dt \tag{21}$$

The control effort, i.e., input voltage applied to the DC motor is measured by using total variation (TV) of the input (u). It quantifies how much sudden change in the control effort the controller demands for faster response, and it is a good measure of smoothness. TV should be as small as possible.

The control effort minimization is assessed using TV as follows:

$$TV = \sum_{i=0}^{\infty} |u_{i+1} - u_i| \tag{22}$$

4.2 Stability Analysis

The maximum sensitivity (M_s) is used for robustness analysis like gain margin (GM) and phase margin (PM). The margin is related to M_s as follows:

$$M_s \leq \frac{GM}{GM - 1} \tag{23}$$

$$M_s \leq \frac{1}{2sin^{-1}(PM/2)} \tag{24}$$

Generally, for robustness, the system should have $GM > 6db$ and $PM > 30^o$. The corresponding M_s value should be less than 2. In the process of IMC controller design, M_s values are plotted against the filter tuning parameter λ. For a stable system, faster response, and good disturbance rejection, it demands a higher value of λ and for control effort minimization, it requires a lesser value of λ. Hence, it is observed that there is a trade-off between faster response and control effort minimization for the selection of λ. *TV* and *IAE* values are plotted against the filter tuning parameter λ. From this plot, λ is selected which provides the global minimum of the convex plot.

4.3 Robustness Analysis

In real time, the plan will have parameter uncertainties. Hence, it is necessary to measure the robustness analysis of the controller in the presence of the model uncertainties. Here, in the DC motor model, the uncertainty can be seen in gain (K) and time constant (τ). According to the small gain theorem, the closed loop system is robustly stable if and only if

$$||\Delta(j\omega)T(j\omega)||_\infty < 1, , \forall\omega = (-\infty, \infty) \tag{25}$$

where $T(j\omega)$ is the complementary sensitivity function and $\Delta(j\omega)$ is the process multiplicative uncertainty. It can be represented as follows:

$$\Delta(j\omega) = \left|\frac{G_p(j\omega) - G_m(j\omega)}{G_m(j\omega)}\right| \tag{26}$$

where $G_m(j\omega)$ is the nominal plant model and $G_p(j\omega)$ is the uncertain plant.

The tuning parameter λ can be selected in such a way that the controller satisfies the following criteria:

$$||T(j\omega)||_\infty < \frac{1}{\Delta(j\omega)} \tag{27}$$

If there is uncertainty in the gain, then Eq. (27) becomes

$$||T(j\omega)||_\infty < \left|\frac{1}{\Delta k/K}\right| \tag{28}$$

If there is uncertainty in the time constant, then Eq. (27) becomes

$$||T(j\omega)||_\infty < \left|\frac{s(\tau+\Delta\tau)+1}{\Delta\tau s}\right| \tag{29}$$

Thus, the tuning parameter has to be selected such that the controller satisfies the performance, control effort minimization, and robustness criteria.

5 Simulation Results and Discussion

The effectiveness of the integer filter IMC-PID controller design for various IMC architectures is compared with Proportional Velocity (PV) Controller whose proportional gain as 5.85 and velocity gain as 0.233 [17] for nominal tracking and perturbation scenarios are as follows.

Nominal Tracking scenario: Reference tracking for stair input is considered, i.e., step increase from 0° to 90° at 1 s and step decrease from 90° to 0°.

Perturbation Tracking scenario: The tracking with constant step disturbance applied at 3 s.

The optimum controller is selected based on minimum IAE and TV values for different filter structures as follows (Table 1).

***Case 1*: Performance Analysis with Proposed Method 1** It can be observed that the tuning parameter λ represents the closed loop time constant. The tuning parameter λ is chosen based on the global minimum for IAE and TV for a particular filter structure as shown in Fig. 4. For Type 1 filter of order 2, λ is chosen as 0.047 as the global minimum, respectively. Figures 5 and 6 illustrate the nominal reference tracking performance and its corresponding control effort. It is observed that the

Table 1 Filter structure in proposed methods

Controller	IMC filter structure
Proposed method 1: IMC with Type 1 Filter structure of order 2	$\frac{1}{(\lambda s+1)^2}$
Proposed method 2: IMC with Type 1 Filter structure of order 3	$\frac{1}{(\lambda s+1)^3}$
Proposed method 3: IMC with Type 2 Filter structure of order 3	$\frac{\beta s+1}{(\lambda s+1)^3}$
Proposed method 4: 2-DOF IMC with Type 2 Filter structure of order 3	$\frac{\beta s+1}{(\lambda s+1)^3}$

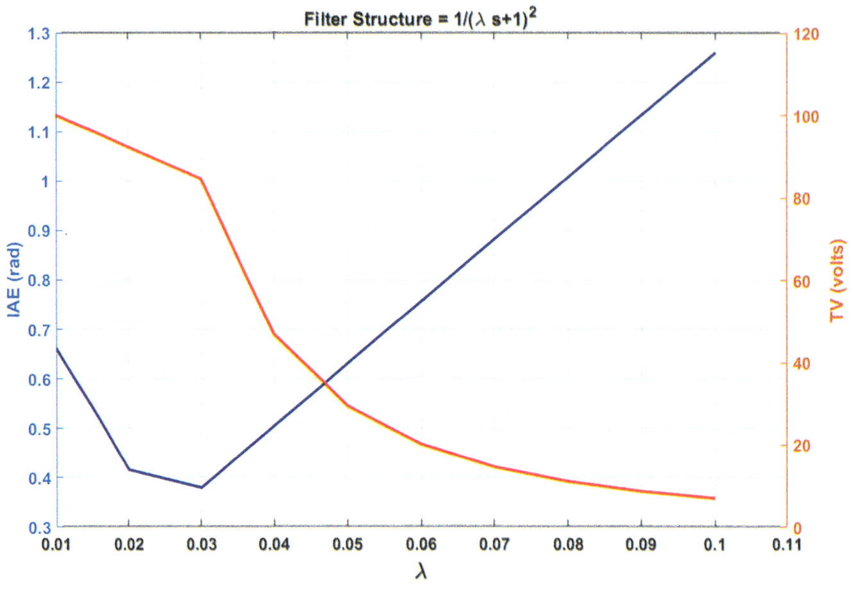

Fig. 4 Comparison between TV, IAE and λ for proposed 1

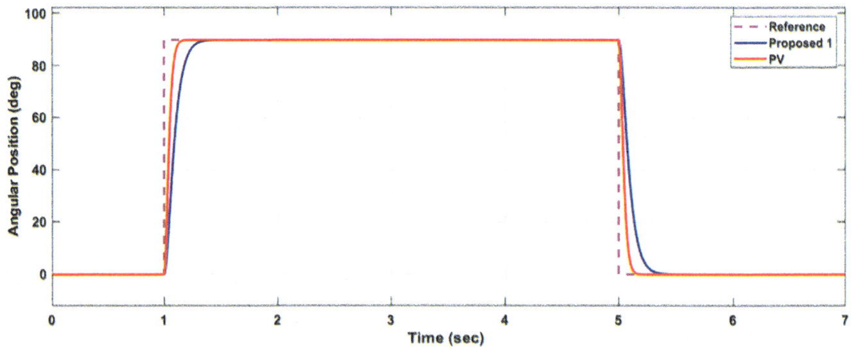

Fig. 5 Comparison of step response between proposed 1 and PV controller for nominal case

response of the proposed 1 lies within the tracking limits, and it also demands only half the control effort required by the PV controller. Figures 7 and 8 depict the process response and control effort demanded in the presence of output disturbance scenario. The disturbance rejection of the proposed 1 is slightly sluggish when compared to the PV controller, but the rejection happens within safer time limits with less demanded control effort.

***Case 2*: Performance Analysis with Proposed Method 2**

The tuning parameter λ is chosen based on the global minimum for IAE and TV for a particular filter structure as shown in Fig. 9. For Type 1 filter of order 3, λ is chosen as

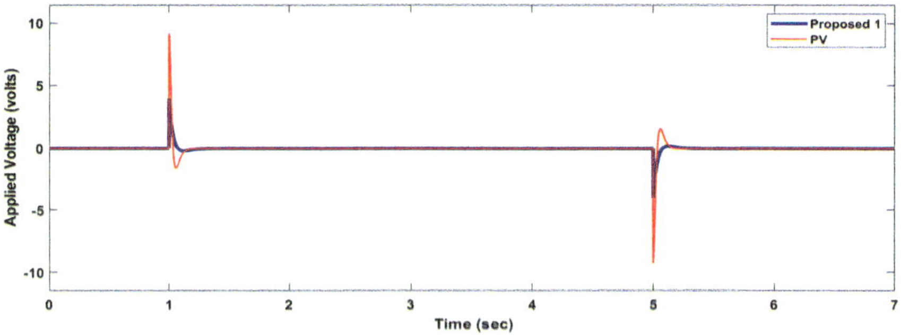

Fig. 6 Comparison of control effort between proposed 1 and PV controller for nominal case

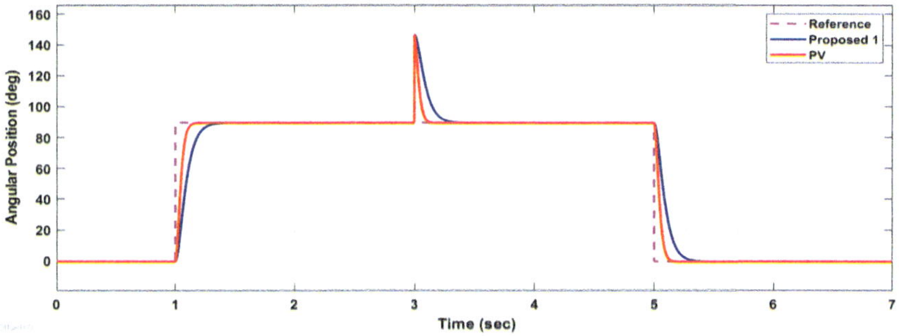

Fig. 7 Comparison of step response between proposed 1 and PV controllers for perturbation case

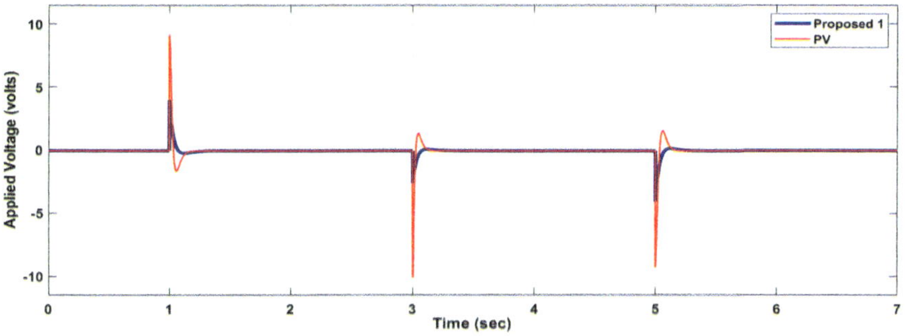

Fig. 8 Comparison of between proposed 1 and PV controllers for perturbation case

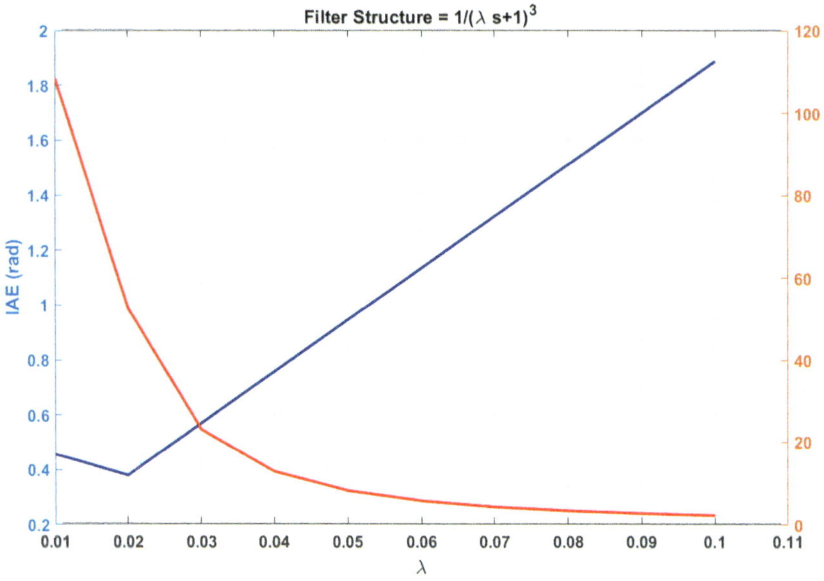

Fig. 9 Comparison between TV, IAE, and λ for proposed 2

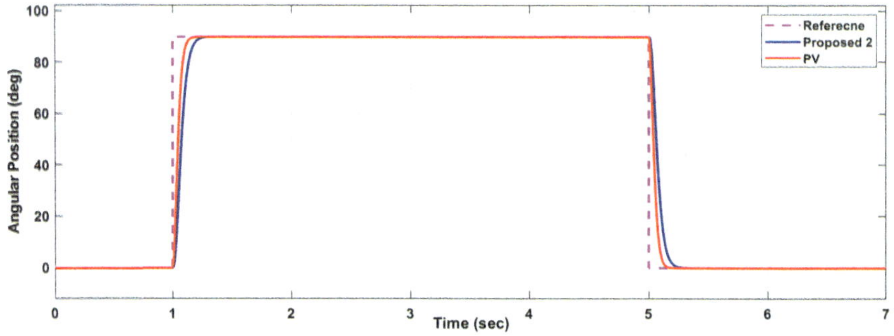

Fig. 10 Comparison of step response between proposed 2 and PV controllers for nominal case

0.026 as the global minimum, respectively. Figures 10 and 11 illustrate the nominal reference tracking performance and its corresponding control effort. It is observed that proposed 2 demands slightly more control effort to have faster response and has improved performance metrics, i.e., IAE. Figures 12 and 13 depict the process response in the presence of output disturbance, and it is also evident that proposed 2 has faster disturbance rejection.

***Case 3*: Performance Analysis with Proposed method 3**

It can be observed from previous analysis that lesser the λ value, faster the system response. Hence for this case, the parameter λ is chosen as 0.015. The speed of the response and disturbance rejection are improved further by adding additional

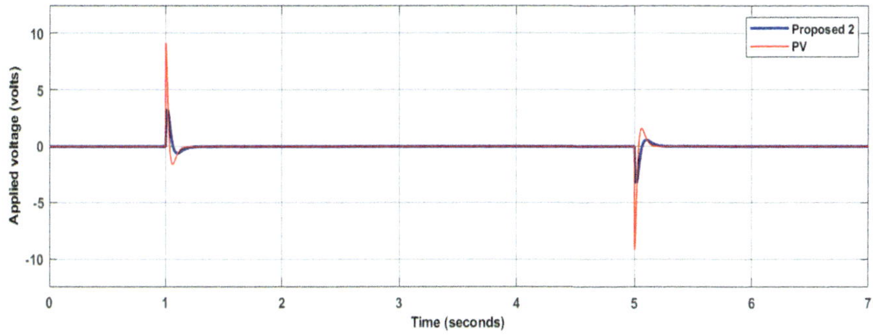

Fig. 11 Comparison of control effort between proposed 2 and PV controllers for nominal case

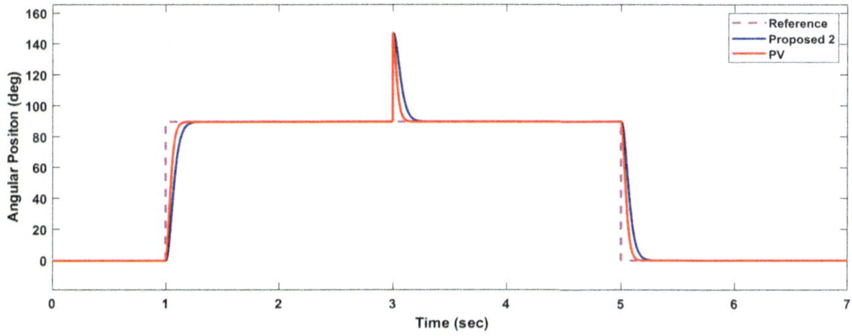

Fig. 12 Comparison of step response between proposed 2 and PV controllers for perturbation case

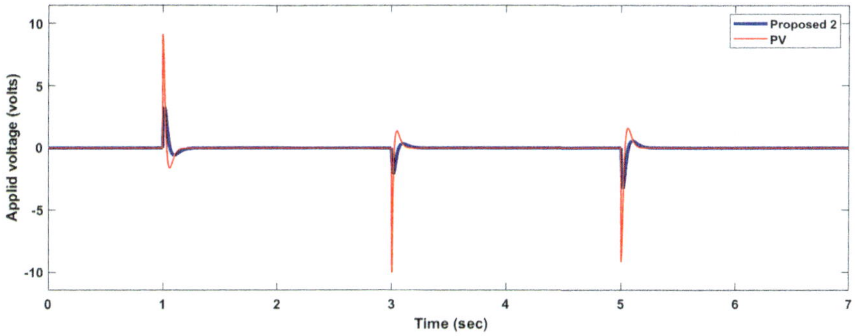

Fig. 13 Comparison of control effort between proposed 2 and PV controllers for perturbation case

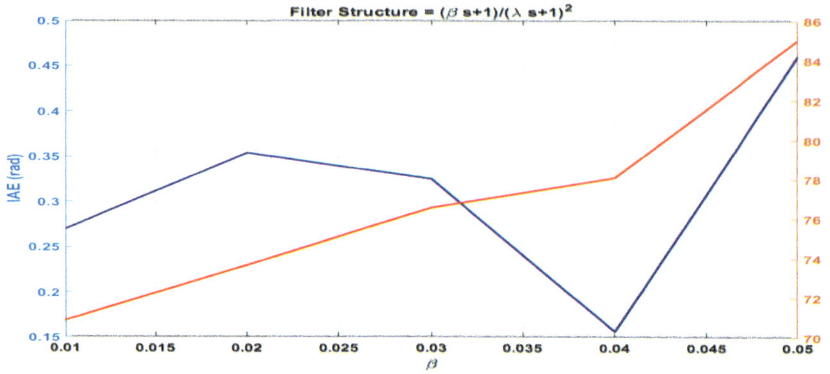

Fig. 14 Comparison between TV, IAE, and λ for proposed 3

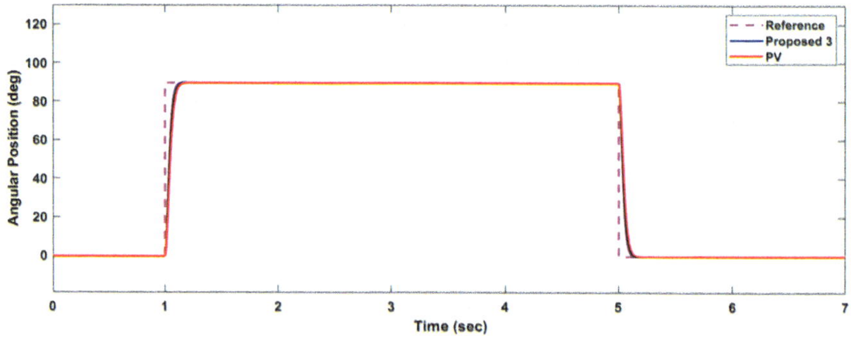

Fig. 15 Comparison of step response between proposed 3 and PV controllers for nominal case

parameter flexibility by adding one LHP zero. This tuning parameter β is selected based on the global minimum for IAE and TV, and it is chosen as 0.04 as shown in Fig. 14. The nominal reference tracking performance and its corresponding control effort are shown in Figs. 15 and 16. It is observed that proposed controller 3 demands more control effort to have faster response and improved performance metrics, i.e., IAE. Figures 17 and 18 depict the process response and control effort in the presence of output disturbance, and it is also evident that proposed 3 has faster disturbance rejection.

***Case 4*: Performance Analysis with Proposed Method 4**

For this case, the parameter λ is chosen as 0.019 and β is chosen as 0.045. To achieve faster system response with reduced control effort, i.e., TV, the controller is implemented with additional DOF-Prefilter. A Prefilter $F(s)$ is shaped such that the closed loop tracking system lies between the desired tracking limits (T_U and T_L) as shown in Fig. 19. T_U and T_L are selected based on the guidelines given in [18] for the following time domain specifications:

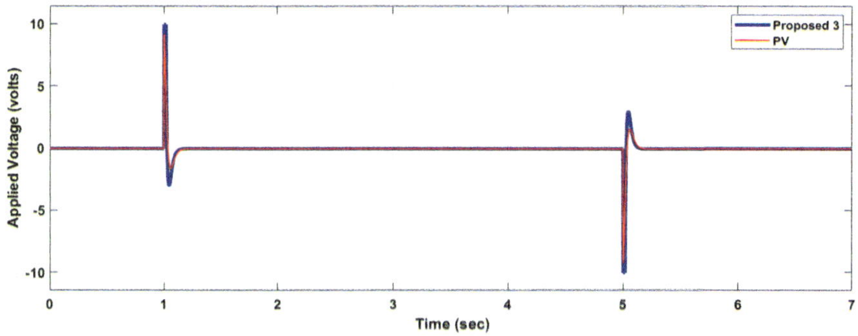

Fig. 16 Comparison of control effort between proposed 3 and PV controllers for nominal case

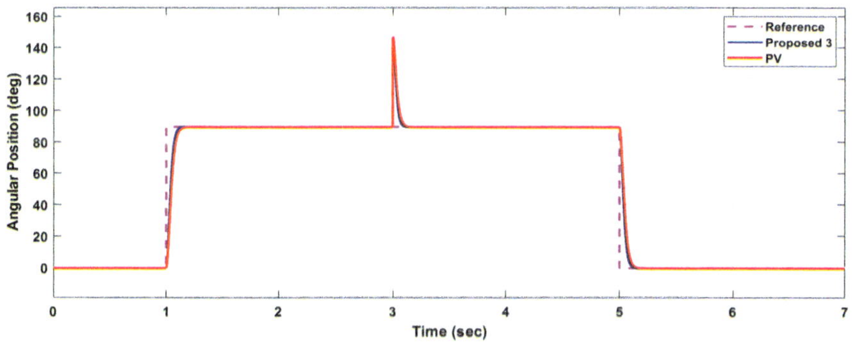

Fig. 17 Comparison of step response between proposed 3 and PV controllers for perturbation case

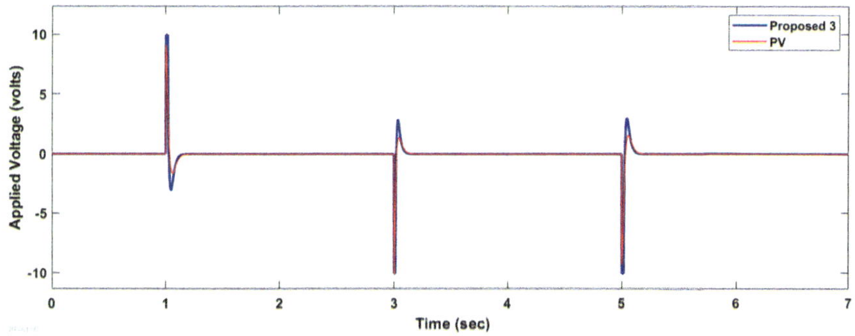

Fig. 18 Comparison of control effort between proposed 3 and PV controllers for perturbation case

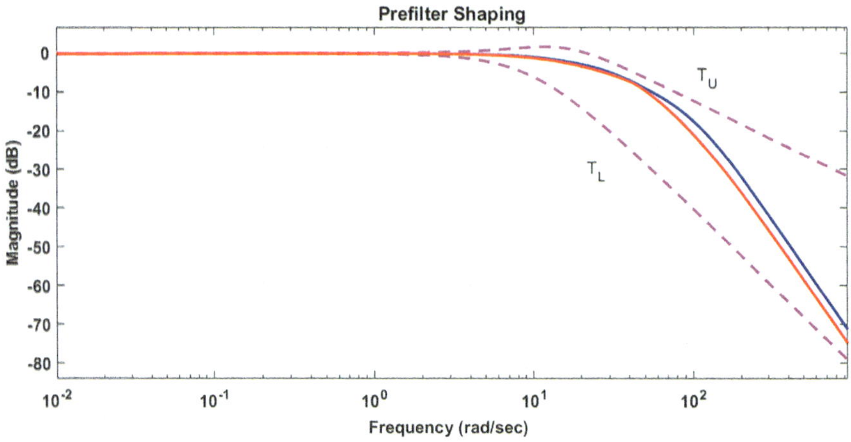

Fig. 19 Prefilter shaping

$$0.0533 \leq t_r \leq 0.336 \text{ s} \quad \text{and} \quad 0.323 \leq t_s \leq 0.583 \text{ s} \tag{30}$$

$$T_L \leq \left| F(j\omega)\frac{G_m(j\omega)C(j\omega}{1 + G_m(j\omega)C(j\omega)} \right| \leq T_U \tag{31}$$

where

$$T_U = \frac{0.99}{[(s/10) + 1]^2} \quad \text{and} \quad T_L = \frac{1.01[(s/10) + 1]}{[(s/15.625)^2 + (2 * 0.8s)/15.625) + 1]} \tag{32}$$

The prefilter designed is shown in Fig. 19, and it is given as

$$F(s) = \frac{1}{\frac{s}{22} + 1} \tag{33}$$

Figures 20 and 21 illustrate the nominal reference tracking performance and its corresponding control effort. It is observed that proposed controller 4 has faster response and improved performance metrics, i.e., IAE with reduced control effort. Figures 22 and 23 depict the process response and demanded control effort in the presence of output disturbance.

The controller structures used for all the cases is shown in Table 2.

It is observed from Table 3 that with proposed 1 structure, i.e., simple PD controller with simple LPF , tracking error is more with less control effort and sluggish disturbance rejection when compared to that of PV controller. Thus with proposed 2, i.e., PD with complex LPF filter, tracking error is improved slightly with more control effort and slightly faster disturbance rejection when compared to proposed 1. Hence, it is observed that increasing the order of the filter does not make much difference in the performance. Proposed 3 controller introduces the additional parameter

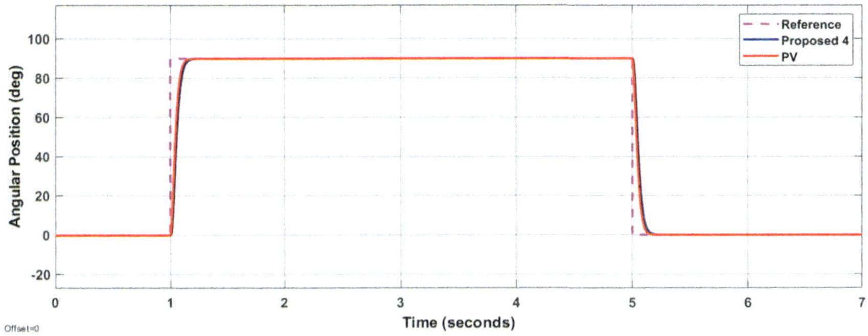

Fig. 20 Comparison of step response between proposed 4 and PV controllers for nominal case

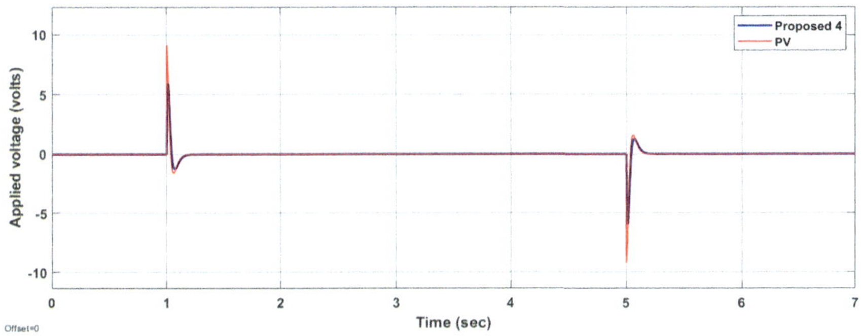

Fig. 21 Comparison of step response between proposed 4 and PV controllers for nominal case

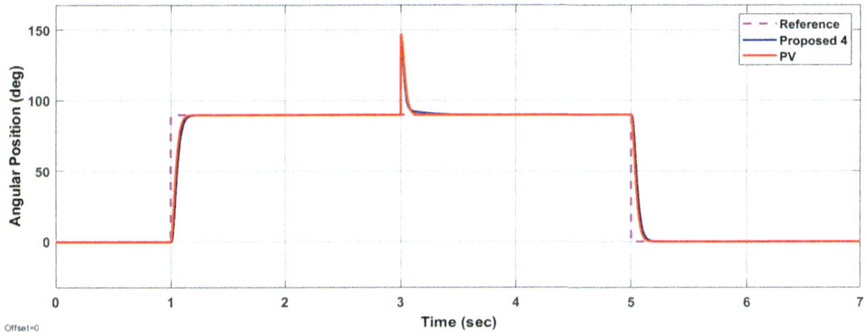

Fig. 22 Comparison of step response between proposed 4 and PV controllers for perturbation case

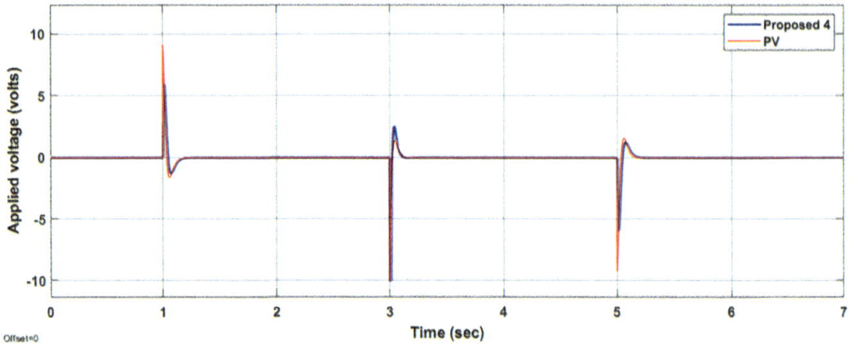

Fig. 23 Comparison of control effort between proposed 4 and PV controllers for perturbation case

Table 2 Controller structure in proposed methods

Proposed controller	Controller structure (Integer Filter)(PD)
Proposed 1	$(\frac{2.55}{s+42.55})(1+7.776s)$
Proposed 2	$(\frac{320.49}{s^2+115.45s+4438})(1+7.776s)$
Proposed 3	$(\frac{66.761(s+25)}{(s+192.3)(s+7.704)})(1+7.776s)$
Proposed 4	$(\frac{36.956(s+22.22)}{(s+145.9)(s+11.99)})(1+7.776s)$

Table 3 Performance metrics comparison between PV and proposed controllers

Controller	Nominal case		Perturbation case		M_s
	IAE	TV	IAE	TV	
PV	0.1487	28.67	0.1886	47.69	1.2380
Proposed 1	0.2953	10.61	0.3893	13.99	1.1542
Proposed 2	0.245	15.62	0.323	20.6	1.2836
Proposed 3	0.1225	36.8	0.1545	54.9	1.2634
Proposed 4	0.1821	28.96	0.2241	46.83	1.1619

by improving the controller flexibility; tracking error is less and faster disturbance rejection when compared to that of PV controller, but it demands more control effort to have faster response. Thus, proposed 4 which has the same controller structure with additional degree of freedom, i.e., prefilter, control effort is reduced when compared to that of proposed 3 and comparable response w.r.t. PV controller is observed. When there is a sudden change in the reference command, it can also be observed that with PV controller, the actuator goes into saturation whereas with IMC controller, it does not go into saturation because the control effort minimization constraint is considered during the design phase of the IMC controller.

6 Conclusions

In this article, an IMC-based robust motion controller was designed for DC servo systems with IMC controllers of different structures of integer filters. Here, the selection of a tuning parameter of IMC filters is done by the global minimum for both IAE and TV. Detailed comparison analysis between proposed structures is done. Simulation results show that the performance of the proposed method 4 is better in tracking as well as control effort minimization for both nominal and perturbation models when compared to that of controllers in the literature. It should be noted that IMC controller does not go into saturation even if there is a sudden change in the reference command. This reduces the wear and tear and improves the life of the actuator.

References

1. Ziegler JG, Nichols NB (1942) Optimum settings for automatic controllers. J Fluids Eng
2. Fruehauf PS, Chien I-L, Lauritsen MD (1994) Simplified imc-pid tuning rules. ISA Trans 33(1):43–59
3. Rivera DE, Morari M, Skogestad S (1986) Internal model control: Pid controller design. Ind Eng Chem Process Design Develop 25(1):252–265
4. Brosilow C, Joseph B (2002) Techniques of model-based control. Prentice Hall Professional
5. Horn IG, Arulandu JR, Gombas CJ, VanAntwerp JG, Braatz RD (1996) Improved filter design in internal model control. Ind Eng Chem Res 35(10):3437–3441
6. Kaya I (2004) Two-degree-of-freedom imc structure and controller design for integrating processes based on gain and phase-margin specifications. IEE Proceed Control Theory Appl 151(4):481–487
7. Ali A, Majhi S (2009) Pi/pid controller design based on imc and percentage overshoot specification to controller setpoint change. ISA Trans 48(1):10–15
8. Skogestad S (2003) Simple analytic rules for model reduction and pid controller tuning. J Process Control 13(4):291–309
9. Arya PP (2022) A modified imc design for second order plus time delayed processes. IFAC-PapersOnLine 55(1):843–847
10. Kumar DS, Sree RP (2016) Tuning of imc based pid controllers for integrating systems with time delay. ISA Trans 63:242–255
11. Rao CN, Sree RP (2010) Imc based controller design for integrating systems with time delay. Indian Chem Eng 52(3):194–218
12. Trivedi R, Padhy PK (2020) Design of indirect fractional order imc controller for fractional order processes. IEEE Trans Circ Syst II: Express Briefs 68(3):968–972
13. Yadav AK, Gaur P (2017) Speed control of an uncertain heavy-duty vehicle using improved imc technique. Arabian J Sci Eng 42(7):2981–2991
14. Li N, Yang H, Mu A (2019) Improved grey particle swarm optimization and new luus-jaakola hybrid algorithm optimized imc-pid controller for diverse wing vibration systems. Complexity 2019:1–21

15. Ranganayakulu R, Seshagiri Rao A, Uday Bhaskar Babu G (2020) Analytical design of fractional imc filter-pid control strategy for performance enhancement of cascade control systems. Int J Syst Sci 51(10):1699–1713
16. Wang Q, Lu C, Pan W (2016) Imc pid controller tuning for stable and unstable processes with time delay. Chem Eng Res Design 105:120–129
17. Servo UMQSR (2013) Experiment set up and configuration
18. Garcia-Sanz M (2017) Robust control engineering: practical QFT solutions. CRC Press

Design and Development of Immersive Vision Drone

Meda Narendra, Seemakurthi Tharun, Rolla Shiva Goud, Duggempudi Bala Krishna, and Manoj Sindhwani

Abstract An unmanned vehicle that achieves vertical flying stability and used to collect the data of aerial views to get immersive sceneries out of them by interfacing vital peripherals to low power embedded microcontrollers/processors. This paper discloses an innovative approach from the initial step i.e., from making the drone to the last step of achieving the precise look at immersive aerial views of places captured with unmanned aerial vehicle and enables the possibility of providing the aerial perspective of captured views through web application to end users. A technical approach for capturing views through camera attached to the drone that achieves various angles of capturing with the pitch and yaw rotations. The drone captures are of 2D pictures, and the captured pictures are further transfers to a separate device, which process the captured pictures/images into 3-D visions. Processing the 3-D visions into server application experiences immersive views of places/spots and further used in developing various next-gen applications.

Keywords Unmanned vehicle · Capturing aerial views · Image processing · Immersive vision · Webserver

1 Introduction

An invention of flying the unmanned vehicles of multirotor drones into air recreated the tech world in wide range of applications ranging from agriculture [1], photography, healthcare, construction, surveillance to high standard defence combat missions [2] of finding and destroying the targets. Thanks to the great engineers behind innovative designs of flying drones with understanding and analysing the aerodynamics behind.

Depending on application, the drone customizations [3] will be done in the following aspects: Flight controller, Motors, propellers and frame (size), Battery

M. Narendra (✉) · S. Tharun · R. S. Goud · D. B. Krishna · M. Sindhwani
School of Electronics and Communication Engineering, Lovely Professional University, Jalandhar 144411, India
e-mail: meda.narendra987@gmail.com

S. Kumar et al. (eds.), *Control Applications in Modern Power Systems*, Lecture Notes in Electrical Engineering 1128, https://doi.org/10.1007/978-981-99-9054-2_17

(voltage, capacity and energy density), Camera (ensure high-resolution), Global positioning system (GPS) [4] and sensors.

We have latest innovations of satellite image mapping technique that captures 2-D images and process into wide applications mainly in navigation sector. Even though, we have maps that renders 3-D representation, which does not feel the real 3-D view.

Whereas, coming to the application of immersive vision drone, the agenda is to create a new era of 3-D visuals experience through unmanned vehicle. The primary task is to achieve a smooth vertical take-off and landing (VTOL) ensuring less vibration, more balance conditions and effective communication between the drone and remote devices.

A high-resolution camera mounted to the frame part of the drone with a dedicated mechanical rotation design to capture the aerial views of sceneries that a man cannot. It was interfaced to an embedded processor for capturing & seamless data storing of the captures into storage chip. The mechanical rotation design can perform operations: Left–right rotation and Top–bottom rotation.

Covering larger area was explored in this paper by separating the area into different blocks and captures the aerial views from all directions and every corner of all blocks with the unmanned.

Few techniques of image alignment and mesh generation were used that results in processing the captured 2-D images into 3-D, ensuring precise immersive looks at sceneries. The same 3-D visuals of all blocks were merged and enabled the experience of real 3-D view to the public through the server.

2 Unmanned Aerial Vehicle Design

Drones are remotely operated unmanned aerial vehicles (UAVs) running with a dedicated hardware for enabling various modes of flying. As mentioned earlier, the drone customizations are designed based on the requirements of application.

2.1 Flight Controller

It is a dedicated hardware of processor running over software with few on-board, off-board sensors and electronic modules embedded in a board.

In this paper, working with APM-2.8 flight controller [5] internally embedded with main microcontroller chips of ATMEGA2560, ATMEGA32U2 responsible for processing and USB functioning (Fig. 1).

It has on-board 3-axis gyro, accelerometer, high-performance barometer [6], and a 4-Megabytes of data flash chip which enables off-board connectivity for GPS module [7] and uBlox LEA-6H modules of inbuilt compass. As the chips are of AVR, it is more compatible with the Arduino IDE and is open-source platform with ready to upload firmware into the flight controller.

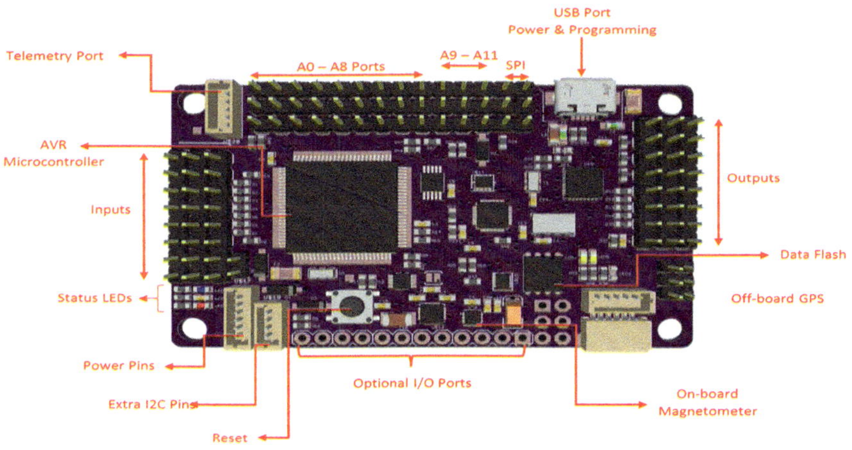

Fig. 1 Flight controller APM 2.8

2.2 *Brushless DC Motor and Electronic Speed Controllers*

For high-performance applications, the BLDC motors were used because of their high speed with low voltage due to their brushless design. They require less maintenance and have an exceptionally long lifespan. Can achieve precise control of the drone through electronic speed controller (ESC). The A2212/13t 1000 kV brushless motors [3] are used for designing the drone.

The electronic speed controller (ESC-30A) [3] is a compact and efficient speed controller designed to use in the drones and multi rotor aircrafts, the number 30A indicates that the maximum current rating is up to 30 amperes, these ESC's supports multiple protocols, it includes many safety features like over-current protection and temperature protection and these ESCs ensures the safe and reliable operations of the drone.

2.3 *Propellers and Frame*

Propellers are the crucial components which play a significant role in flying and hovering of the drone. The selection of the perfect propellers has a major impact on the overall flying performance of the drone. The more propellers attached, the more efficiently the drone works. All the propellers collectively lift the drone by rotating and producing the air lift through the airflow.

The drone frames are the foundation in achieving stability and superior performance during airborne. It typically consists of four arms extending from a central body, which holds the motors and propellers [8]. These frames are made up of lightweight materials like plastic or carbon fiber. A well-designed drone frame supports and secures components such as cameras, batteries and processors that control the overall drone.

2.4 Wireless Receiver (Rx) and Transmitter (Tx)

For long distance control of the drone during airborne was achieved through radio frequency signals transmitting from remote end to drone end receiver. The "Fly Sky FS-i6" [9] radio control system (6 active channels) was used in making the drone. It's very small in size, light in weight and supports multiple protocols and is highly recommended for designing the drone because of advanced features it provides to techies (Fig. 2).

For take-off, a drone into air, all the components mentioned above are necessary as the input signals will be sent to the flight controller [10] with the receiver and transmitter channels through radio frequency signals. The flight controller processes the input and controls the brushless motor through analog signals sent to the ESCs.

BEC circuit was implemented to reduce the voltage of 12 to 5 V for powering the flight controller, raspberry pi and modem modules as their max's operating voltage was 5 V. A GPS module was attached to the off-board port of the flight microcontroller [11].

The same functioning can be done in many ways of sending inputs to the flight controller through cloud by simply interfacing a separate module of SIM7600 modem with latest Wi-Fi processors [12] of raspberry pi or ESP32.

3 Camera Design, Interfacing and Capturing

Whereas, coming to the design and interfacing part of the immersive vision drone, a high-resolution camera mounted below to the frame part of drone with a dedicated mechanical rotation design enabling left–right rotation and top–bottom rotations.

Fig. 2 Complete drone parts and shows alignment of motors, propellers, lander, and camera on frame

3.1 Camera Design

A Raspberry Pi high quality camera "Sony IMX477" sensor of 12.3 megapixel was interfaced to an embedded processor. It comes up with the features of 7.9 mm diagonal image size, 1.55 μm × 1.55 μm pixel size, interchangeable lenses with adjustable back focus (M12 & CS Mount variant). It was mounted to the bottom board of the frame with 2 servo motors controlled primarily by the remote inputs or alternatively from cloud (Raspberry Pi-IoT). Out of which, first servo motor (pitch) is responsible for top–bottom rotations [13] and the other (yaw motor) is for left–right rotations.

3.2 Camera Interfacing and Capturing

Raspberry Pi high quality "Sony IMX477" cameras sensor can be interfaced to the raspberry pi through MIPI camera serial interface protocol. This camera is one of the ready to use camera modules and works with all Raspberry Pi's those of having camera connectors by default.

For interfacing the camera, need a ribbon cable (connector), raspberry pi board and lenses (if needed). Screw in the lens, attach the one end of the ribbon cable to the rear of the camera and the other end to the Raspberry Pi's camera CSI port. Power up the board and enable the camera in the raspberry pi preferences then the board requests for re-boot.

Once the re-boot was done, just type "raspistill" tool in the terminal to check whether the camera was correctly installed or any issue with the cable/camera module.

Immersive aerial views will be captured through radio frequency signals or over cloud through sending a high signal (5v) to a particular GPIO pin of Raspberry Pi board Captures get stored in the SD card, which was initially inserted into the board for booting the board with Raspbian operating system.

To cover a larger area, the best route is to divide the area into plots then capture the aerial views from all directions and every corner of the plot with the unmanned vehicle (see Fig. 3).

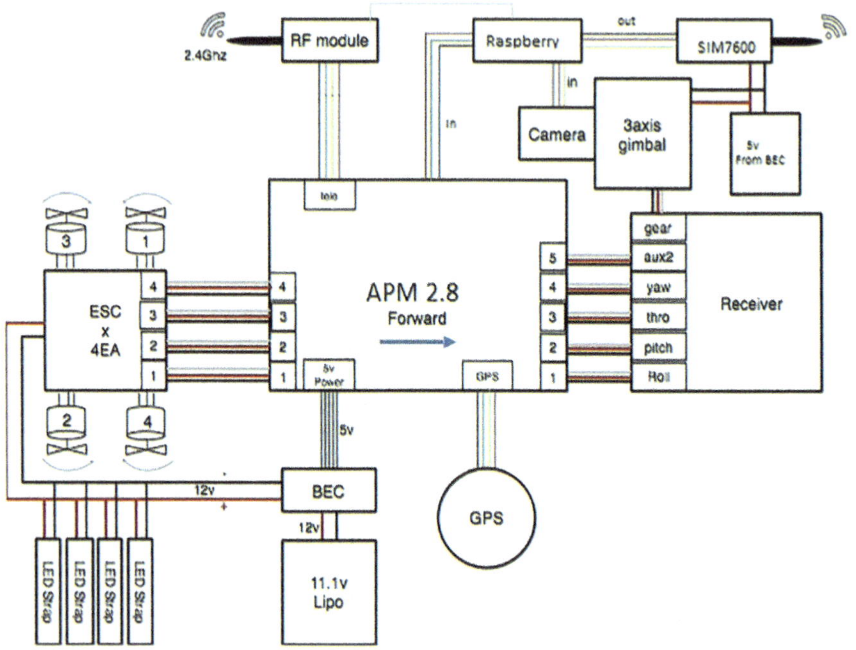

Fig. 3 Quadcopter component connections

4 Image Processing

Prior to converting the raw 2D captures to 3D vision, it is important to review those captures and eliminate the images which are not that useful using the techniques of image elimination.

4.1 Image Elimination

Make sure that there will be no moving things in those captures. Pick the images that don't have their own shadow. Make sure that you have captured every single area very clearly.

4.2 Image Conversion

There are several steps to convert these 2D captures to 3D, by processing these steps can achieve the precise 3D look at blocks.

Photogrammetry. This step involves the generation of primary conversion of 2D to 3D using the series of images to analyze and determine the 3D structure of the objects in the scene [14].

Image Alignment. The first step in photogrammetry is to align the images so that they are all in the same coordinate system. This is done using various algorithms to identify common features in the images and align them to the correct positions, if they are misplaced due to any common features in multiple positions align them manually in the software (Fig. 4).

Point Cloud Generation. Once the images are aligned, these software's uses them to generate a point cloud, where all the objects in the scene are clearly visible from the point, it means it was a collection of points in 3D space representing the objects in the scene (Fig. 5).

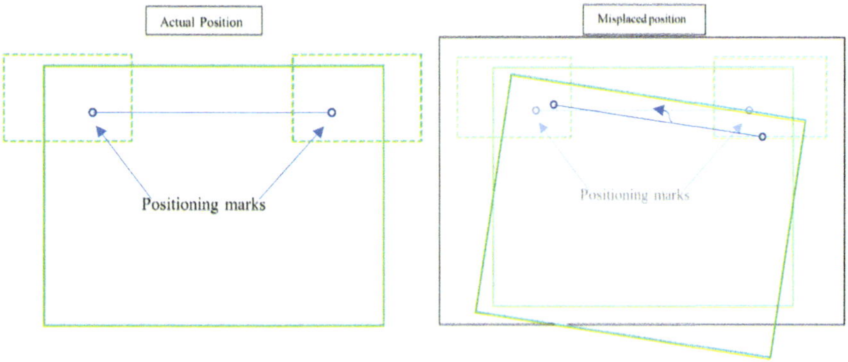

Fig. 4 Image position alignment

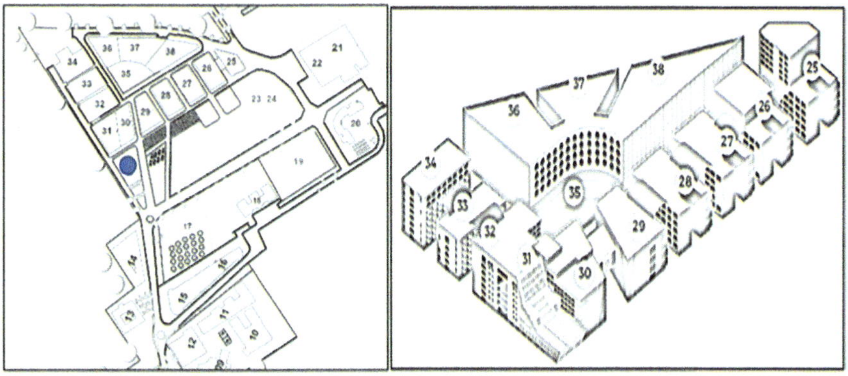

Fig. 5 Point cloud generation

Mesh Generation. The point cloud is then used to generate a 3D mesh, which is a surface representation of the objects in the scene.

Texturing: The ultimate step is to texture the 3D mesh, which involves mapping the original 2D images onto the surface of the mesh [14] to give it a more realistic appearance.

Will get the final precise 3D immersive vision of our places, colleges, and our desired locations of the cities by performing these five simple steps. The final 3D model can then be exported to a variety of file formats, such as OBJ, FBX or 3D PDF for use in other 3D software or for 3D printing (Fig. 6).

5 Conclusion

In this tech-world the existing applications show and fulfil all our requirements from searching of places to navigation support for those places with real-time tracking. With the latest innovations, currently there are satellite image mapping techniques that help in detection and depiction of 2D structures/images but are not exactly precise (Fig. 7).

Further, there are existing maps that render 3D representation of earth based on satellite imagery. Hence in view of emerging technologies, the present paper disclosed an aerial drone to capture 3-D immersive images to provide immersive vision of famous spots to the end user. The drone is equipped with a camera to capture the 2-D images and is further processed to 3-D visions through image processing. The 3-D visions, processed from 2-D images of unique slots are combined to form a frame and further can process into server in specific applications installed on their mobile phones or webs to experience the immersive views of famous tourist spots.

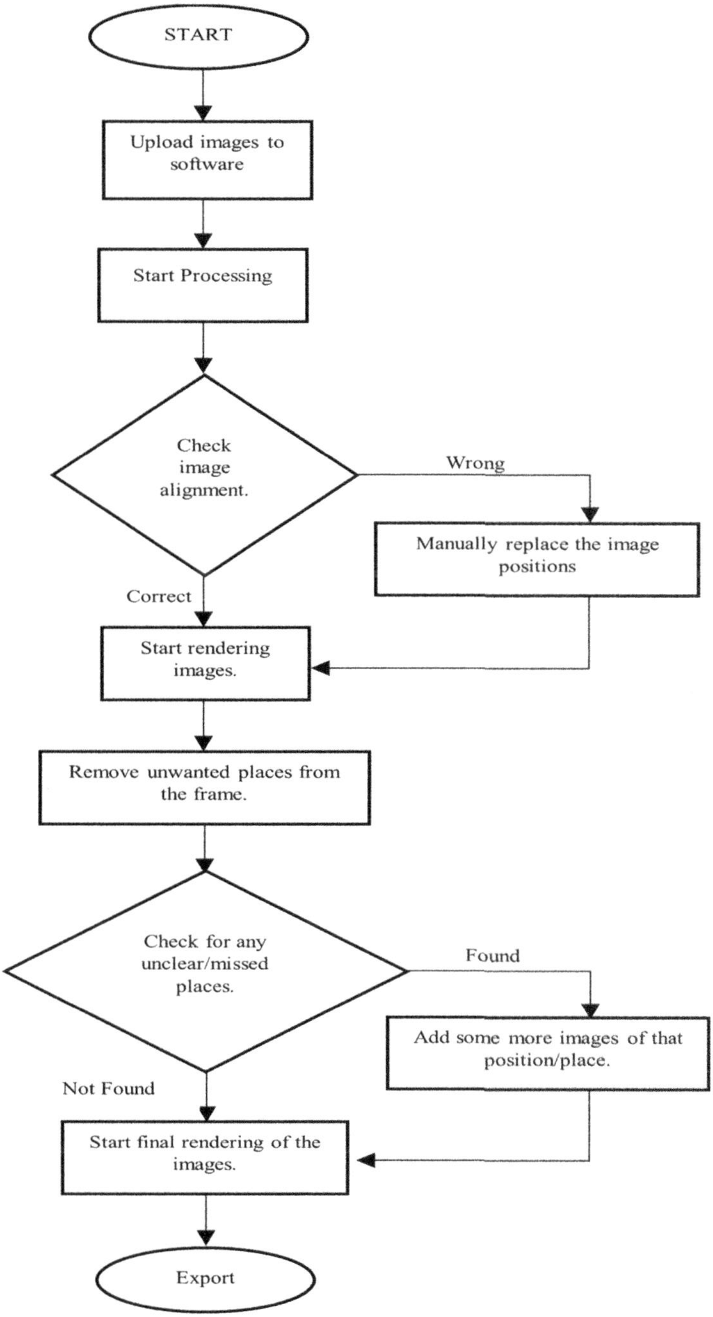

Fig. 6 Entire process of image processing

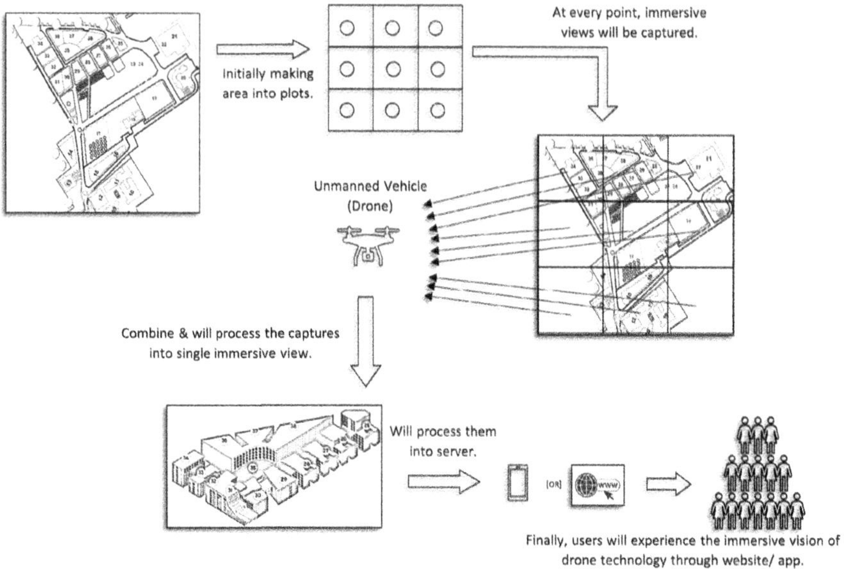

Fig. 7 Diagrammatic representation of entire process

References

1. Ahirwar S, Swarnkar R, Bhukya S, Namwade G (2019) Application of drone in agriculture. Int J Curr Microbiol App Sci 8(01):2500–2505
2. Nanomi K (2007) Prospect and recent research & development for civil use autonomous unmanned aircraft as UAV & MAV. J Syst Des Dyn 1(2)
3. Mogili UR, Deepak B (2020) An intelligent drone for agriculture applications with the aid of the MAVlink protocol. In: Deepak B, Parhi D, Jena P (eds) Innovative product design and intelligent manufacturing systems. Lecture notes in mechanical engineering. Springer, Singapore. https://doi.org/10.1007/978-981-15-2696-1_19
4. Park S, Kim HT, Lee S, Joo H, Kim H (2021) Survey on anti-drone systems: components, designs, and challenges. IEEE Access 9:42635–42659
5. Ebeid E, Skriver M, Terkildsen KH, Jensen K, Schultz UP (2018) A survey of open-source UAV flight controllers and flight simulators. Microprocess Microsyst 61:11–20
6. Lim H, Park J, Lee D, Kim HJ (2012) Build your own quadrotor: opensource projects on unmanned aerial vehicles. IEEE Robot Autom Mag 19(3):33–45
7. Saha A, Kumar A, Sahu AK (2017) FPV drone with GPS used for surveillance in remote areas. In: 2017 third international conference on research in computational intelligence and communication networks (ICRCICN), Kolkata, India, pp 62–67. https://doi.org/10.1109/ICRCICN.2017.8234482
8. Ebeid ESM, Skriver M, Jin J (2017) A survey on open-source flight control platforms of unmanned aerial vehicle.
9. Chao H, Cao Y, Chen YQ (2007) [IEEE 2007 international conference on mechatronics and automation - Harbin, China (2007.08.5–2007.08.8)] 2007 international conference on mechatronics and automation - autopilots for small fixed-wing unmanned air vehicles: a survey, pp. 3144–3149
10. Chao HY, Cao YC, Chen YQ (2010) Autopilots for small unmanned aerial vehicles: a survey 8(1):36–44

11. Mészáros J (2011) Aerial surveying UAV based on open-source hardware and software. Int Arch Photogram Remote Sens Spat Inf Sci 37
12. Nam C, Danielson S (2011) Development of a small UAV with real-time video surveillance In: Paper presented at 2011 ASEE annual conference and exposition, Vancouver, BC. https://doi.org/10.18260/1-2-17762
13. Zaheer Z, Usmani A, Khan E, Qadeer MA (2016) Aerial surveillance system using UAV. In: 2016 thirteenth international conference on wireless and optical communications networks (WOCN), Hyderabad, India, 2016, pp 1–7
14. Wahbeh W, Ammann M, Nebiker S, van Eggermond M, Erath A (2021) Image-based reality-capturing and 3D modelling for the creation of VR cycling simulations. In: ISPRS annals of the photogrammetry, remote sensing and spatial information sciences.

LFC of Three Area Power System with hyHS-RS Tuned T2-Fuzzy Controller in Constraint to Eoconomic Dispatch

Ashok Kumar Mohapatra, Srikanta Mohapatra, Prakash Chandra Sahu, and Padarabinda Samal

Abstract This article demonstrates a research work on load frequency control (LFC) of a multi generating station based three area power system in concern to the Economic load dispatch (ELD). Both the action of LFC and ELD are obtained in a common platform with employing a noble fuzzy type-2 controller under a disturbance of step load perturbation of 2%. The controller has an advantages of highly accessing area since the concerned membership function creates a three dimensional structure as per its process. Further, this controller handles the huge uncertainty of power system smoothly due to its 3D structure. This enables to this controller for providing improved control performances as compared to the standard type-1 fuzzy controller and conventional PID controller. Since the outcomes of the controller largely depends upon the selection of proper parameter, this study has applied a novel hybridized harmony search and random search (hyHA-RS) technique for optimal gain formulation of this type-2 fuzzy controller. The optimal property of the suggested hyHS-RS techniques also has been examined over original random search and harmony search algorithms by simulated responses and results. Finally, it is concluded that applied hyHS-RS tuned T2-Fuzzy controller is much effective to obtain LFC and ELD in a common power system scenario.

Keywords Load frequency control · Automatic generation control · hyHS-RS optimization technique · Fuzzy T2 controller · Settling time · Overshoot

A. K. Mohapatra · S. Mohapatra · P. Samal
School of Electrical Engineering, KIIT Deemed to be University, Bhubaneswar 751024, India

P. C. Sahu (✉)
Electrical Department, Aryan Institute of Engineering and Technology, Bhubaneswar 752050, India
e-mail: prakashsahu.iter@gmail.com

S. Kumar et al. (eds.), *Control Applications in Modern Power Systems*, Lecture Notes in Electrical Engineering 1128, https://doi.org/10.1007/978-981-99-9054-2_18

1 Introduction

The electrical power is an essential asset to electrify various loads as per the requirement. The power needed for the load is not available directly at the load end. The entire scheme for the electrical power to meet load demand gas been undergone different stages. These stages are power generation, power transmission and final stage is power distribution. The system which varies all these three stage pf electrical is called power system. The main objective of the power system is to generate power reliable to fulfill the net demand exactly followed by loss less power transmission and finally meet demand with good power quality. In electrical power system scenario, the loads are most dynamic in nature and varies throughout. It is the objective of the power system to update net power generation as per the new demand [1, 2]. It indicates net power generation should be such that it should cover the net demand including loss in the system. However, in loss less system the net power generation should be equivalent with the new demand. This phenomenon of the power system is termed as the power balance in the power system [3, 4]. The power balance in the power system is able to make nominal voltage and frequency in the system. So at this level the loads get nominal electrical quantity from the system and operates at its maximum efficiency level. If the power generation deviates from the net demand the system frequency and voltage starts to oscillates and could not able to maintain stable and nominal parameter in the system. Under such scenario, the loads undergo maloperation and doesn't operate successfully. It is the responsibility of a power engineer to take necessary action over the power generation as per the new demand. This could be only possible if a highly control action will be involved in the governing system and controls the fuel injection as per new demand. This phenomenon is called load frequency control (LFC) of the power system [5–9]. The LFC in the power system needs a controller to act as secondary control loop in system. The idea and theoretical concept of the LFC has been represented through various research papers and journals. In the beginning of the LFC analysis, single and two area power systems were modelled to control frequency and tie-line power of the system under various load dynamics. Now research has been developed in multi area power systems to progress LFC study of the power system. Since the controller is the heart of the LFC model, different basic controllers such as PI, PID have been employed in the LFC model to carry out successful research of the system [10–12]. These above said controllers very simple and could obtain improved outcomes only in linear system. Then various fuzzy rules controllers have been implementing in the LFC model of different power system [13–16]. It is great question for the power engineer, whether the generated power is economic or not. As per the power system reliable concern, the generation must meet the new demand and consequently afford minimum cost for their power generation. If the power generation will be done with at minimum cost to meet the new demand such concept is called economic load dispatch (ELD) [17, 18]. The obtain ELD in the power system also needs huge control action in the system. The analysis on ELD and LFC suggests requirement of good control actions for their process of execution. Various research papers on ELD as well as combined

LFC and ELD are published in different reputed journals [19, 20]. Sahu and Prusty [21] have addressed on a novel fractional based fuzzy controller principle to adopt ECO-AGC of a multi generation based power system. The controllers are not validated until unless they are designed optimally with applying suitable optimization technique. The concept of salp swam, sine cosine algorithm, grey wolf technique, different hybridized and improved techniques are well addressed in different articles [15, 22–25] to obtain LFC of interconnected power system. The well-known type-1 fuzzy controller only gives better output in low uncertainty based power system. The proposed power system model for this work carries lots of uncertainty in the analysis of LFC and ELD. To obtain good control action, this work has adopted a robust type-2 fuzzy control approach for the study. In optimal structure concern, the work has applied a noble hybridized harmony search and random search algorithm in different load dynamics.

2 System Investigated

The proposed model is a three area model where each area is electrified with installing a thermal power station. The thermal station is modelled completely by its governor and non-reheat type turbine. The structure model is fit both for LFC as well as ELD analysis. To carry out bot LFC and ELD in common power model two different controllers are installed at two different positions. The ED (economic dispatch) controller is termed as primary controller and is responsible to obtain combined ELD and LFC in the system. This controller is also known as primary controller. Another controller called secondary controller is fit to create only LFC in the system. Both the controllers are activated through a switch with proper arrangement. For this work study the secondary controller operates for only LFC study for a time period of 1–15 s however the primary controller acts for both combined LFC-ELD for time duration of 15–30 s. The overall rating of this huge system is 1500 MW to handle smoothly a load of 1200 MW (Fig. 1).

3 Type-2 Fuzzy Approach

The fuzzy ruled controllers are best substitute for the conventional PI and PID controllers. There are basically two types of fuzzy controllers. These controllers are type-1 and type-2 fuzzy controllers. The PI and PID controllers are only suitable for linear system for making necessary control actions in the system. In non-linear systems the fuzzy PID controller is much more effective to produce suitable control actions in the system. This type-1 fuzzy controller brings good results only in little uncertainty oriented power systems. Since the proposed model for this research study holds large uncertainty, this type-1 does not produce improved results in the system. So an advanced fuzzy controller called type-2 fuzzy approach is employed for this

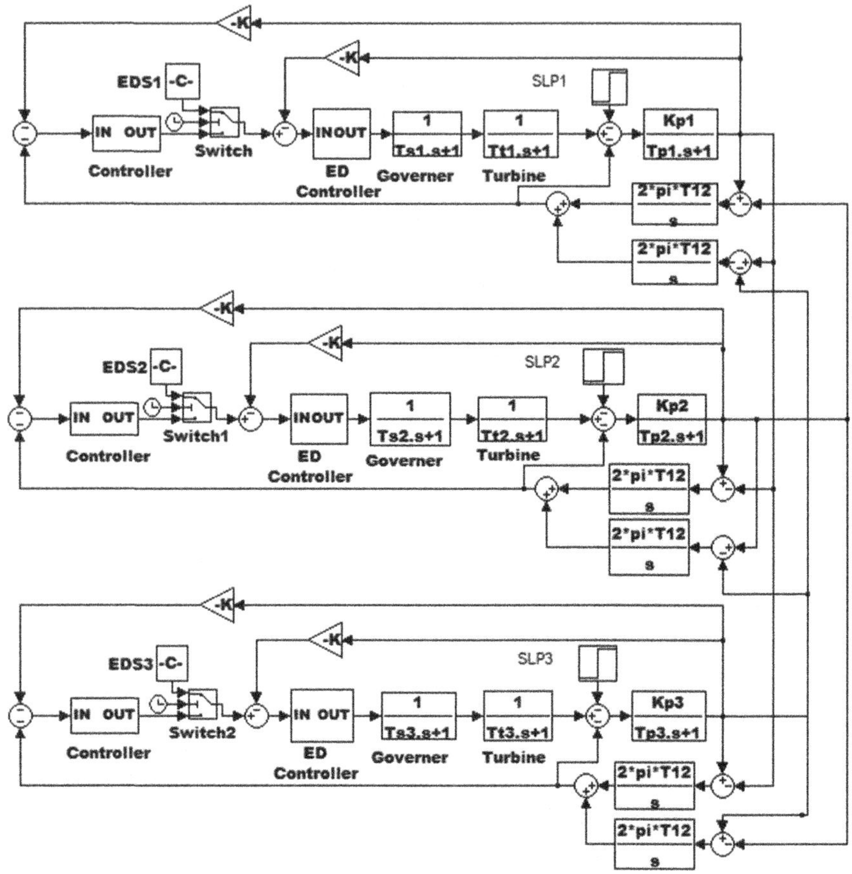

Fig. 1 Proposed system model

study since it empowers the control actions in the system under high uncertainties. Since the area of access is more due to structure of double base valued membership function the controller has high capacity to handle the large uncertainty. The uncertainty associated at the base of the membership function is called foot print of uncertainty. The operation of this novel type-2 fuzzy approach exactly similar with that of type-1 fuzzy controller except there is an action called type reducer which is involved in type-2 fuzzy controller working [26–29]. The block model of the type-2 fuzzy PID approach is deployed in Fig. 2

The work of this research study also has employed an improved objective function named integral of time multiplied absolute error (ITAE) to produce best fir parameters of this fuzzy controller.

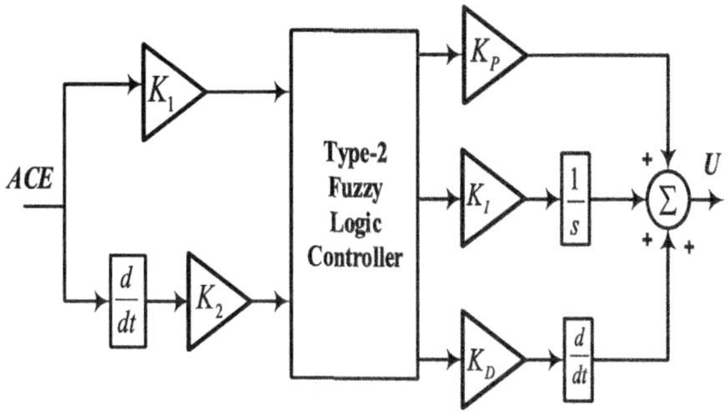

Fig. 2 Model of type-2 fuzzy PID

4 Hybridized Harmony Search and Random Search (hyHS-RS) Technique

This hybridized technique has been inspired by the fundamentals of basic harmony search and random search algorithm [30, 31]. The developed flow chart of this proposed hyHS-RS technique is given in Fig. 3

5 Results and Analysis

This work of research study is conducted in an application software called MATLAB/ SIMULINK software. At first the mathematical model of the suggested LFC dynamic model is designed in the MATLAB forum. The controller structure is also modelled in this software to progress this research activity in effective manner. The LFC as well ELD study has been conducted in concern to a step load variation of 1%. As the controller has great role to manage the system frequency and tie-line power, it is needed to justify the activeness and importance of the proposed type-2 fuzzy controller. The controller performances are also judged by the nature of the optimization technique. So this section also presented the supremacy of suggested hyHS-RS technique in different relative studies and analysis.

Study1: Controller activity study

It is found that the system behavior is highly dependent on the activity of the suggested controller. In order to show most potential behavior of the suggested type-2 fuzzy controller, it is required to make relative of this proposed controller with few standard controllers like fuzzy PID and PID controllers. In this section activity, the leadership of the proposed type-2 fuzzy PID controller has been tested over basic fuzzy PID and PID approaches in different abnormal condition especially step load disturbance.

This validation section of research study justifies the excellency of the employed type-2 fuzzy PID approach over fuzzy PID and PI controllers under the injection of a step load noise of 1%. The above controllers function both as primary controller

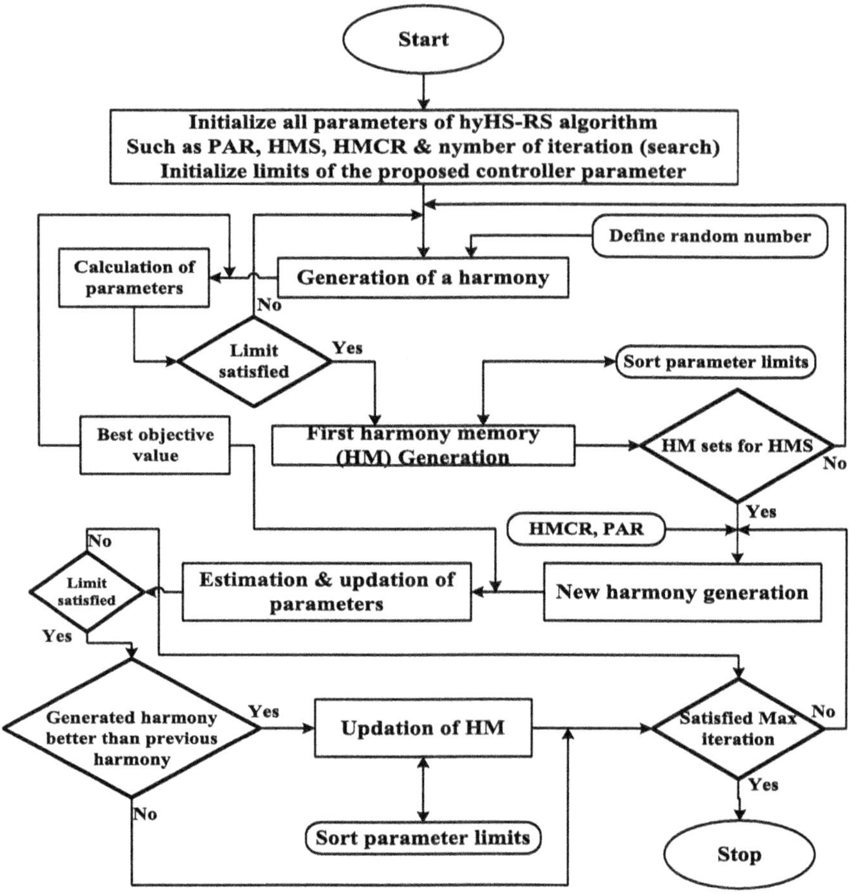

Fig. 3 Algorithm of proposed hyHS-RS technique

and secondary controller. At first all the three employed controller parameters are obtained in MATLAB software with applying an artificial technique oriented soft computing technique called hyHS-RS algorithm. Then the controller parameters are adopted in the whole system model to extract dynamic response of the system.

Frequency Responses

In a common optimal scenario, the response of fluctuation in area1 frequency response and area2 frequency response with all three employed controllers is given in Fig. 4a and b respectively. For area1 frequency response, time band (0–10) s for only LFC study and time (10–20) s is for mutual LFC-ELD study. The responses which are resulted due to type-2 fuzzy approach settles faster with little oscillation. This clearly suggests novel action of the type-2 fuzzy PID approach for frequency monitor of the power system.

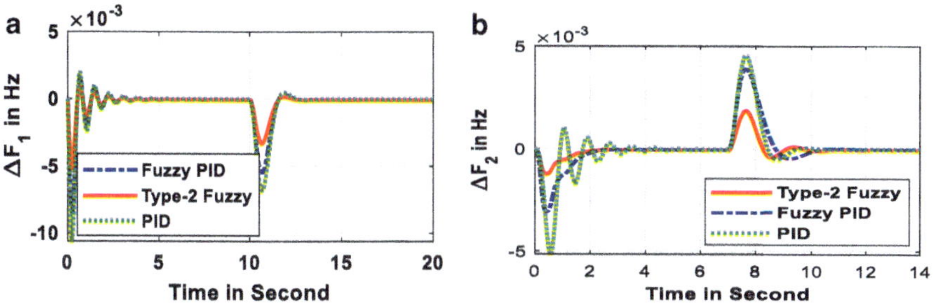

Fig. 4 **a** Area1 frequency error. **b** Area2 frequency error

Tie-Line Power Responses

In the same manner the ability of the suggested type-2 fuzzy approach also has been verified by the dynamic tie-line power responses. In continuation with frequency response, the tie-line power responses between area1 & area2 and area1 area3 are given in Fig. 5a and b respectively. As per the only LFC concern (0–7 s), the error response backs to zero after some time but as per LFC-ELD (7–20 s) the tie-line response tries to settles another stable value as power always flows from the plant of low incremental fuel cost to high incremental fuel cost. Overall study confers outstanding function of the proposed type-2 fuzzy controller.

Power Generation Response

Unlike frequency and tie-line power response, the work also has developed power generation responses of the power plants in regard to ELD and LFC of the multi source power system. With this the power generation of area1 power plant is depicted in Fig. 6a and also the response of dynamic area2 power generation is given in Fig. 6b. Overall analysis concludes excellent performance of the implemented type-2 fuzzy approach in related to the LFC-ELD of the proposed multi source power system. For 1% SLP only generator1 takes up the additional total power.

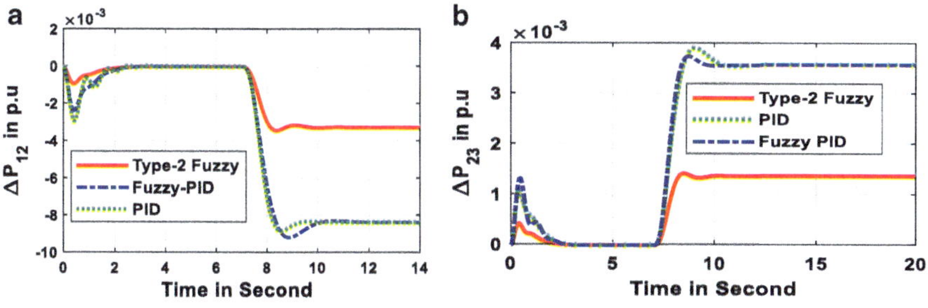

Fig. 5 **a** Area1 & Area2 tie-line power. **b** Area2 & Area3 tie-line power

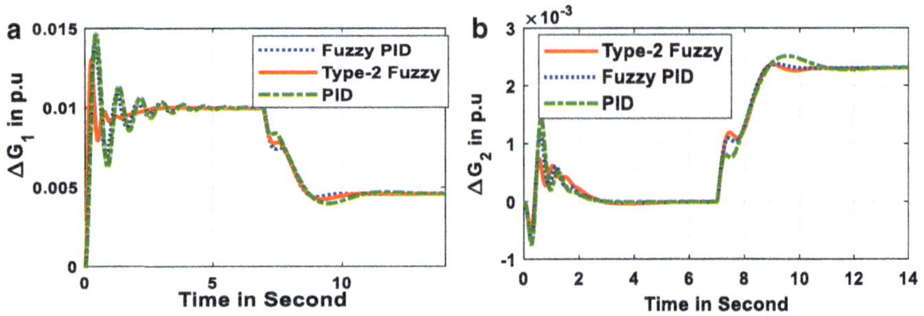

Fig. 6 **a** Power generation of area1. **b** Power generation of area2

6 Conclusion

In technical concern the research work on LFC and ELD is has been carried out successfully in a common power grid model. The outcomes of the entire analysis are well demonstrated in the result outcome section. Focusing on this result section concludes that the additional load is well taken up by the convenient generator and also capable of generating power with economic. In LFC concern, the applied type-2 fuzzy approach maintains stable in system frequency quickly under the injection of a step load dynamic. It is also clearly visualized from the joint contribution of LFC and ELD that still this proposed fyzzy type-2 controller gives brilliant results over type-1 and basic PID controller. The action of the adopted hyHS-RS technique is also found to be more effective over original HS and RS techniques to design to the suggested type-2 fuzzy controller. Overall study made a conclusion of most supreme performances of the hyRS-HS based type-2 fuzzy controller for LFC-ELD of multi source power system.

Appendix

Power regulation parameter of each area (=R1 = R2 = R3) is 2.4 Hz/pu MW; Biasing parameter of each area (B1 = B2 = B3) = 0.425 pu MW /Hz.

Governor time constant of each area ($T_{S1} = T_{S2} = T_{S3}$) = 0.08 s; Turbine time constant of each area ($T_{t1} = T_{t2} = T_{t3}$) = 0.3 s.

Power system gain of each area ($K_{P1} = K_{P2} = K_{P3}$) = 120 Hz/pu MW; Power system time constant of each area ($T_{P1} = T_{P2} = T_{P3}$) = 18 s.

Coefficient of synchronization of each area ($T_{12} = T_{23} = T_{13}$) = 20.06.

References

1. Hahn H, Krautkremer B, Hartmann K, Wachendorf M (2014) Review of concepts for a demand-driven biogas supply for flexible power generation. Renew Sustain Energy Rev 29:383–393
2. Mishra D, Sahu PC, Prusty RC, Panda S (2022) Power generation monitoring of a hybrid power system with I-GWO designed trapezoidal type-II fuzzy controller. Int J Model Simul 42(5):797–813
3. Sahu PC, Prusty RC, Panda S (2021) Active power management in wind/solar farm integrated hybrid power system with AI based 3DOF-FOPID approach. Energy Sources, Part A: Recovery, Util Environ Effects 1–21
4. Reymov KM, Rafikova GR, Nematov LA, Esemuratova S (2020) Existing condition and prospects of making power balance and managing load of electric consumers in Uzbek power system. In: E3S web of conferences, vol 209. EDP Sciences, p 07015
5. Sahu PC, Mohapatra S, Sahoo S, Sahu BK, Debnath MK (2021) Intelligent frequency control of an AC microgrid with HSA designed tilt controller. In: 2021 international conference in advances in power, signal, and information technology (APSIT). IEEE, pp 1–5
6. Sahu PC, Prusty RC (2019) Stability analysis in RECS integrated multi-area AGC system with modified-SOS optimized fuzzy controller. Recent Adv Electr Electron Eng (Former Recent Patents Electr Electron Eng) 12(6):532–542
7. Mahapatra AK, Samal P, Mohapatra S, Sahu PC, Panda S (2021) Analysis of Gaussian fuzzy logic-sliding model control and flexible AC transmission systems controllers for automatic generation control of hybrid power system under chaotic-water cycle algorithm approach. Int Trans Electr Energy Syst 31(12):e13163
8. Sahu PC, Sahoo S, Prusty RC, Sahu BK (2021) Binary dragonfly algorithm-designed fuzzy cascade controller for AGC of multi-area power system with nonlinearities. In: Green technology for smart city and society. Springer, Singapore, pp 445–456
9. Jena NK, Sahoo S, Sahu PC, Sahu BK, Mohanty KB (2021) Automatic generation control of a renewable energy integrated power system in the presence of fractional order 3DOF controller. In: 2021 1st Odisha international conference on electrical power engineering, communication and computing technology (ODICON). IEEE, pp 1–6
10. Sahu PC, Prusty RC, Sahoo BK (2020) Modified sine cosine algorithm-based fuzzy-aided PID controller for automatic generation control of multiarea power systems. Soft Comput 1–18
11. Begum B, Sahu PC, Sahu BK (2021) Comparative performance investigation of fractional order and conventional PID controller implemented for frequency stability. In: Green technology for smart city and society. Springer, Singapore, pp 487–496
12. Sahu PC, Prusty RC, Panda S (2020) Optimal design of a robust FO-multistage controller for the frequency awareness of an islanded AC microgrid under i-SCA algorithm. Int J Ambient Energy 1–13
13. Sahu PC, Prusty RC, Panda S (2020) Frequency regulation of an electric vehicle operated micro grid under WOA tuned fuzzy cascade controller. Int J Ambient Energy 1–18
14. Prusty UC, Sahu PC, Prusty RC, Panda S (2022) TEG Integrated hybrid power system for frequency regulation study under fuzzy controller approach. In: 2022 IEEE 2nd international symposium on sustainable energy, signal processing and cyber security (iSSSC). IEEE, pp 1–6
15. Mohapatra S, Sahu PC, Agrawal R, Prusty RC, Patel NC (2020) Superiority of JAYA optimized 2DOF-IDD control strategy for AGC of hybrid power system. In: 2020 international conference on computational intelligence for smart power system and sustainable energy (CISPSSE). IEEE, pp 1–6
16. Sahu PC, Prusty RC, Panda S (2017) MFO algorithm based fuzzy-PID controller in automatic generation control of multi-area system. In: 2017 international conference on circuit, power and computing technologies (ICCPCT). IEEE, pp 1–6
17. Deb S, Abdelminaam DS, Said M, Houssein EH (2021) Recent methodology-based gradient-based optimizer for economic load dispatch problem. IEEE Access 9:44322–44338
18. Yu J, Kim CH, Rhee SB (2020) Clustering cuckoo search optimization for economic load dispatch problem. Neural Comput Appl 32(22):16951–16969

19. Mishra D, Mohapatra S, Sahu PC, Prusty RC, Baral KK (2021) Combined AGC and ELD analysis of a multi area power system with WCA based PID controller. In: 2021 international conference in advances in power, signal, and information technology (APSIT). IEEE, pp 1–6
20. Baral KK, Sahu PC, Barisal AK, Mohanty B (2022) Combined analysis on AGC and ELD of a hybrid power system with D-WCA designed Gaussian type-2 fuzzy controller. Evol Syst 1–18
21. Sahu PC, Prusty RC (2020) Frequency and tie-line power awareness in eco-AGC of multi-area power system with SSO-based fractional order controller. Int J Power Energy Convers 11(2):200–221
22. Sahu PC, Mishra S, Prusty RC, Panda S (2018) Improved-salp swarm optimized type-II fuzzy controller in load frequency control of multi area islanded AC microgrid. Sustain Energy Grids Netw 16:380–392
23. Sahu PC, Prusty RC, Panda S (2019) A gray wolf optimized FPD plus (1+ PI) multistage controller for AGC of multisource non-linear power system. World J Eng
24. Sahu PC, Prusty RC, Panda S (2020) Approaching hybridized GWO-SCA based type-II fuzzy controller in AGC of diverse energy source multi area power system. J King Saud Univ Eng Sci 32(3):186–197
25. Sahu PC, Bhoi SK, Jena NK, Sahu BK, Prusty RC (2021) A robust multi verse optimized fuzzy aided tilt controller for AGC of hybrid power system. In: 2021 1st Odisha international conference on electrical power engineering, communication and computing technology (ODICON). IEEE, pp 1–5
26. Sahu PC, Prusty RC, Panda S (2020) Improved-GWO designed FO based type-II fuzzy controller for frequency awareness of an AC microgrid under plug in electric vehicle. J Ambient Intell Humaniz Comput 1–18
27. Sahu PC, Baliarsingh R, Prusty RC, Panda S (2020) Automatic generation control of diverse energy source-based multiarea power system under deep Q-network-based fuzzy-T2 controller. Energy Sources Part A: Recovery Util Environ Effects 1–22
28. Mohapatra S, Sahu PC, Baral KK, Bhoi SK, Prusty RC (2021) Performance stability study of a hybrid power system under levy flight CSA based fuzzy controller. In: 2021 international conference in advances in power, signal, and information technology (APSIT). IEEE, pp 1–6
29. Sahu PC, Baliarsingh R, Prusty RC, Panda S (2022) Novel DQN optimised tilt fuzzy cascade controller for frequency stability of a tidal energy-based AC microgrid. Int J Ambient Energy 43(1):3587–3599
30. Kim M, Chun H, Kim J, Kim K, Yu J, Kim T, Han S (2019) Data-efficient parameter identification of electrochemical lithium-ion battery model using deep Bayesian harmony search. Appl Energy 254:113644
31. Bhatta SK, Mohapatra S, Sahu PC, Swain SC, Panda S (2021) Load frequency control of a diverse energy source integrated hybrid power system with a novel hybridized harmony search-random search algorithm designed Fuzzy-3D controller. Energy Sources Part A Recovery Util Environ Effects 1–22
32. Prasad S, Purwar S, Kishor N (2017) Non-linear sliding mode load frequency control in multi-area power system. Control Eng Pract 61:81–92
33. Prasad S (2020) Counteractive control against cyber-attack uncertainties on frequency regulation in the power system. IET Cyber-Phys Syst Theory Appl 5(4):394–408

Comparison of Perturb and Observe and Incremental Conductance MPPT Method in Matlab/Simulink

Tanuj Monu and Shelly Vadhera

Abstract For maximizing power output of a solar photovoltaic system, some power tracking mechanism should be adapted in the system so that our operating point should be near around maximum power point. A study of two most prominent method of maximum power point tracking (mppt) is presented in this paper. These two methods are Perturb and Observe method and Incremental Conductance method. These two methods have been used to maximize output power of the solar pv system. Simulations have been done in matlab and simulink environment by providing same temperature and irradiation level to both method.

Keywords Incremental conductance · Perturb and observe · Matlab/Simulink · Solar pv system

1 Introduction

Solar energy contributes to a major portion of renewable energy [1]. This is due to the fact that solar energy is easily available in most part and it is environmentally friendly. Also solar energy helps in reducing greenhouse gases and effect of global warming [2]. Also its installation and maintenance is quite easy as compared to other renewable sources. For converting solar energy direct into electrical energy, PV cells are used which works on principle of photoelectric effect. The pv arrays generate DC power which can be converted to AC power with the help of power inverters. MPPT is a technique which is used to operate the pv modules in such a way so that maximum output energy from system can be extracted [3]. This MPPT can be achieved by controlling output current/voltage [4]. This method can be useful under extreme weather condition like cloudy, rainy and cold weather.

T. Monu (✉) · S. Vadhera
National Institute of Technology, Kurukshetra, Haryana, India
e-mail: tanuj.kashyap1998@gmail.com; tanuj_32114304@nitkkr.ac.in

S. Vadhera
e-mail: shelly_vadhera@nitkkr.ac.in

S. Kumar et al. (eds.), *Control Applications in Modern Power Systems*, Lecture Notes in Electrical Engineering 1128, https://doi.org/10.1007/978-981-99-9054-2_19

Operating point of the pv system is defined as intersection of I-V curves between pv array and the load. In general scenario, pv arrays do not operate on their maximum power point (MPP). MPP varies mainly due to different level of solar irradiation and temperature and generates fluctuating power output. To overcome this problem maximum power point tracking (MPPT) methods are used. This MPPT technique constantly maintains the operating point of solar pv array near to its MPP. MPP trackers are used to maximize power which is being transferred to load from pv arrays. MPPT tracks maximum power point considering various environmental constraints such as temperature and solar irradiance [5]. At least there are 19 mppt algorithm which can be implemented in various ways to optimize the output energy performance of solar pv systems [6].

MPPT system is installed between pv system and load. The solar pv system is connected with arrangement of converter and battery system. This converter can be either buck/boost or both. This converter system receives control signal from the MPPT controller which continuously measures voltage and current of solar pv module. Many researchers has worked on this topic. Gow [7] has made a generalized simulation of pv model on PSPICE platform. El-Shatter [8] has calculated maximum power using fuzzy logic control unit for a solar pv model. Tsai [9] has also implemented a model of insulation oriented solar pv model in matlab/simulink environment.

Two of the most commonly applied MPPT algorithm include incremental conductance method and perturb and observe (P&O) method [10]. Despite of some drawback like oscillations in power of system in perturb and observe method, it is the most generally used mppt technique as its implementation is quite easy and its cost effectiveness [11] (Fig. 1).

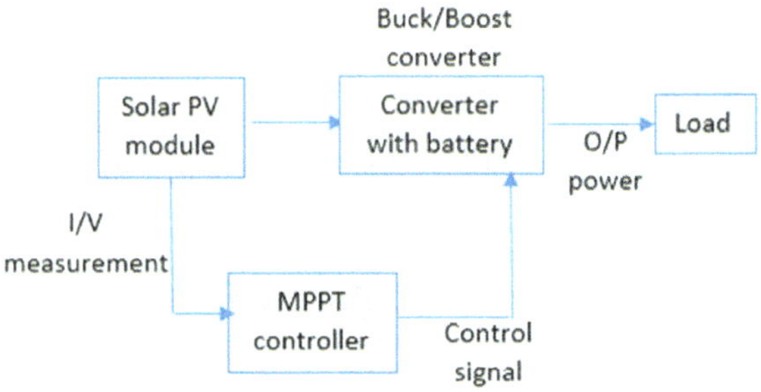

Fig. 1 Solar pv system having mppt controller

2 Methodology

2.1 Incremental Conductance Method

This method basically uses two sensors which continuously monitors the output current and voltage of pv modules simultaneously. The converter duty cycle is directly controlled by the output pv power [12]. The power derivative with respect to voltage curve is considered in this incremental conductance method. If the slope of power voltage curve is considered then.

$\frac{dP}{dV} = 0$, at maximum power point(mpp)

$\frac{dP}{dV} > 0$, at left side of mpp and $V < V_{mpp}$

$\frac{dP}{dV} < 0$, at right side of mpp and $V > V_{mpp}$

The maximum power point is tracked by making a comparison between incremental conductance and instantaneous conductance [13].

$$\frac{dP}{dV} = \frac{d(VI)}{dV} = I + V\frac{dI}{dV} \tag{1}$$

MPP of the system is observed when this derivative becomes zero.

$$\frac{dP}{dV} = 0$$

$$\mathrm{I} + \mathrm{V}\frac{dI}{dV} = 0 \Rightarrow \frac{dI}{dV} = -\frac{I}{V} \tag{2}$$

where this dI/dV is defined as incremental conductance and I/V is defined as instantaneous conductance. On comparing these two, a relation can be defined for maximum power point in power voltage curve [14].

$\frac{dI}{dV} =-\frac{I}{V}$, at mpp

$\frac{dI}{dV} >-\frac{I}{V}$ left of mpp point

$\frac{dI}{dV} <-\frac{I}{V}$, right of mpp point.

Flowchart of incremental conductance method can be stated as (Fig. 2).

2.2 Perturb and Observe Method

This method is the most commonly used MPPT method. In this method, the operating voltage or current of pv array is modified until maximum power is obtained from it. The output power of solar pv array is constantly monitored and is compared with previous value, and based on this there is increment or decrement in current or voltage until the maximum power point is obtained [15]. The system perturbs current

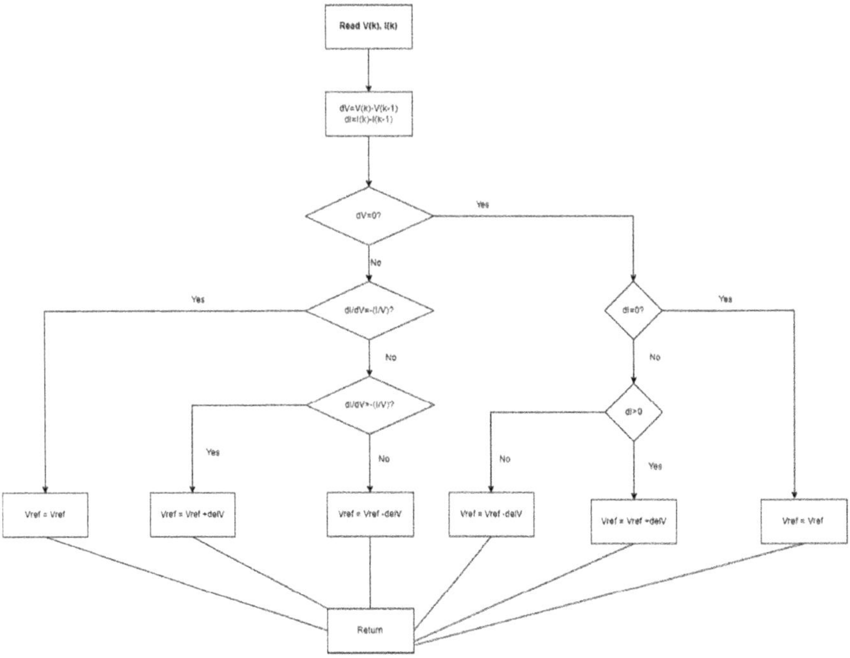

Fig. 2 Flowchart of incremental conductance method

or voltage based on the sign of derivative of power. If dP/dV > 0 and this sign of derivative matches with previous value of comparison then this perturb and observe algorithm will make the system move in the same specific direction by increasing or decreasing current/voltage whatever it has applied in previous observation. If dP/dV comes out to be opposite of previous observation, then it means that maximum power point has been crossed and this algorithm will reverse direction of perturb voltage or current. Although this method is very easy to apply but it has some drawback like power loss at maximum power point due to constant oscillation and stability issue when there is a rapid change in irradiance or temperature levels [16] (Fig. 3).

3 Model of MPPT Based Solar pv System in Simulink

A general model of MPPT is developed in which pv system is connected with a resistive load with the help of a dc-dc boost converter. MATLAB function block is used which consists the code of Perturb and Observe method and incremental conductance method. Irradiation and temperature are inputs of solar pv system. Irradiation is taken as 1000W/m^2 and standard temperature is considered as 25°C. The solar pv array is taken of Suntech Power of each module of 230Wp and arranged in combination of 10 series model and 47 parallel strings (Figs. 4, 5 and 6, Table 1).

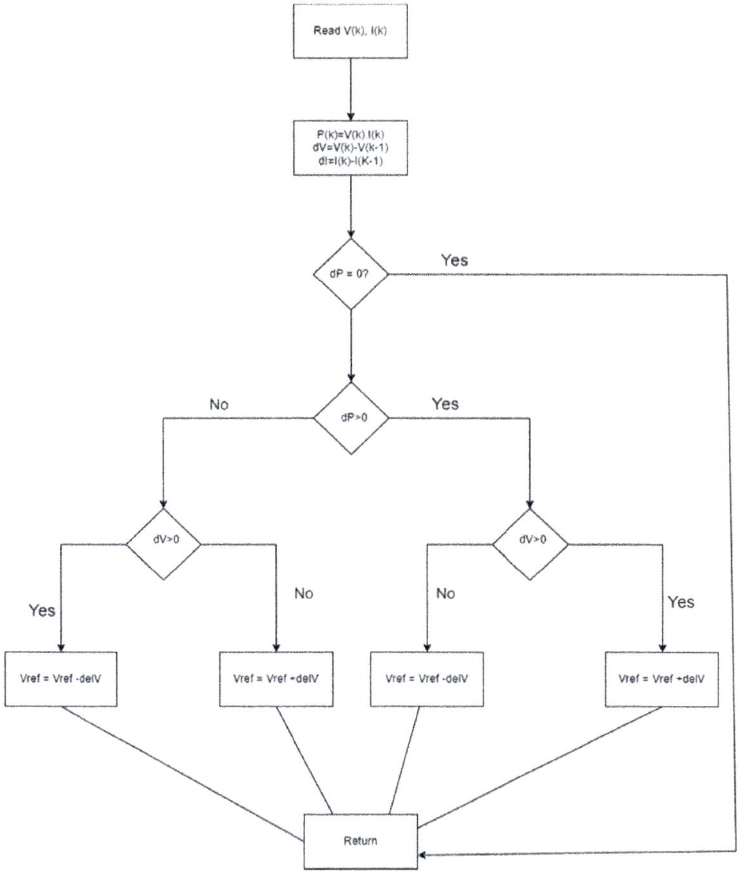

Fig. 3 Flowchart of perturb and observe method

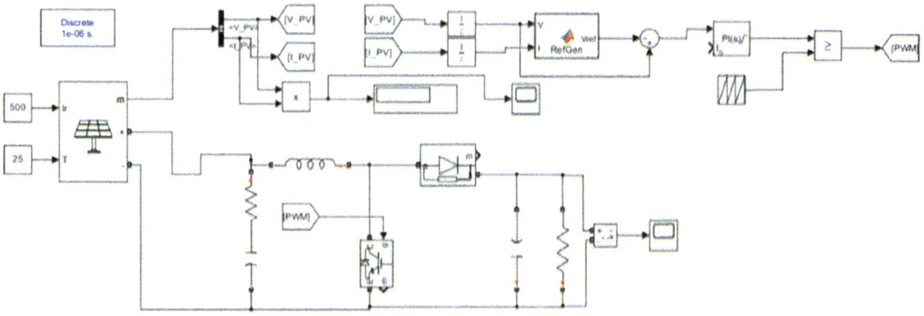

Fig. 4 Model of pv system in matlab/simulink

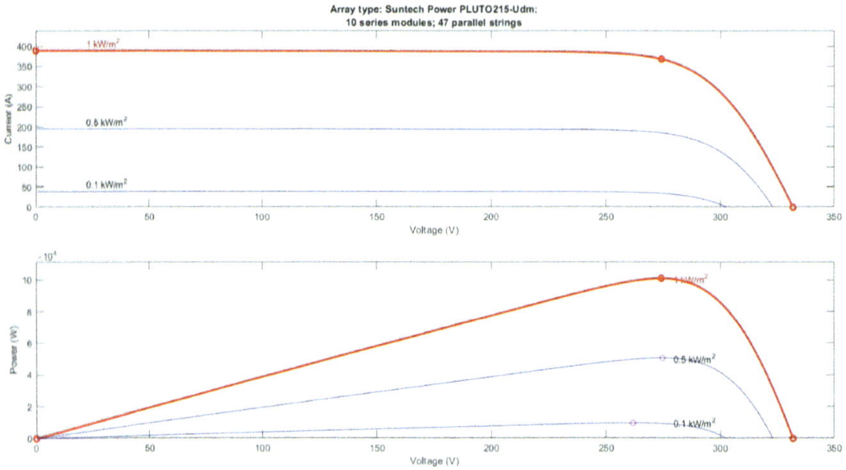

Fig. 5 Variation of pv array current versus voltage and power versus voltage at different irradiance level and constant temperature

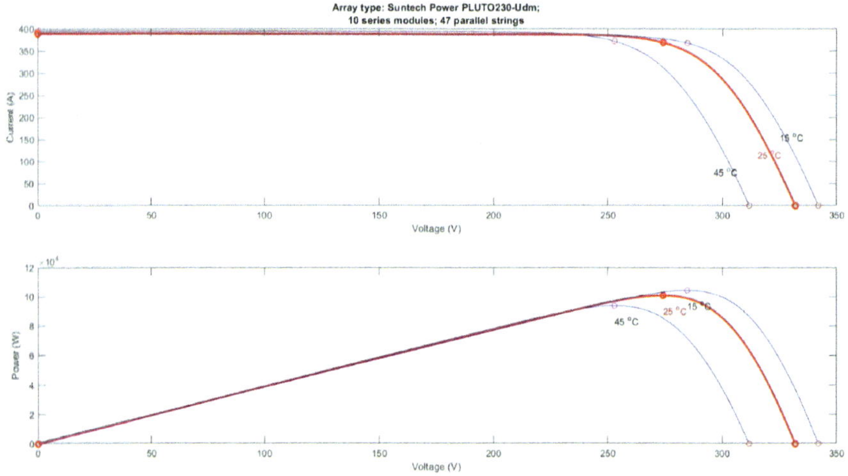

Fig. 6 Variation of pv array current versus voltage and power versus voltage at different temperature level and constant irradiation

Table 1 Specification parameters of boost converter

Boost converter specification		
1	Series inductance(L)	1.45mH
2	Capacitance	3.227mF
3	Load resistance	2 ohms

Fig. 7 Output results of both mppt method

4 Results

The simulation has been performed for a standard irradiation level of 1000W/m^2 and a temperature of 25 °C. On the x coordinate simulation time is taken and on the y coordinate output power is taken. A separate subsystem has been created for both the method and then a common scope has been taken to monitor the output power of both method. The yellow line represents the incremental conductance method while the blue line represents perturb and observe method. The results of simulation of both incremental conductance method and perturb and observe method are shown in same scope for same input parameters. The incremental conductance method is found to achieve a better result. Perturb and observe method is found to give large oscillations as compared to that of other one. Whenever there is sudden change in irradiation or temperature, the perturb and observe method is found to give more oscillations and it settles down in more time as compared to incremental conductance method (Fig. 7).

5 Conclusion

A comparison between incremental conductance method and perturb and observe method, which are two of the most commonly used MPPT method is presented in this paper. Matlab/Simulink environment has been used to create the model, which consists of solar pv module, boost converter and other required controller for MPPT. The results have been obtained for both method for step input and by considering the same irradiation level and temperature. The results for incremental conductance method are found to be better than perturb and observe method in term of power obtained.

References

1. Ibrahim O, Yahaya NZ, Saad N, Umar MW (2015) Matlab/simulink model of solar PV array with perturb and observe MPPT for maximising PV array efficiency. In: 2015 IEEE conference on energy conversion (CENCON). Johor Bahru, Malaysia, pp 254–258. https://doi.org/10.1109/CENCON.2015.7409549.
2. Abu-Rub H, Ellabban O, Blaabjerg F (2014) Renewable energy resources: current status, future prospects & their enabling technology. Renew Sustain Energy Rev 39:748–764
3. Rekioua D, Matagne E (2012) Optimization of photovoltaic power systems, modelization, simulation and control. Springer
4. Camacho EF, Berenguel M, Rubio FR, Martínez D (2012) Control of solar energy systems. In: Advances in industrial control. Springer, p 52
5. Jena PK, Mohapatra A, Srikanth, Choudhary P (2016) Comparative study of solar PV MPPT by perturbation and observation and fuzzy method. In: 2016 IEEE Uttar Pradesh section international conference on electrical, computer and electronics engineering (UPCON). Varanasi, India, pp 515–518. https://doi.org/10.1109/UPCON.2016.7894707
6. Banu IV, Beniugă R, Istrate M (2013) Comparative analysis of the perturb-and-observe and incremental conductance MPPT methods. In: 2013 8th International symposium on advanced topics in electrical engineering (ATEE). Bucharest, Romania, pp 1–4. https://doi.org/10.1109/ATEE.2013.6563483
7. Gow JA, Manning CD (1999) Development of a photovoltaic array model for use in power-electronics simulation studies. IEE Proc-Electr Power Appl 146:193–200
8. El-Shatter TF, Eskander MN, El-Hagry MT (2006) Energy flow and management of a hybrid wind/PV/fuel cell generation system. Energy Convers Manage 47:1264–1280
9. Tsai HL (2010) Insolation-oriented model of photovoltaic module using Matlab/Simulink. Sol Energy 84:1318–1326
10. Yu T-C, Chien T-S (2009) Analysis and simulation of characteristics and maximum power point tracking for photovoltaic systems. In: International conference on power electronics and drive systems. pp 1339–1344
11. Kumar M, Kachhwaya M, Kumar B (2016) Development of MATLAB/Simulink based model of PV system with MPPT. In: 2016 7th India international conference on power electronics (IICPE). Patiala, India, pp 1–4. https://doi.org/10.1109/IICPE.2016.8079336
12. Kumar H, Tripathi RK (2012) Simulation of variable incremental conductance method with direct control method using boost converter. In: 2012 students conference on engineering and systems. Allahabad, India, pp 1–5. https://doi.org/10.1109/SCES.2012.6199050
13. Esram T, Chapman PL (2007) Comparison of photovoltaic array maximum power point tracking techniques. IEEE Trans Energy Convers 22(2):439–449
14. Teulings WJA, Marpinard JC, Capel A (1993) A maximum power point tracker for a regulated power bus. In: Power electronics specialists conference, PESC '93 record, 24th annual IEEE. pp 833–838
15. Villalva MG, Ruppert F E, Gazoli JR (2009) Analysis and simulation of the P&O MPPT algorithm using a linearized FV array model. In: Industrial electronics, 2009. IECON '09. 35th annual conference of IEEE. pp 231–236
16. Sera D, Kerekes T, Teodorescu R, Blaabjerg F (2006) Improved MPPT algorithms for rapidly changing environmental conditions. In: Power electronics and motion control conference. pp 1614–1619

Intelligent Modelling of Smart Grid by Fractal Analysis: Towards Scale Invariant Networking

Swati Devabhaktuni

Abstract Deep learning approaches are supposed to improve prediction accuracy by being stochastic and allowing bi-directional connections between neurons as a development of neural network-based prediction methods. This paper is an effort to model smart grid through pragmatic approach focusing on fractal analysis as a potential tool to mathematically formulate and assess the dataset and come with a solution which makes development of smart grids scale invariant. Additionally, this paper analyses solar energy generation in India and creates a mechanism to predict it with little uncertainty because solar-based power generation is a significant alternative for developing nations like India that can increase energy security, decrease global warming, and improve supply. All in all, this paper takes you through demand side management and also reducing uncertainty in renewable energy generation and help assist engineers in scaling devices to existing networks.

Keywords Deep learning · Solar energy · Fractal approach · Smart grid · Energy management

1 Introduction

Things are more intricate than they appear to us.' Here I make use of this very statement and have developed a mechanism which enables us to add devices, sources of energy to the grid (preferably renewable) and make it scale invariant. As the electric grid transitions to a sustainable, efficient and adaptable network, more and more complicated procedures are required. Although urbanization and electrification trends point to a rise in global electrical consumption as well as the use of renewable sources of energy over time, these trends show that total energy demand will increase in the future.

A comprehensive review of the literature mostly on load forecasting reveals a number of unsolved research problems. The most crucial and challenging goal of

S. Devabhaktuni (✉)
National Institute of Technology, Warangal, India
e-mail: swatikjm@nitw.ac.in

S. Kumar et al. (eds.), *Control Applications in Modern Power Systems*, Lecture Notes in Electrical Engineering 1128, https://doi.org/10.1007/978-981-99-9054-2_20

providing a ground-breaking energy prediction technique is to attain exactness in predicting accuracy. The execution of the established algorithm across the edge nodes, which results in successful communication between connected devices in an Internet of Things (IoT) network for energy utilization, is another significant topic that has not gotten enough attention in the literature [1]. An improved Model Predictive Control (MPC) scheme is created and linked with a hybrid energy storage system in [2] for optimal power dispatch in a smart grid.. A genetic algorithm (GA) is paired with a state space model (SSM) in this study to create MPC control, which seeks the smallest energy exchange between the power grid and the hybrid renewable energy storage system. The most important and difficult task when presenting a new method of energy prediction is to achieve precision in the accuracy of the forecast [3]. One of the hardest problems to solve in an Internet of Things (IoT) network for energy utilization, this significant topic has not received enough attention in the literature. Recently, resource-constrained IoT devices have showed considerable potential in a variety of industries, including video analytics, healthcare, and more [4].

Additionally, while addressing the issue of short-term load forecasting (STLF), a significant concern is the reduced temporal complexity of an energy forecasting system [5]. Anomaly energy demand prediction, which relies on cloud and fog computing paradigms, is underutilized in the literature on energy forecasting, despite their value as reliable platforms for big data analysis and real-time decision making [5].

The main aim of this paper is to develop a fractal approach to find a solution to the smart grid energy management in order to help assist optimal demand side management and enable feedback for renewable energy source generation. The overall objectives met in this paper are:

i. The mathematical foundations of fractal geometry and decipher the fractal dimensions.
ii. The energy prediction techniques, achieving exactness in forecasting accuracy.
iii. Between the grid and a hybrid renewable energy storage system, the genetic algorithm seeks the least amount of energy exchange.
iv. Enabling Tracking of Random Trends by Load Forecasting using Efficient Machine Learning Algorithms.
v. Decipher the purpose behind the use of Deep Learning Algorithms
vi. Recognize the Long Short Term Memory Model and Recurrent Neural Networks, then adapt fresh iterations of these in accordance with the issue.
vii. Work on Extreme dataset with extreme climatic conditions to ensure model's robustness.

2 Demand Side Management

The generation of electrical energy is rapidly being bolstered by the use of renewable energy sources. Renewable energy sources have benefited from countries' efforts to reduce fossil fuel use. New issues arise when wind and solar energy are integrated into power distribution systems, which are becoming more commonplace [6]. Distribution and demand side management are two concepts that play a significant part in the smart grid technologies because of the intertwining of production and consumption, monitoring and control. Flexible, dependable, efficient, and environmentally friendly, smart grids connect producers and consumers in both directions via two-way communication. With an ever-increasing demand for electricity and a rapid advancement in communication technology, power management is necessary. An energy management system's goals include system stability, higher power quality, and energy optimization.

Model Summary:

It's crucial to understand the error metrics utilized in the results before analyzing them. There isn't a single indicator that can be used to determine how accurate (or inaccurate) a forecast is. The only way to find the ideal Key Performance Indicator (KPI) for you is through experimentation. You'll see that every indicator has a certain set of issues. Consider the difference between the predicted and actual demand as the error to start. If the demand forecast is higher than the forecast according to this criterion, the forecast error will be positive. If the demand projection is lower than the actual demand, the error will be negative.

$$e_t = f_t - d_t \tag{1}$$

So, here e_t is the error.
Now for bias, since it being average error, we have,

$$bias = \frac{1}{n}\sum_{n} e_t \tag{2}$$

'n' stands for the number of previous periods when you have both a forecast and a demand. Due to the fact that a positive error on one item can cancel out a negative error on another, a forecast model can attain extremely low bias while still being imprecise. Naturally, bias alone will not be sufficient to evaluate the accuracy of your prognosis. However, a forecast that is significantly skewed indicates that something is amiss with the model. The Mean Absolute Percentage Error (MAPE), which measures forecast accuracy, is one of the most popular KPIs. MAPE is the total of all absolute errors divided by the demand (each period separately). It's the sum of all percentage errors.

$$MAPE = \frac{1}{n}\sum \frac{|e_t|}{d_t} \tag{3}$$

The Mean Absolute Error (MAE) is also a metric for determining forecast accuracy. It is the mean of the absolute error, as the name implies.

$$MAE = \frac{1}{n}\sum |e_t| \tag{4}$$

$$\%MAE = \frac{\frac{1}{n}\sum |e_t|}{\frac{1}{n}\sum d_t} = \frac{\sum |e_t|}{\sum d_t} \tag{5}$$

The Root Mean Squared Error (RMSE) is an unusual KPI but one that is extremely useful, as we will see later. The square root of the average squared error is what it's called.

$$RMSE = \sqrt{\frac{1}{n}\sum e_t^2} \tag{6}$$

$$\%RMSE = \sqrt{\frac{\frac{1}{n}\sum e_t^2}{\sum d_n}} \tag{7}$$

Many techniques are based on the Mean Squared Error (MSE), which is closely connected to the Root Mean Square Error (RMSE) and is especially useful for machine learning. Since MSE is quicker to compute and manipulate than RMSE, it is frequently employed in algorithms. The error is squared, therefore it is not scaled to the original error, leading to a KPI that we are unable to link to the initial demand scale. Therefore, we won't use it to evaluate our statistical forecasting models.

$$MSE = \frac{1}{n}\sum e_t^2 \tag{8}$$

Implementation Process in Brief for Demand Side Management:

The flow diagram in Fig. 1 summarizes the implementation process followed for demand side management using LSTM Networks. There are several datasets available, but to make a model robust, we need to choose datasets that are under extreme conditions to test model's robustness and compatibility. Choosing right loss function is extremely important to reach the consistent result. Setting parameters and hyper parameter tuning is also required to tune the model for better performance. Thus, in next chapter we shall see the model results and summary and draw inference from the same.

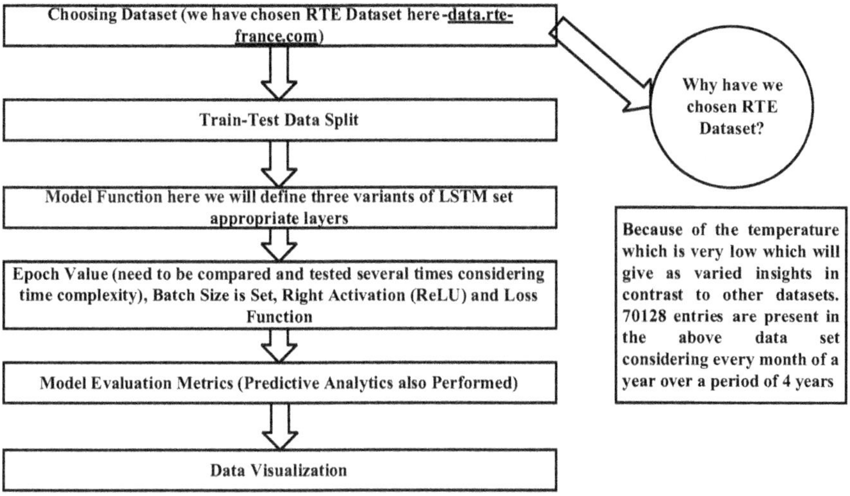

Fig. 1 Flow diagram

3 Experimental Results and Analysis

Now let's infer the results obtained by the three models.

a **Experimental Results for LSTM Model-1**

i **Model Summary**:

(See Fig. 2).

Figure 3 shows the plot results for the LSTM Model-1 with respect to actual values.

```
Layer (type)                 Output Shape              Param #
=================================================================
lstm_2 (LSTM)                (None, 60)                15840

dense_2 (Dense)              (None, 1)                 61

=================================================================
Total params: 15,901
Trainable params: 15,901
Non-trainable params: 0
_________________________________________________________________
None
Compilation Time w.r.t start :  0.012639284133911133
```

Fig. 2 Shows the model summary for the LSTM Model-1

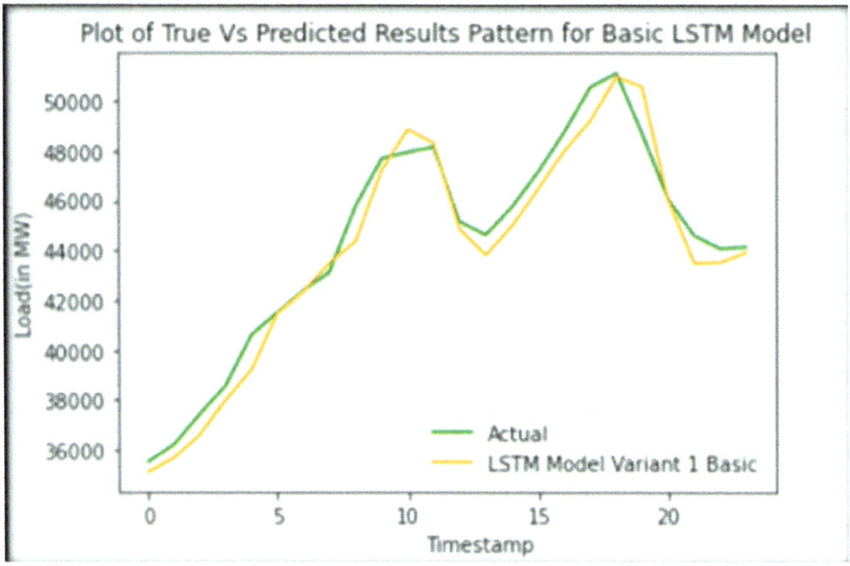

Fig. 3 Plot results

ii. **Plot Results (True versus Predicted)**

iii. **Errors**

Figure 4 displays the errors obtained for LSTM Model-1; the type of errors has already been explained in the introduction of this section.

Note: The above model testing and simulation were performed on Jupyter Notebook. Coding was done in Python (Epochs = 120, Batch Size = 1024).

b **Experimental Results for LSTM Model-2**

i **Model Summary**:

Figure 5 shows the model summary for the LSTM Model-2.

Figure 6 shows the plot results for the LSTM Model-2 with respect to actual values.

```
Finally Saved our model to disk
MAPE1(Test): 5.283
MSE1(Test): 562.926
```

Fig. 4 Errors

```
Layer (type)                 Output Shape              Param #
=================================================================
lstm_14 (LSTM)               (None, 5, 60)             15840

lstm_15 (LSTM)               (None, 60)                29040

dense_6 (Dense)              (None, 1)                 61

=================================================================
Total params: 44,941
Trainable params: 44,941
Non-trainable params: 0
```

Fig. 5 Model summary for LSTM-2

ii. **Plot Results (True versus Predicted)**
iii. **Errors**

Figure 7 displays the errors obtained for LSTM Model-2; the type of errors has already been explained in the introduction.

Note: The above model testing and simulation were performed on Jupyter Notebook. Coding was done in Python (Epochs = 70, Batch Size = 1024).

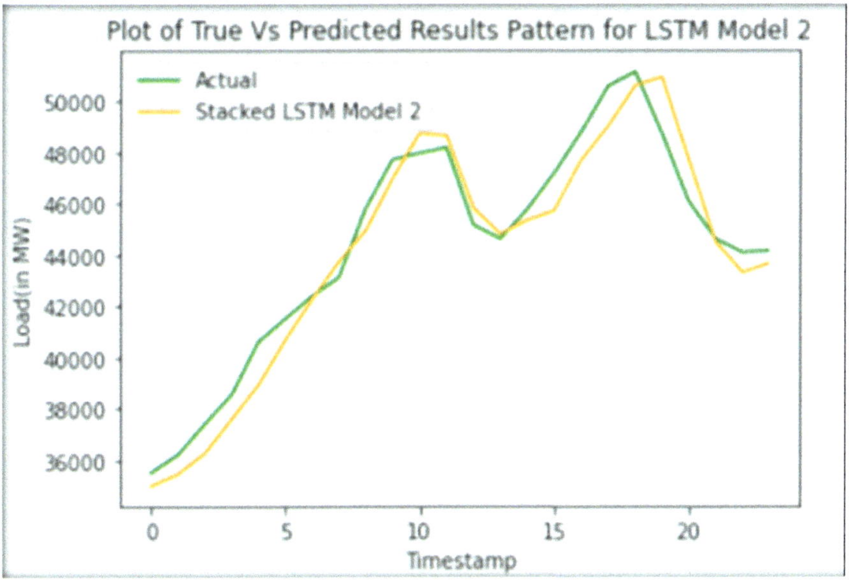

Fig. 6 Plot results

```
Finally Saved our model to disk
MAPE2(Test): 5.304
MSE2(Test): 671.746
```

Fig. 7 Errors

c **Experimental Results for LSTM Model-3**

i **Model Summary**

Figure 8 shows the model summary for the LSTM Model-3.

ii **Plot Results (True versus Predicted) and Errors**

Figure 9 displays the errors obtained for LSTM Model-3; the type of errors has already been explained in the introduction of this section.

Layer (type)	Output Shape	Param #
lstm (LSTM)	(None, 5, 128)	68608
dropout (Dropout)	(None, 5, 128)	0
lstm_1 (LSTM)	(None, 64)	49408
dropout_1 (Dropout)	(None, 64)	0
dense (Dense)	(None, 16)	1040
dense_1 (Dense)	(None, 1)	17

Total params: 119,073
Trainable params: 119,073
Non-trainable params: 0

Fig. 8 Model summary for LSTM-3

```
Finally Saved our model to disk
MAPE3(Test): 5.300
MSE3(Test): 580.472
```

Fig. 9 Errors

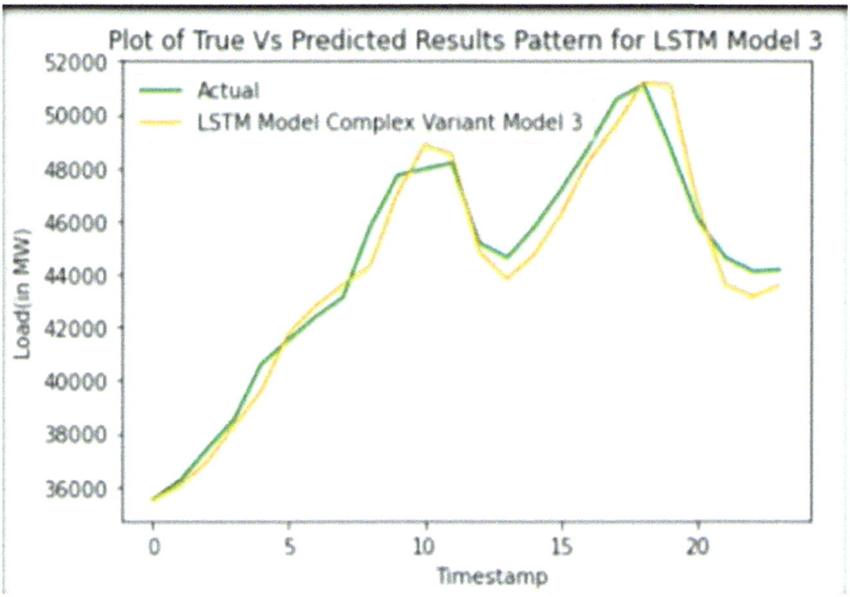

Fig. 10 Plot results

Figure 10 shows the plot results for the LSTM Model-3 with respect to actual values.

Note: The above model testing and simulation were performed on Jupyter Notebook. Coding was done in Python (Epochs = 60, Batch Size = 1024).

d. **Inference and Model Compatibility**

Let's first compare the three models and infer the plots simultaneously. Figure 11 shows the comparison plot of the three models.

i **Plot (True vs Predicted for All Three Models**
ii **Errors Obtained**:

Figure 12 shows the errors obtained in the three models developed.

e. **Understanding the Q-Q Plot**

Statistics employs Quantile–quantile (Q-Q) charts to visually examine and contrast two probability distributions by contrasting their quantiles. The points on the Q-Q plot will precisely sit on a straight line y = x if the two distributions we're comparing are exactly equal. The type of distribution for a random variable, such as a Gaussian, Uniform, Exponential, or even Pareto distribution, can be identified using Q-Q plots. It is possible to identify the sort of distribution represented by the Q-Q plot only by looking at it. We're talking about normal distributions in general because we have a lovely concept called the 68–95–99.7 rule that fits

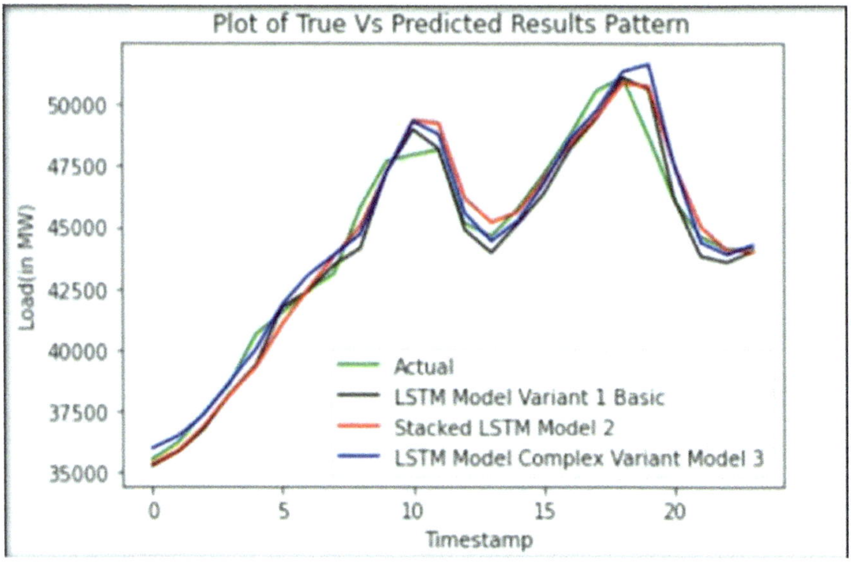

Fig. 11 Plot results

Fig. 12 Errors

MAPE1(Test): 5.312
MAPE2(Test): 5.287
MAPE3(Test): 5.271
MSE1(Test): 599.571
MSE2(Test): 562.857
MSE3(Test): 705.637

well into the normal distribution. So, we know how much of the data falls inside the first, second, and third standard deviations from the mean. Knowing whether a distribution is Normal allows us to readily experiment with the data. Second, Normal Distributions appear frequently in most large-scale natural phenomena. Because it is evenly aligned with the typical normal variate, or Q-Q plot, we may state categorically that this distribution is normally distributed if all of the points represented on the graph fall on a straight line.

Figure 13 shows the Q-Q Plot for our model and thus as the points are closer to the line y = x, we can safely say that the distribution is normal. R squared value of 0.985 shows a high coefficient of determination. The best fit line crosses

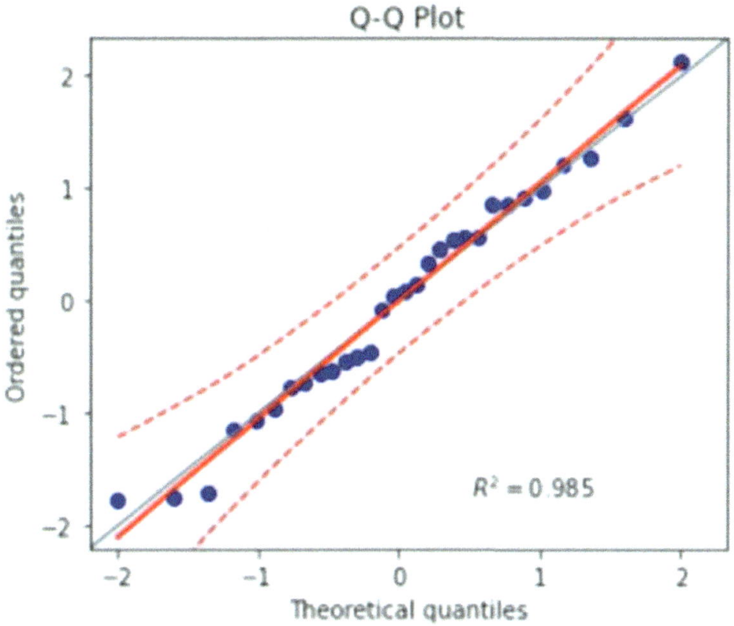

Fig. 13 Q-Q plot

over many of the data points in the model if the r-squared value is high and thus it proves our model's compatibility.

4 Conclusion

When it comes to energy management, it is also very important to understand and detect drifts that help us to solve issues pertaining to sudden requirement of demand and help us prepare for sudden fluctuations and enable control over scaling. Performing energy disaggregation is one technique to quantify building adaptability in real time. Flexibility is the capacity to increase or decrease building energy consumption at a particular point in time. Disaggregation is the separation of appliance-level energy consumption signals from an aggregated energy consumption signal for the entire building or family. Only this aggregated signal is generally made available to the grid operator via the smart meter technology due to end user privacy concerns. In LSTMs, the concept of cell states allows the gradient to easily go backward in time, making it more resistant to the vanishing gradient problem. Defining Prediction Intervals enables us to be prepared for minimum and maximum variation that can happen and thus gives better preparedness. This work specifically has covered extreme datasets and thus enabled robust models for demand side management. The concept used in this work is broad and can be applied in diverse areas and thus enlarging the scope to several domains. For the optimization, we have incorporated several variants both for demand side management so as it fit in several use cases and reduce errors. Also, current issue of energy management has been understood especially with respect to scaling and thus modelling has been done intelligently to cater the growing demand. The verification methods adopted have been very instrumental in making the model relevant and robust.

References

1. Han T, Muhammad K, Hussain T, Lloret J, Baik SW (2021) An efficient deep learning framework for intelligent energy management in IoT networks. IEEE Internet Things J 8(5):3170–3179. https://doi.org/10.1109/JIOT.2020.3013306
2. Chen X, Cao W, Zhang Q, Hu S, Zhang J (2020) Artificial intelligence aided model predictive control for a grid-tied wind-hydrogen-fuel cell system. IEEE Access 8:92418–92430
3. Freire VA, De Arruda LVR, Bordons C, Márquez JJ (2020) Optimal demand response management of a residential microgrid using model predictive control. IEEE Access 8:228264–228276. https://doi.org/10.1109/ACCESS.2020.3045459
4. Kong W, Dong ZY, Jia Y, Hill DJ, Xu Y, Zhang Y (2019) Short-term residential load forecasting based on LSTM recurrent neural network. IEEE Trans Smart Grid 10(1):841–851. https://doi.org/10.1109/TSG.2017
5. Mamun AA, Sohel M, Mohammad N, Haque Sunny MS, Dipta DR, Hossain E (2020) A comprehensive review of the load forecasting techniques using single and hybrid predictive models. IEEE Access 8:134911–134939. https://doi.org/10.1109/ACCESS.2020.3010702
6. Akhtar I, Kirmani S, Jameel M, Alam F (2021) Feasibility analysis of solar technology implementation in restructured power sector with reduced carbon footprints. IEEE Access 9:30306–30320. https://doi.org/10.1109/ACCESS.2021.3059297M. J. Duran and F.

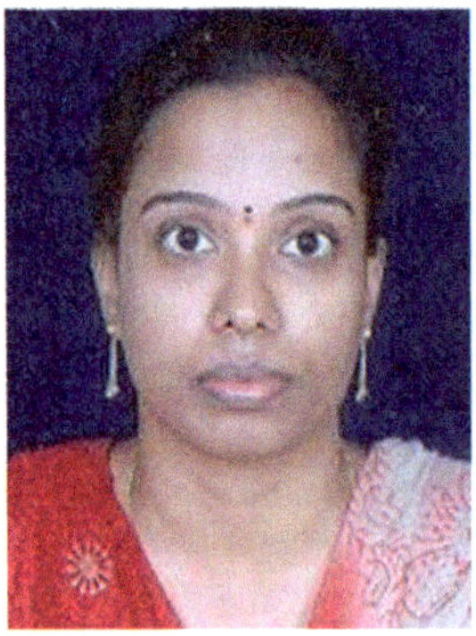

Swati Devabhaktuni received a Ph.D. degree in electrical engineering from the Jawaharlal Nehru Technological University, Hyderabad in 2014. Currently, she is working as an Assistant professor at the National Institute of Technology, Warangal. Her research interests are power electronics, AC motor drives, and control systems.

Battery Monitoring System for Electric Vehicles

P. Sobha Rani, B. Jyothi, S. K. Nagoor, and S. Meghanadh

Abstract The selection of the battery system has a significant impact on the state of electric vehicle (EV) technology. To access a battery's functional utilization, its capacity and the method utilized to determine its operating time is to be known. Based on load requirements, cell voltage, current, temperature readings, as well as anticipated battery state of charge, capacity, and impedance, battery monitoring system decides how efficiently to charge and discharge batteries. In this paper, a battery monitoring system is proposed to continuously track performance in comparison to the estimated battery capacity. This study suggests using the Raspberry Pi-Pico, a processor that can explore computation and access programming through interfaces, to implement a battery monitoring system. It can be combined with the framework for monitoring data, which can manage, watch, and log the data into an online database and send to user that can be monitored through the display (or android app).

Keywords Battery management system (BMS) · Raspberry Pi-Pico · Android app

1 Introduction

Electric Vehicles (EV) are an alternative solution for vehicles which run with gas/diesel/petrol. The advantages being less pollution, energy saving and environmental safety [1]. Electric vehicles are classified as Plug-in electric vehicles, Hybrid Electric vehicles and battery electric vehicles. In EV, cells are connected in series/parallel to provide high voltage or high capacity battery packs. Lithium ion batteries are used in electric vehicles at present. They are rechargeable and have high energy efficiency, best high temperature performance, low self-discharge compared to nickel–cadmium or lead-acid batteries. Charging of batteries in minimum time is the main issue in electric vehicles. There are various methods of EV charging like constant

P. Sobha Rani (✉) · B. Jyothi · S. K. Nagoor · S. Meghanadh
Lakireddy Bali Reddy College of Engineering (Autonomous), Jawaharlal Nehru Technological University, Mylavaram, Kakinada (JNTUK), Andhra Pradesh, India
e-mail: sobhareveru@gmail.com

S. Kumar et al. (eds.), *Control Applications in Modern Power Systems*, Lecture Notes in Electrical Engineering 1128, https://doi.org/10.1007/978-981-99-9054-2_21

current, constant voltage, constant power, constant current constant voltage and multi charging [2]. Normally charging takes 8 h' time from a 7 kW charger [2].

Battery models helps to design state of charge (SoC), state of health (SoH) of batteries [3]. The SoC and SoH of battery are defined as per Eqs. (1) and (2)

$$SoC = \frac{Remaining\ capacity\ of\ battery}{Total\ capacity\ of\ battery} \tag{1}$$

$$SoH = \frac{Maximum\ charging\ capacity\ of\ battery}{Total\ capacity\ of\ battery} \tag{2}$$

Battery models in general are classified as physical, thermal, electrochemical and equivalent electrical circuit models [3]. The equivalent electrical circuit models (EECM) consists of electrical elements resistors, voltage source and capacitors. There are various EECM models available in literature. EECM models are basically classified as Impedance based EECM and voltage-current based EECM models. The battery VI relationship empirically is defined as per Eq. (3) [4]:

$$v_k = E_0 - RXi_k - \frac{\mu}{(1 - SoC_K)Xi_k} \tag{3}$$

Battery management system (BMS) facilitates safe and reliable operation of batteries [5]. BMS senses individual cell voltage, current and temperature [1]. The main function of BMS is to provide (i) over voltage protection, (ii) over current protection (iii) estimation of state of charge and state of health of each cell, (iv) cell balancing.

The main parts of electric vehicle is shown in Fig. 1 [6]. Battery management system (BMS) is bi-directionally connected to battery and electric control unit. It monitors the speed of vehicle, battery capacity, (SoC) and battery health [7]. It can notify the user regarding over charging or over discharging of battery there by preventing fire accidents. BMS also monitor the speed of vehicle and gives appropriate signal to electric control unit. BMS plays a significant role in increasing the life time of battery and performance of electric vehicle.

A well-designed BMS helps protect the battery from damage, extends the battery's lifespan, and ensures that the vehicle has sufficient range for its intended use.

Due to advanced technology in wireless communication, BMS can notify the battery status to the users or manufacturers remotely also [8]. The technologies useful in BMS are global system for mobile communication (GSM), Zigbee, Android and Blue tooth communication [9]. In [10], Arduino based BMS was implemented and Google Apps script was used to monitor battery details data base. In [11], a passive control method was used for differential control of lithium–ion battery cells. In [12], IoT based BMS was implemented to notify the battery performance. This system consisted of micro controller which receives data from voltage, current, temperature sensors and sends the data to the cloud like Thingspeak. An alert mail could be sent if data goes beyond threshold level.

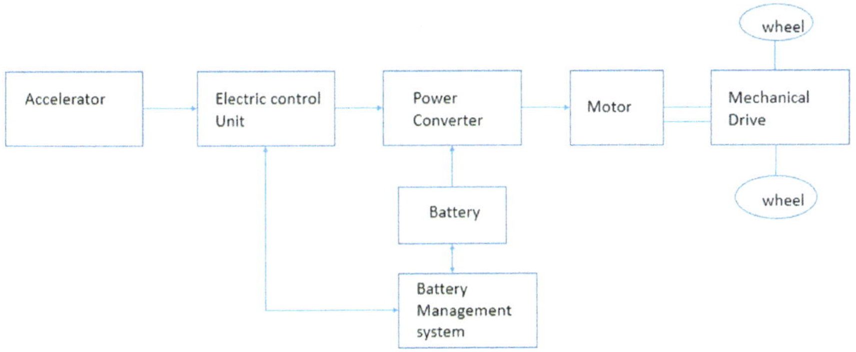

Fig. 1 Block diagram of electric vehicle

In this paper Raspberry Pi–Pico controller based BMS is proposed for monitoring data, which can manage, watch, and log the data into an online database and send to user and can be monitor through the display (or android app). Raspberry Pi–Pico is an easy to use device which consumes less power. It supports micro python /C^{++} programming language.

2 Proposed System

The proposed system is represented in block diagram as shown in Fig. 2 consists of:

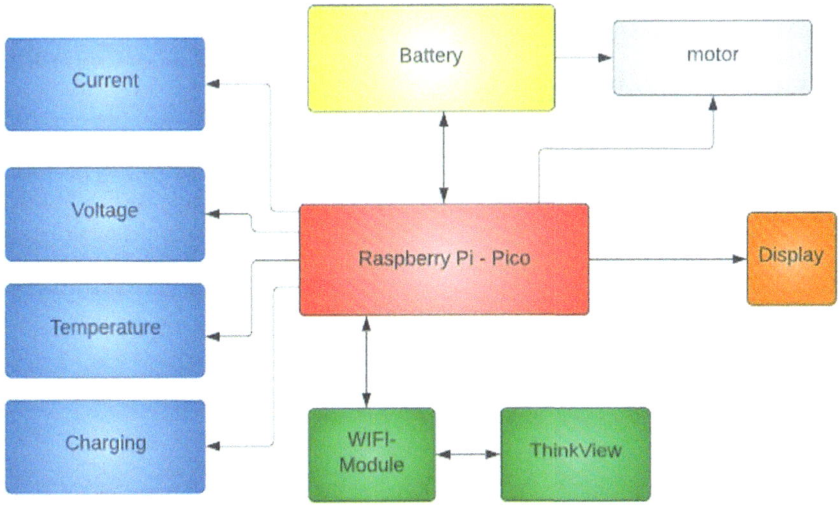

Fig. 2 Block diagram

- Power Supply Unit.
- Raspberry Pi–Pico.
- WIFI- Module.
- LCD Display.
- DC Motor.
- Battery.
- Current Sensor.
- Voltage Sensor.
- Temperature Sensor.
- Switch.

The block diagram consists of a power supply connected to the micro controller (Raspberry Pi- Pico) which can operate in with the micro-Python code. The controller uses around 38 mA and can be capable of delivering a minimum of 1.8 V and maximum of 5.5 V. A battery pack with USB to micro-USB cable can also power the controller. The controller is connected to the 16X2 LCD display for output reading, and a Wi-Fi module is inserted into the Raspberry Pi, which can give information of battery state via IoT with an android app known as ThinkView and monitor the voltage, current, charging, discharging, and temperature parameters to stabilize the risks that may occur during the operation of the electric vehicle with an estimation to the user.

Hardware configuration of Raspberry Pi- Pico controller is shown in Fig. 3. It has a total of 40 pins arranged in a 2×20 grid. These pins are split into different types of functions, including General-Purpose Input / Output (GPIO), Analog-to-Digital Converter (ADC), Pulse Width Modulation (PWM), and communication interfaces. The Pico's pins provide a wide range of options for interfacing with sensors, actuators, and other components, making it an adaptable and powerful microcontroller board.

3 Software Implementation

Thonny Ide software is used to accumulate the required parameters with a raspberry pi-pico, and micro-Python is used to implement the programming, taking advantage of its robustness, versatility, and easy implementation. Following installation and coding, the application is dumped to the Pico using a USB connection linked to the system and the Pico board's source pin. As a result, the software installation is complete, and the application is successfully saved in the Raspberry pi-Pico. Snapshot of coding is shown in Fig. 4.

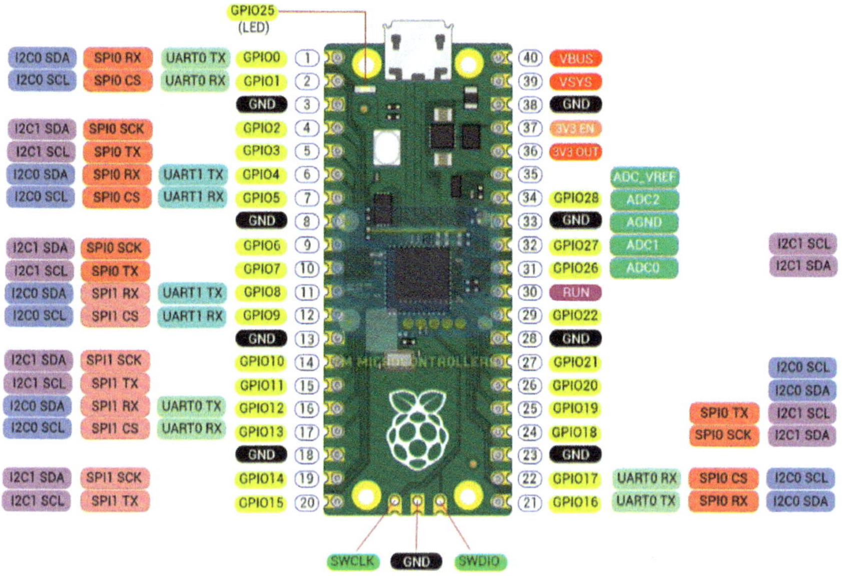

Fig. 3 Hardware configuration

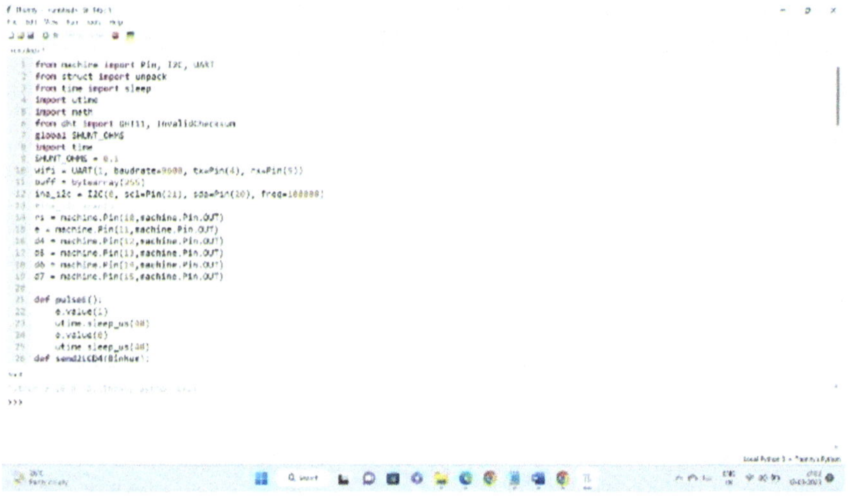

Fig. 4 Snapshot of codind with Raspberry pi-Pico

Fig. 5 Hardware setup

4 Results

Figure 5 depicts the prototype model of the proposed battery management system. The data monitoring such as the nominal voltage of battery and the current, further power developed is estimated during the operation. The BMS also depicts the Charging Time\present percentage of charge available and the temperature over the system when connected in the vehicle.

Battery current, voltage, power, charging, discharging, and temperature are monitored and transmitted to the server's primary database. The app can subsequently be used to check the battery's health and charge level after receiving these details.

The software results are obtained in the Thing-view app which are connected to the server through ESP8266 and Raspberry pi-Pico. The results show the individual parameters of battery under different time periods. The graphs as shown in Fig. 6.9, 6.10, 6.11, 6.12, and 6.13 represent voltage, current, charging, temperature and power readings (Figs. 6, 7, 8, 9 and 10).

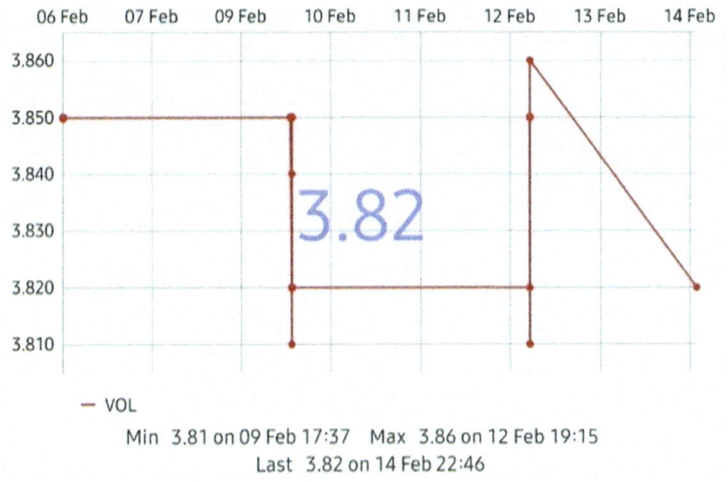

Fig. 6 Voltage readings

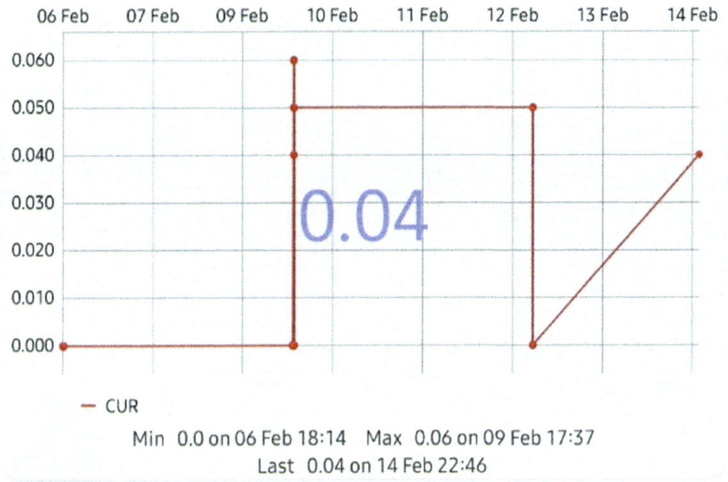

Fig. 7 Current readings

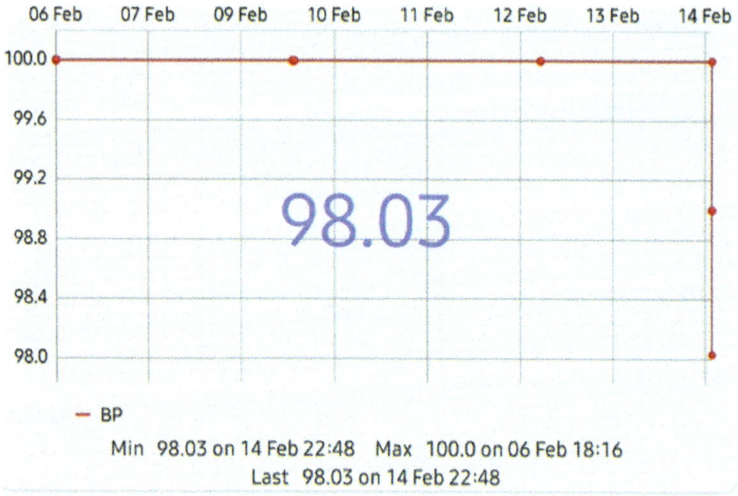

Fig. 8 Charging

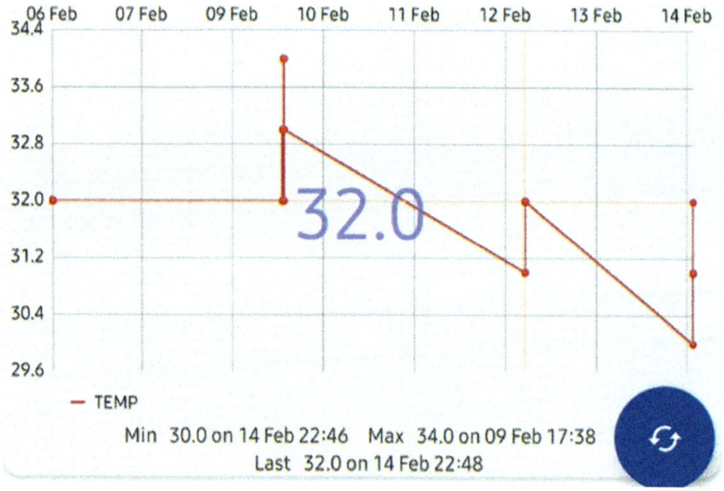

Fig. 9 Temperature reading

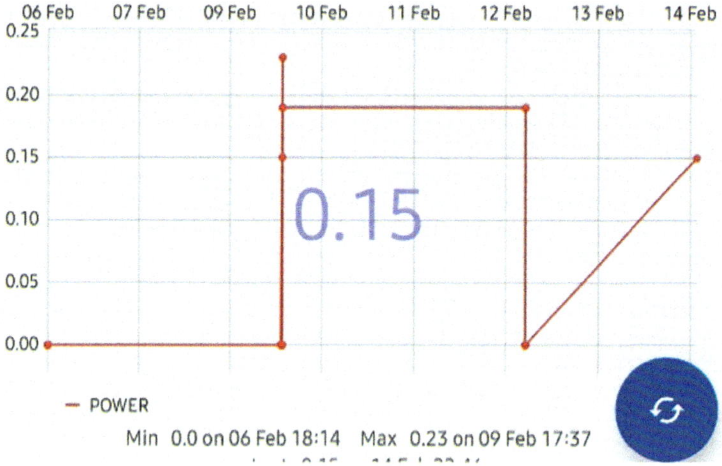

Fig. 10 Power reading

5 Conclusion

This battery monitoring system can be used to keep track of the current condition of the batteries in electric automobiles such as cars, trucks, and bikes. By collecting data from batteries and other similar sources in one location for analysis and applications, it allows for advancement. To predict and enhance battery life, this can also be applied to businesses and houses that use huge batteries for backup power and other commercial applications. Knowing the life significantly earlier would make it possible to replace the components promptly without interfering with a person's or an industry's regular operations.

References

1. Mathew SA, Prakash R, John PC (2012) A smart wireless battery monitoring system for electric vehicles. In: 12th international conference on intelligent systems design and applications (ISDA). Kochi, India, pp 189–193
2. Pareek S, Sujil A, Ratra S, Kumar R (2020) Electric vehicle charging station challenges and opportunities: a future perspective. In: 2020 international conference on emerging trends in communication, control and computing (ICONC3). Lakshmangarh, India, pp 1–6
3. Singirikonda S, Obulesu YP (2020) Battery modelling and state of charge estimation methods for energy management in electric vehicle-a review. In: IOP conference series, Material science Engineering, vol 937
4. Zhang C, Li K, Mcloone S, Yang Z (2014) Battery modelling methods for electric vehicles-a review. In: 2014 European control conference (ECC). pp 2673–2678
5. Liu W, Placke T, Chau KT (2022) Overview of batteries and battery management for electric vehicles. Energy Rep (8), 4058–4084

6. Ramkumar MS, Subba Rami Reddy C, Ramakrishnan A, Raja K, Pushpa S, Jose S, Jayakumar M (2022) Review on Li-Ion battery with battery management system in electrical vehicle. Adv Mater Sci Eng (2022):1–8
7. Vaideeswaran V, Bhuvanesh S, Devasena M (2019) Battery management systems for electric vehicles using lithium ion batteries. In: 2019 innovations in power and advanced computing technologies (i-PACT). Vellore, India, pp 1−9
8. Le Gall G, Montavont N, Papadopoulos GZ (2022) IoT network management within the electric vehicle battery management system. J Sign Process Syst 94:27–44
9. Mishra S, Swain SC, Samantaray RK (2021) A review on battery management system and its application in electric vehicle. In: 2021 international conference on advances in computing and communications (ICACC). Kochi, Kakkanad, India, pp 1–6
10. Senthilkumar M, Suresh KP, Sekar TG, Pazhanimuthu C (2021) Efficient battery monitoring system for E-Vehicles. In: 2021 7th international conference on advanced computing and communication systems (ICACCS). Coimbatore, India, pp 833–836
11. Kivrak S, Özer T, Oğuz Y (2018) Battery management system implementation with pasive control method. In: 2018 IV international conference on information technologies in engineering education (Inforino). Moscow, Russia, pp 1–4
12. Krishnakumar S et al (2022) IoT-based battery management system for E-Vehicles. In: 2022 3rd international conference on smart electronics and communication (ICOSEC). Trichy, India, pp 428–434

On the Improved Storage Capacity in Power Generation Using Piezoelectric Sensors

M. Keerthana, E. Gopi Chand, B. Sai Prasad, and Ambar Bajpai

Abstract At present, the whole world is sustained because of energy, which is derived from natural resources. As the population is increasing day by day, the energy resources are depleting which produce electricity. Therefore, in order to get electrical energy for our existence, we can make use of the energy waste of human locomotion. By using piezoelectric sensors, we can convert this kinetic energy of human locomotion into electricity. To implement this system, we have used six piezoelectric sensors, a voltage booster, an Arduino Nano board, I2C converter. In this paper, we have discussed a detailed description and processing of the developed prototype. The performance and analysis of the proposed system is presented in a table. The power that can be transferred to the rural villages is shown with the LED indication. This paper gives the analysis of the obtained output voltages from the footstep power-generation model and shows the functionality of the piezoelectric model.

Keywords Electricity · Force · Piezoelectric sensor · Pressure · Voltage booster

1 Introduction

Natural resources produce electricity and are depleting day by day, and there is need for alternative energy resources and in the rural areas, which do not have access to electricity need to be accessed with electricity. Power that is used for our sustainability is generated using natural resources like coal, water, wind, geothermal energy etc. Renewable energies are reasonable, eco-friendly in nature with various forms like solar, wind, tidal, biomass, and geothermal. In the present situation we are wasting energy in various forms. Therefore, the major solution is to reform this

A. Bajpai (✉)
Department of EECE, GITAM University, Bengaluru, India
e-mail: ambarbajpai@ieee.org

M. Keerthana · E. Gopi Chand · B. Sai Prasad
Department of ECE, Atria Institute of Technology, Bangalore, India

S. Kumar et al. (eds.), *Control Applications in Modern Power Systems*, Lecture Notes in Electrical Engineering 1128, https://doi.org/10.1007/978-981-99-9054-2_22

energy wastage into usable form. Present industries also developing many gadgets, electronic devices that consumes power. As technology increases day by day the gadgets or electronic devices that run on electricity also increases. Therefore, there is a necessity for an alternative power generation method. One of the alternative power generative method is to extract the power from human locomotion that is converting kinetic energy into electrical energy. This can be achieved by using piezoelectric sensor, which converts pressure into voltage. This piezoelectric model is used to generate voltage using human footstep force. This proposed system is useful to generate power using pressure. This piezoelectric model in large scale can be implemented in public areas such as railway stations, shopping malls, colleges, schools, etc. [1–4].

1.1 Challenges

Some of the cities and villages have several power cuts because of the increasing demand for electricity and lack of resources. One of the methods to supply electricity to rural areas is by extracting energy from human footsteps or from wastage pressure. To design a footstep model using piezoelectric sensors to generate and transfer the electricity to rural areas [5].

1.2 Literature Survey

In the earlier piezoelectric models, the voltage produced is small due to small strain sensing devices. Those models require additional power supply to operate the appliances. Later with advent of piezoelectric tales, the voltage level is increased. With the advent of piezoelectric sensors with in-built spring produces higher voltage levels. This is implemented in areas where locomotion or movement is more and is given to appliances, which need lower voltage to operate. The obtained voltage is less and hence given to nearby appliances. There were many losses in supplying the power from system model to appliances [2, 3, 6].

2 Preliminaries

2.1 Piezoelectricity

Piezoelectricity is defined as the electricity that is generated when mechanical energy is converted into electrical energy by piezo sensors. A piezoelectric sensor consists of quartz, when this quartz is squeezed and stretched it generates a slight electrical potential difference between its ends. Through this squeezing and stretching phenomena voltage is generated, this effect is called piezoelectric effect [3, 5] (Fig. 1).

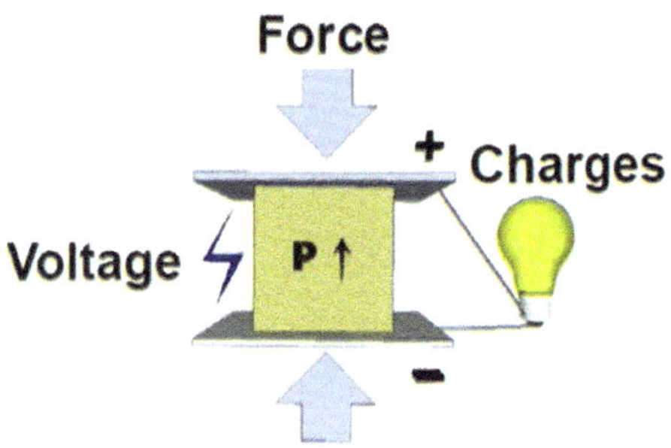

Fig. 1 Piezoelectric effect [3]

Fig. 2 Voltage booster [1]

2.2 Voltage Booster

A voltage booster is a boost converter, a DC-to-DC power converter. Voltage booster is used to step up the voltage that is generated from piezo sensors input to the output. It is helpful in boosting the voltage than the obtained voltage. Voltage booster is used in places where higher voltage is required than the original voltage [1] (Fig. 2).

2.3 Piezoelectric Sensor

A piezoelectric sensor is a device which has both positive and negative terminals. It senses the pressure or force applied on it and converts that force into electricity. It uses the piezoelectric effect principle that generates voltage. Piezo sensors are robust in nature and used in many applications [7] (Fig. 3).

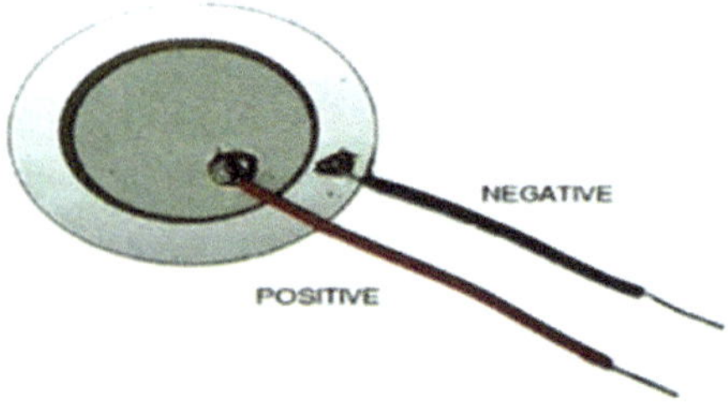

Fig. 3 Piezoelectric sensor [7]

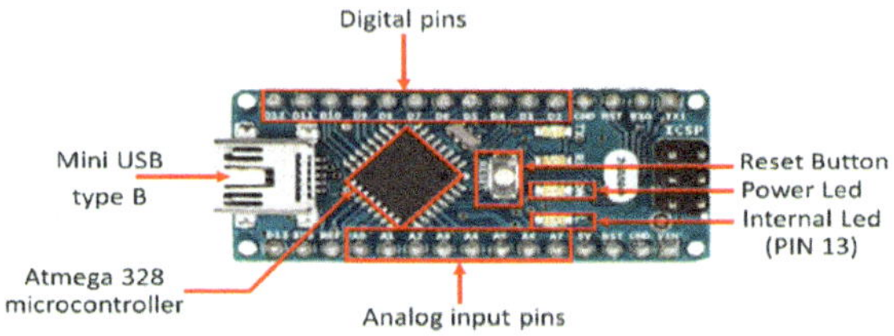

Fig. 4 Arduino nano [7]

2.4 Arduino Nano

The Arduino Nano is type of microcontroller's board, designed by Arduino. This microcontroller is small, breadboard friendly, which is based on the ATmega328P. It is more flexible and smaller as compare to Arduino Uno. The operating voltage of the Nano board is in the range of 5 to 12 V. Arduino Nano has 22 I/O pins, 14 digital pins and 8 Analog pins. The applications of Arduino Nano include robotics, control systems, automations, etc. [7] (Fig. 4).

2.5 I2C Converter

I2C converter is an interface adapter module, which is used for 16 to 2 LCD display. It can make the display easier. This converter uses the PCF8574T ic chip that converts serial data to parallel data for the LCD display [4] (Fig. 5).

Fig. 5 I2C converter [4]

3 Proposed Solution

3.1 Block Diagram

See (Fig. 6).

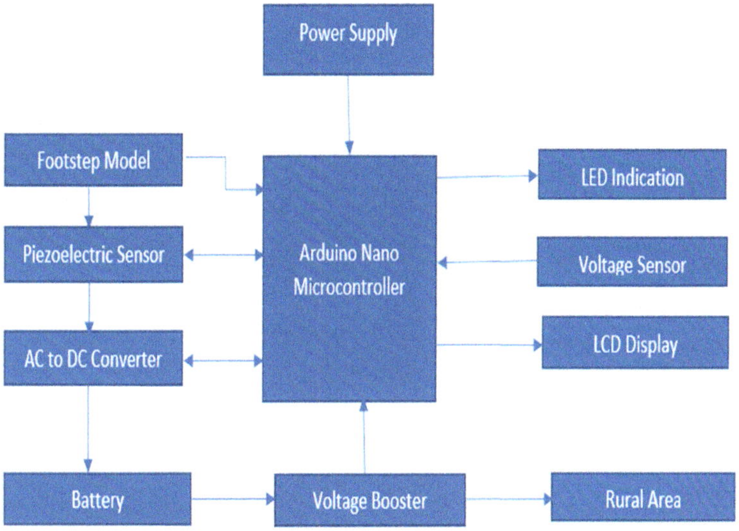

Fig. 6 Proposed block diagram

3.2 Involved Coding

```
#include<Wire.h>
#include<LiquidCrystal_I2C.h>LiquidCrystal_I2C lcd(0*27,16,2);
Const int voltsensor = A0;
float Voltageout = 0.0;
float Voltagein = 0.0;
float S1 = 3000.0;
float S2 = 750.0;
void setup()
{
lcd.init();
//lcd.init();
lcd.backlight();
lcd.print("Measure>25V");
delay(2000);
}
Void loop()
{
value = analogRead(voltageSensor);
Voltageout = (value * 5.0) / 1024.0;
V = Voltageout / (S2/(S1+S2));
//Serial.print("Input =  ");
//Serial.println(Voltagein);
Lcd.setCursor(0,0);
Lcd.print("Input =          ");
Lcd.setCursor (9,0);
Lcd.print(Voltagein);
delay(500);
}
```

3.3 Working Model

This piezoelectric model is based on the principle of piezoelectric effect. When a force is applied on the piezo sensors which is by human locomotion, this pressure is sensed by the piezoelectric sensors. Piezoelectric sensors consist of quartz material that produces potential when it is strained. Thus, piezo sensors produce voltage. This generated voltage is transferred to voltage booster that boosts the original voltage. It acts as a step-up voltage converter. This is further connected with Arduino Nano board. Arduino Nano board is also connected with I2C converter, which is used for LCD display. The final voltage is stored in the batteries, which can store the power for future use. This stored power can be transferred to rural villages for various home applications (Fig. 7).

4 Result and Analysis

The output voltage for our project model is 10 V. After stepping up the voltage using voltage booster it is 12 V. Before the voltage is supplied to LED's, it is stepped down and is stored in the battery for the future use. The table below shows analysis of the project model, Fig. [8] shows the voltage that is stored in the battery, Fig. [9] shows the led indication that is supplied to the rural villages, and LED's glows as per the voltage that is stored in the battery.

In our piezoelectric model, we are able to boost and produce the voltage of about two volts from each piezoelectric sensor. In this project, we used six piezoelectric

Fig. 7 The system model

sensors that is able to produce the voltage of twelve, which is of about 10.36 V in the existing system. In our project, we are able to produce about 1.24 V more than the existing system [6] (Figs. 8 and 9, Table 1).

Fig. 8 Voltage output from the charged battery

Fig. 9 The LED's indication of system model

Table 1 Analysis of system model

Weight (kg)	Time (sec)			
	5 s	10 s	15 s	20 s
45	1.98 V	2.15 V	2.80 V	3.78 V
50	0.83 V	1.23 V	2.38 V	3.12 V
55	1.76 V	2.73 V	4.66 V	5.65 V
60	2.75 V	4.59 V	5.31 V	9.06 V

4.1 Applications

The advantages of power generated through piezoelectric sensors are zero side effect on human body, and it can be easily installed in shopping malls, roadways or areas with frequent locomotion. The output is proportional to number of steps and energy can be easily extracted. The system is environmentally friendly, easy-to-go utility [7–12].

5 Conclusion

This power generation model using piezoelectric sensors is an innovative method to produce electricity. This method of generating electricity is an alternative way that utilizes non-conventional resources, which are used for producing electricity. This model has a voltage booster, which is used for boosting the obtained voltage from piezoelectric sensors. The generated power is displayed on LCD using I2C converter. This proposed system improves the storage capacity of the footstep model with 12 V. This boosted voltage can transfer to rural areas, which do not have access to power. Thus, we can be able to supply electricity to rural areas without depleting natural resources.

References

1. Somalaraju K, Singh JG (2020) Enhancement of power generation from electromagnetic scavenging title. In: 2020 international conference on power electronics & IOT applications in renewable energy and its control (PARC). Uttar Pradesh, India, pp 405–410
2. Hole P, Gophane P (2020) Footstep power generation using Piezo electric sensor. Int Res J Eng Technol 7(3):270–272
3. Ruman MR, Das M, Istiaque Mahmud SM (2019) Human footsteps for energy generation by using piezoelectric tiles. In: Innovation in power and advanced omputing technologies (i-PACT). Vellore, India, pp 1–6
4. Prabaharan R et al (2013) Power harvesting by using human foot step. Int J Innov Res Sci, Eng Technol 2:3001–3009

5. Prasad PR, Bhanuja A, Bhavani L, Bhoomika N, Srinivas B (2019) Power generation through footsteps using piezoelectric sensors along with GPS tracking. In: 2019 4th international conference on recent trends on electronics, information, communication & technology (RTEICT). Bangalore, India, pp 1499–1504. https://doi.org/10.1109/RTEICT46194. 9016865.
6. Patil SS, Sushmita MK, Baliga V, Vishwanath SL (2021) Footstep power generation. Int Res J Eng Technol 8(3):2835–2837
7. https://www.ijraset.com/research-paper/footstep-power-generation-using-piezoelectric-sensor
8. Kamboj A, Haque A, Kumar A, Sharma VK (2017) Design of footstep power generator using piezoelectric sensor. In: 2017 international conference on innovations in information. Embedded and communication systems (ICIIECS). Coimbatore, India, pp 1–3. https://doi.org/10.1109/ICIIECS,8275890
9. Methane NV (2015) Foot step power generation using piezoelectric material. Int J Adv Res Electron Commun Eng (IJARECE) 4(10)
10. Aravind A, Joy J, Sreekalpa S, Sujith S, Rama Krishnan R (2016) Power generation through human locomotion. J Electron Commun Syst 1(1):1–9
11. Shekhar Panda S An investigation on generation of electricity using foot step. Sci J Impact Factor: 3.449 (ISRA), Impact Factor: 1.852. ISSN: 2277–9655
12. Veena RM, Reddy BH, Shyni SM (2016) Maximum energy harvesting from electromagnetic micro generators by footsteps using photo sensor. In: International conference on computation of power, energy information and commuincation (ICCPEIC). Chennai, pp 757–761. https://doi.org/10.1109/ICCPEIC.2016.7557321

Analysis and Comparison of the P&O and INC MPPT Techniques for Solar Energy Systems When Compared to Various Atmospheric Temperatures

Sneha Bharti, Ravikant Kumar, Monika, and U. K. Sinha

Abstract The results of the perturb and observe (P&O) and incremental conductance (INC) MPPT algorithms for an autonomous photovoltaic (PV) system are compared in this research. Under defined temperature, irradiance, and ambient conditions, the INC tracking technique is difficult but better than P&O of PV system. This paper's goal has to provide implementation and performance ratings for the INC and P&O MPPT algorithms. The compare demonstrates that the INC method is more efficient and offers faster tracking speed for the MPPT of a PV array. The MATLAB/ SIMULINK programmer is used to simulate the system.

Keywords Photovoltaic system (PV) · Boost converter · Maximum power point tracking (MPPT) · P & O and INC algorithm

1 Introduction

PV power plants are one of the most efficient, ecologically sound, and clean renewable energy sources available. They offer an outstanding chance to generate electricity. One of the greatest challenges with PV offering is that the output power can be affected by Weather conditions such as irradiance and temperature. Using the maximum power point tracking technique, the efficiency of PV generating can be enhanced [1].

MPPPT methods of various kinds are available, Hill climbing, perturb & observe, INC, and algorithms that use artificial intelligence are examples. The two most frequently used MPPT algorithms are P&O and INC. Because of their ease of implementation and independence from the system, these methods are frequently used. At steady state, the P&O approach has more oscillation around MPP, causing higher power loss in the PV system [2]. To address the issue with the P&O approach and enhance tracking accuracy and dynamic performance under situations that change

S. Bharti (✉) · R. Kumar · Monika · U. K. Sinha
Department, of Electrical Engineering National Institute of Technology, Jamshedpur, India
e-mail: sneha.bharti9999@gmail.com; 2022rsee001@nitjsr.ac.in

S. Kumar et al. (eds.), *Control Applications in Modern Power Systems*, Lecture Notes in Electrical Engineering 1128, https://doi.org/10.1007/978-981-99-9054-2_23

rapidly. The INC method can track rising and declining temperatures with better precision than the P&O method. And it can determine the place of the actual running point with respect to the MPP as well as distance to it. When the MPP has been attained, it can also stop perturbing. We can determine that INC is superior to P&O.

The main objective of this paper is to provide a comparative evaluation of the P&O and INC procedures, and to make an obvious distinction between these two techniques. This paper also provides a detailed analysis of these two techniques to help the industry and researchers in selecting the right tracker for a given application. MPPT approaches differ in terms of algorithm complexity, implementation cost, and sensors utilized. Parameters such as (a) duty cycle, (b) current, and (c) voltage may be employed depending on the technique of control [3].

However, the incremental conductance (INC) at the MPP, which is defined by the slope function of the PV power curve and voltage, is zero. The INC commonly performs iteration in a set Step size. The fixed step size iteration manage has a limit efficiency [4].

To address the shortcomings of traditional MPPT algorithms in terms of speed, stability, and precision. Based on P&O and INC logic, a unique maximum power point tracking controller has been designed. The major objectives of this work are to create a boost converter and a solar panel, as well as to extract the most power from a PV panel through various MPPT algorithms in order to create a more efficient system, and analyses different MPPT techniques applied to a circumstances, a solar PV panel is used. Both the P&O and INC are currently used on commercial PV inverters, debating the widely held idea that higher INC Performance implies better performance.

The performance of these two approaches was compared by varying the amplitudes of the perturbations and the frequency of operation. The results obtained "suggest that the two algorithms are similar" [5], however the author is unable to clarify why the two techniques perform so similarly.

Description of the Structure of the System

The system's block diagram is given in the figure below. (1) Because it is made up of a PV screen, a boost converter, and an MPPT controller, and a load, the power that comes from a single solar cell is relatively low. (1–1.5w).

In order to produce the required power, solar cells can be participated in series to make panels, which can then be connected in parallel or series to form photovoltaic arrays.

The MPP adjustments with the surrounding atmosphere, necessitating the installation of a controller capable of controlling the PV load as close to the MPP as possible (Fig. 1).

$$\frac{dS}{dV_{gpv}} < 0$$

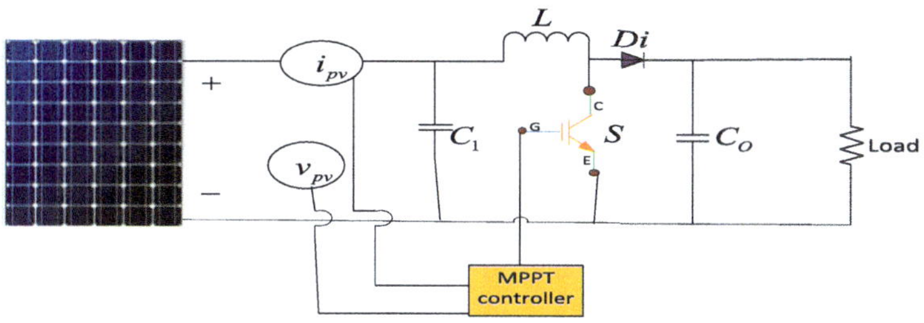

Fig. 1 The diagram schematic for implementing MPPT in a PV system

$$\frac{dS}{dV_{gpv}} < 0 \ (At\ MPP)$$

$\frac{dS}{dV_{gpv}} > 0$ (At left hand side (LHS) of MPP)
$\frac{dS}{dV_{gpv}} < 0$ (At right hand side (RHS) of MPP) Equations.

2 PV Modeling System and its Feature

A PV array is a collection of PV cells that are linked together to generate electricity from a PV module. PV array contains of different PV cells connected in series and parallel due to series connection we get increased voltage and due to parallel connection increased current [6].

The equivalent circuit of a PV array comprises of a diode (d), a current source I_{gout}, series resistance R_{Se} caused by the electrode, and parallel resistance R_{Sh} caused by P–N junction leakage [7] (Fig. 2).

Applying KCL

$$I_{gout} = I_{phirr} - I_{di} - I_{shnt} \tag{1}$$

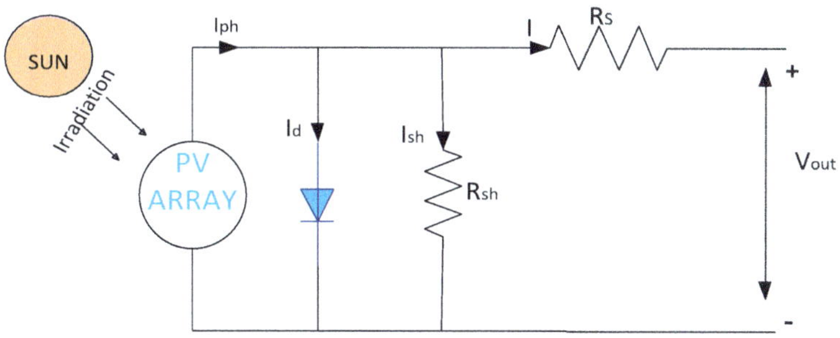

Fig. 2 Equivalent circuit diagram of solar cell

I_{gout} → O/P current of solar cell.
I_{shnt} → photo-electric current generated due to irradiation.
I_{di} → Diode current.
I_{shnt} → Shunt current.
By applying schokely equation. By using KVL

$$I_{di} = I_{st}\left(e^{\left[\frac{q_r(V+IR_S)}{mKT_C}\right]} - 1\right) \tag{2}$$

$$V_{shnt} = V + IR_S \tag{3}$$

$$I_{shnt} = \frac{V + IR_S}{R_{sh}} \tag{4}$$

Putting the value of I_{di} and I_{shnt} in Eq. (1)

$$I_{gout} = I_{phirr} - \left(I_{st}e^{\left[\frac{2(V+IR_S)}{mKT_C}\right]} - 1\right) - \frac{V + IR_S}{R_{sh}} \tag{5}$$

I_{st} → Diode saturation current
$q_r \rightarrow \left(1.6 \times 10^{-19}c\right)$
$K \rightarrow$ Boltzmann const. $\left(1.38 \times 10^{-23}j/k\right)$

$T \rightarrow$ Tem.
So, current at maximum power point given below.

$$I_{\max p} = I_{ph} - \left(e^{\left[\frac{q(V_{\max p}+I_{\max p}R_s)}{mKT_C}\right]} - 1\right) - \frac{\left(V_{\max p} + I_{\max p}\right)}{R_{sh}} \tag{6}$$

Corresponding power generation by PV array

$$P_{main} = V_{\max p}\left\{\left(I_{ph} - I_0(e^{\left[\frac{q(V_{\max p}+I_{\max p}R_S}{mKT_c}\right]} - 1)\right) - \left(\frac{V_{\max p} + I_{\max p}R_s}{R_{sh}}\right)\right\} \tag{7}$$

Where
$I_{\max p}$ → Maximum panel current., $V_{\max p}$ → Maximum panel voltage (Fig. 3).

3 Boost Converter Design and Operation

The basic principle of boost converter is depending on the energy storage element inductor and capacitor. Voltage drop across the rate of change of current flowing through an inductor is directly proportional to it [8].

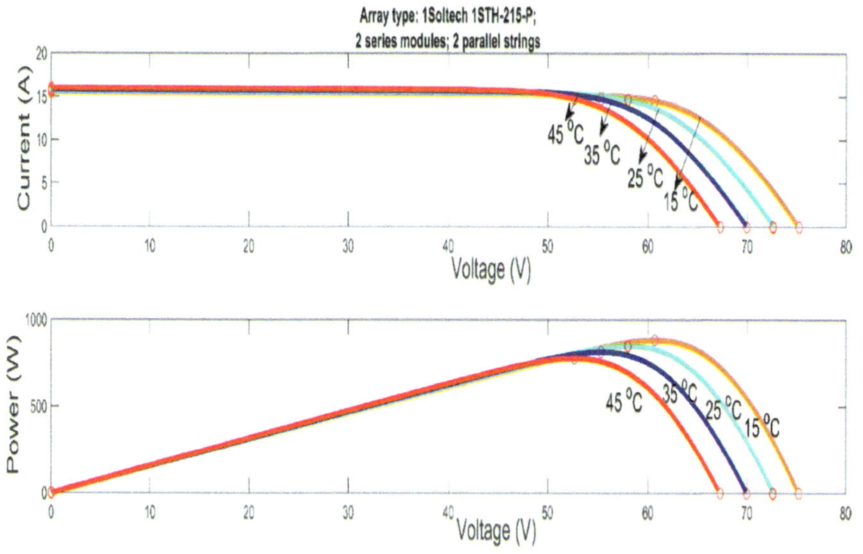

Fig. 3 The current–voltage and power–voltage graphs obtained at various temperature with 1000 W/m^2

In this Fig. 2 IGBT used as switch(s) and connected in series with the source. A diode(d) is used as one more switch(Sw) which acts as a freewheeling diode (d) and provides a discharging path for capacitor capacitors and load (R_{Ld}) are connected in parallel. Inductor reduce the current ripple and capacitors reduce the voltage ripple [9] (Fig. 4).

$$V_{gpv} = I_{gpv}R_{Ld}(1 - d_r)^2 \tag{8}$$

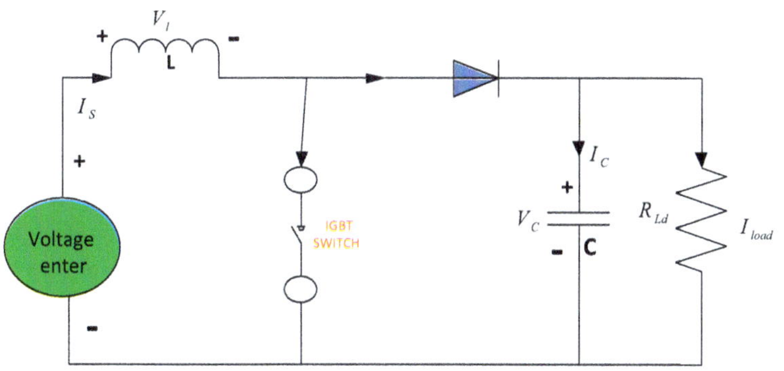

Fig. 4 Equivalent circuit diagram of boost converter

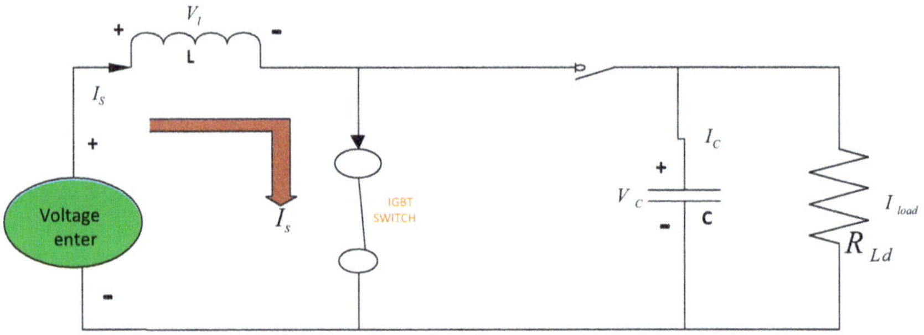

Fig. 5 Boost converter mode 1 operation

where the current and voltage of the PV array are indicated by I_{gpv}& V_{gpv} his load resistance is denoted as R_{Ld}, d_r—Duty ratio.

The boost converter has two ways of operation.

Operation 1 → Sw is On and diode (d) is Off.

Operation 2 → Sw is Off and d is On.

Operation1

In this operation, Sw is in the closed situation, i.e. on state, while diode (d) is in the open situation, i.e. off state, allowing current to flow through switch (Sw.) it. The entire current will travel through the closed loop path. Switch (Sw), inductor L, and back in to the source (Fig. 5).

In this case, the inductor's sign will change depending on the path of current flow. This mode of operation, the diode (d) is reverse biased, which inhibits current from flowing through load in the circuit. We can say that the voltage across the switch S equal to the voltage across the R_{Ld}, resulting in the output voltage.

Operation2

Switch (Sw) is in the open state, and diode (d) is in the on state, in this manner of operation. Current can pass through switch diode (d), but current cannot flow through switch diode Sw [10].

We know that, inductive element in retains energy in the form of a field of magnets, act as source of power when the switch (Sw) is turned on As a result, diode (d) shuts down, and the inductor (L) transfers the energy and collected in the earlier state when Sw was closed The polarity of the inductor changes during release of energy contained in inductors, causing the diode (d) to become forward biased. As a consequence, current can flow through the diode (d) in the circuit.

The release energy gradually dissipates in the load resistance, assisting in keeping the current flowing in the same way through the load while increasing the voltage. Current passing through the inductor is exponentially decreasing and will stop (Fig. 6).

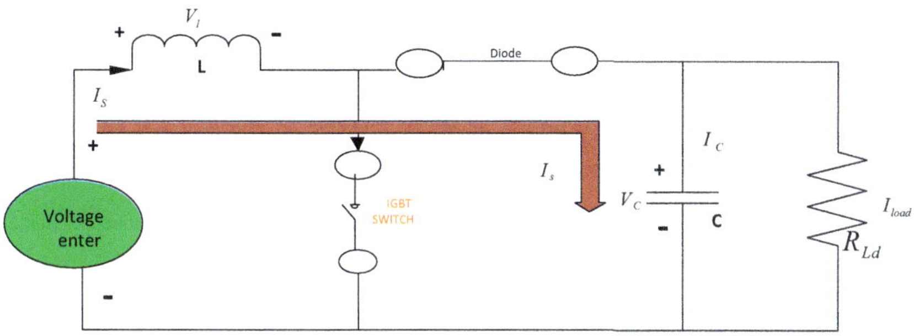

Fig. 6 Boost converter mode 2 operation

MPPT Technique

Because of the non-linear nature of data PV arrays, The MPPT method ensures that the PV array always delivers the maximum available power. P&O, INC, and incremental resistance are the most famous conventional MPPT techniques due to their simple structure and ease of application, In circumstances of swift changes in solar irradiation or load resistance, the reaction of the P&O and INC algorithms with fixed mark size is load, so that P&O and INC algorithm proposal [11, 12].

INC MPPT Approach

The INC algorithm evaluates the instantaneous PV array at the incremental conductance to determine the MPP. This method desires more computation in the controller, but it is faster at monitoring changing conditions than the P&O algorithm of the PV array power v/s voltage curve is zero at the MPP. In this technique make use of the INC $\left(\frac{\Delta I}{\Delta V}\right)$ of the PV array to calculate the sign of the variation in power with respect to voltage $\left(\frac{\Delta P}{\Delta V}\right)$. The INC Algorithm calculate the maximum power by comparison of the incremental conductance with the array conductance [14].

If $\frac{\Delta I}{\Delta V} = \frac{I}{V}$ The output voltage is the MPP voltage.

$\frac{dI}{dV} > \frac{-I}{V}$ On the LHS of MPP, $\frac{dI}{dV} < \frac{-I}{V}$ On the RHS of MPP.

In INC algorithm flowchart measures the current & voltage of PV array (Fig. 7).

P&O MPPT Approach

The P&O algorithm is widely used in the command of algorithmic methods for MPPT for PV energy generation since it is a simple structure, low in cost, and small no. of parameters. The possible to introduce improvement and $d_r = d_r + \Delta d_r$ [15].

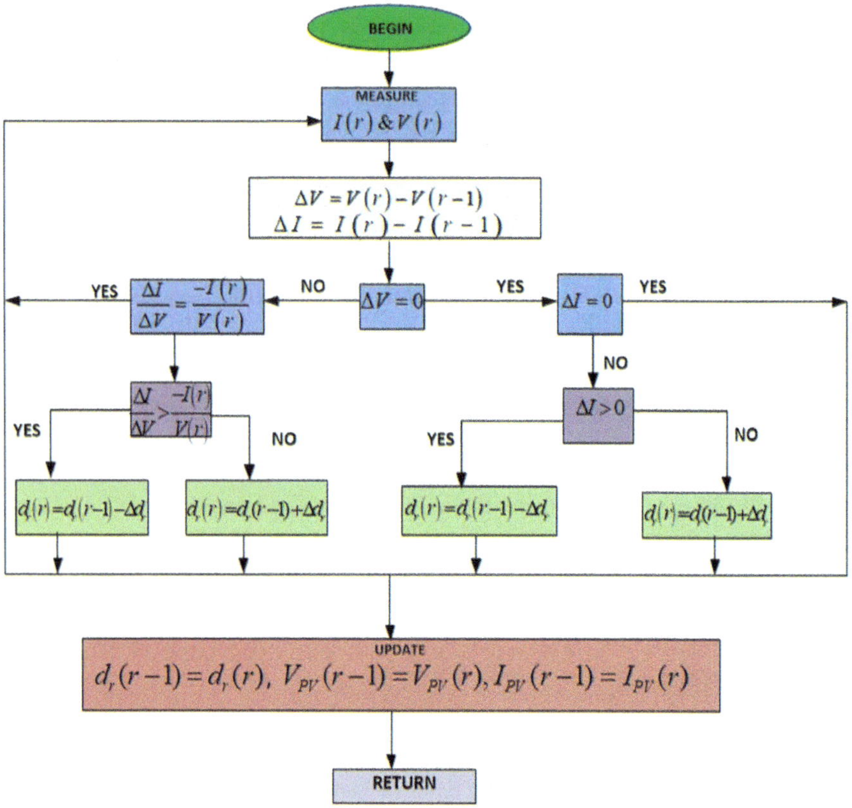

Fig. 7 The flow chart of INC algorithm

To accomplish this, the duty cycle is started and the MPPT sampling frequency is selected. At the MPPT sampling frequency, the input current and voltage signals are detected and sampled [16, 17].

$V_{PV}(r-1) \rightarrow$ Voltage at previous cycle, $I_{PV}(r-1) \rightarrow$ Current at previous cycle.

$V_{pv}(r), I_{pv}(r) \rightarrow$ Voltage, current at present cycle.

The PV array power is multiplying by $I_{PV}(r)$ and. $V_{PV}(r)$.

S(r) $\rightarrow$ power at present cycle., S(r−1) $\rightarrow$ power at previous cycle (Fig. 8 and Table 1).

Experimental Result and Compare with INC and P&O

P&O and INC MPPT methods for photovoltaic PV systems are contrasted here for different tem of 15, 25, and 35°C, as well as fixed irradiation of 1000 W/m$_2$. The figure depicts simulation modelling of a system through P&O and the INC MPPT algorithm with varying temperature.

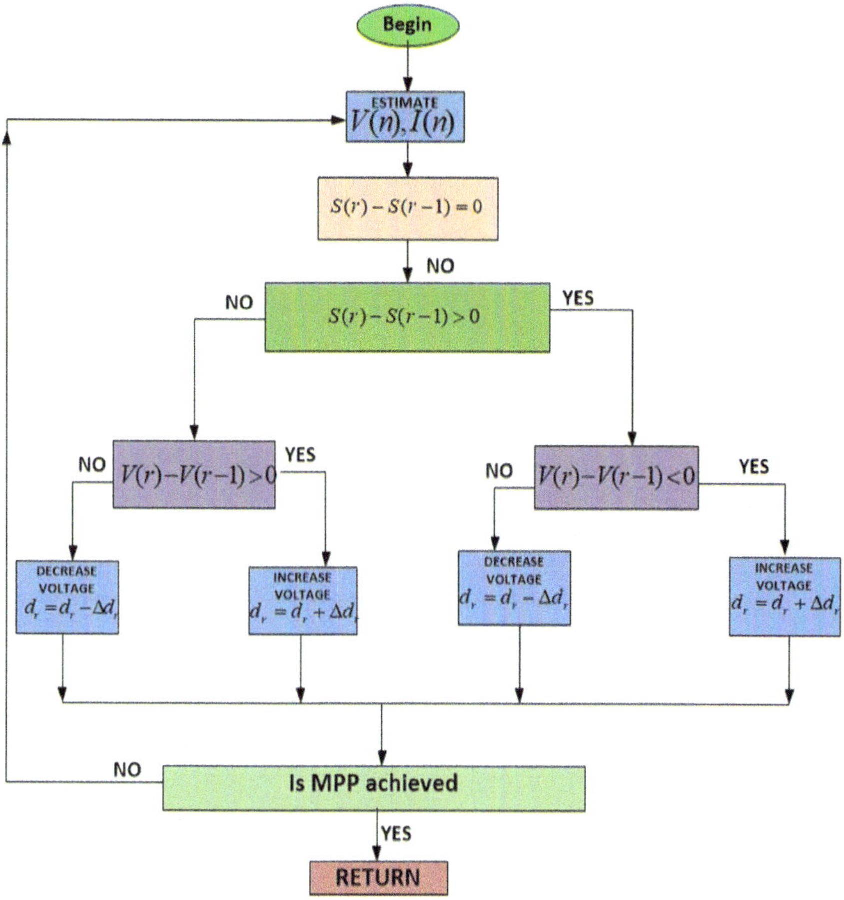

Fig. 8 flow chart of P&O algorithm

At 15°C INC algorithm model behaviour (Figs. 9, 10, 11 and 12, Table 2).
At 15 °C P&O algorithm model behavior.
At 25°C INC algorithm model behaviour (Figs. 13, 14 and 15, Table 3).
At 25 °C P&O algorithm model behavior
At 35°C INC algorithm model behaviour (Figs. 16, 17, 18, 19 and 20, Table 4).
At 35 °C P&O algorithm model behavior.

Table 1 Metlab simulink model specification

PV array	
Parallel string	2
Series connected modules per string	2
Maximum power	211.15W
Cell per module	60
Voltage at maximum power	29V
Voltage at open circuit	36.3V
Short-circuit current Isc	7.84A
Temp coeff	0.36099
Shunt resistance Rsh	313.3991
Series resistance Rse	0.3983
BOOST CONVERTER L = 5mH C = 100μF Fs = 25 kHz	

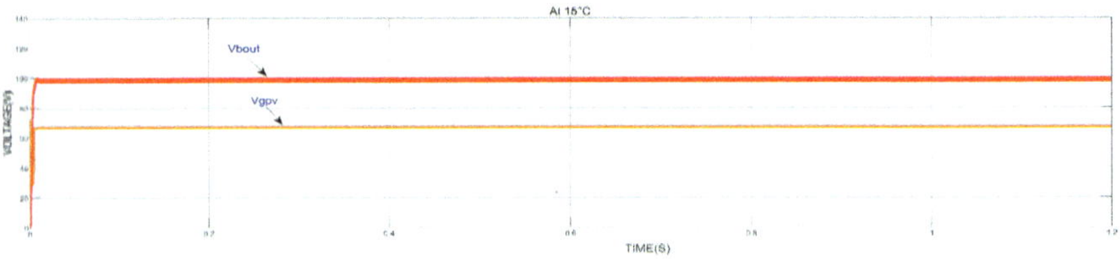

Fig. 9 Voltage across PV array and voltage across boost converter output

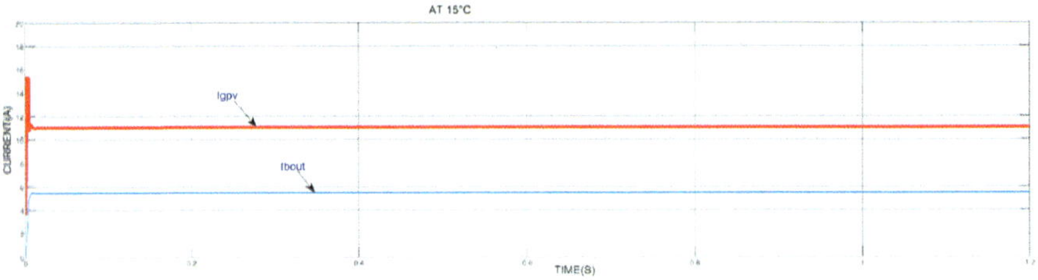

Fig. 10 Current on PV array side and current on boost converter side

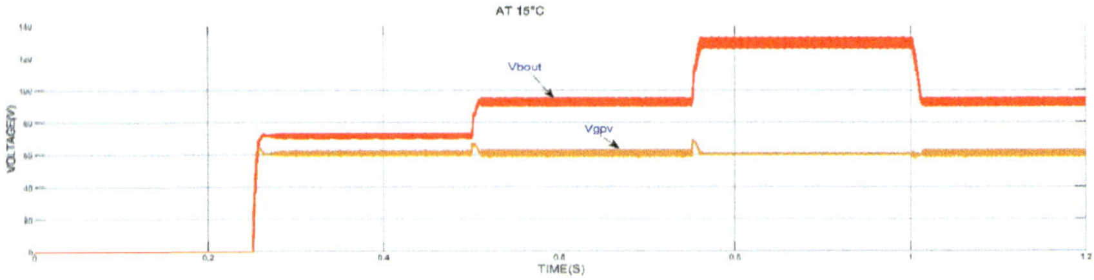

Fig. 11 Voltage across PV array and voltage across boost converter output

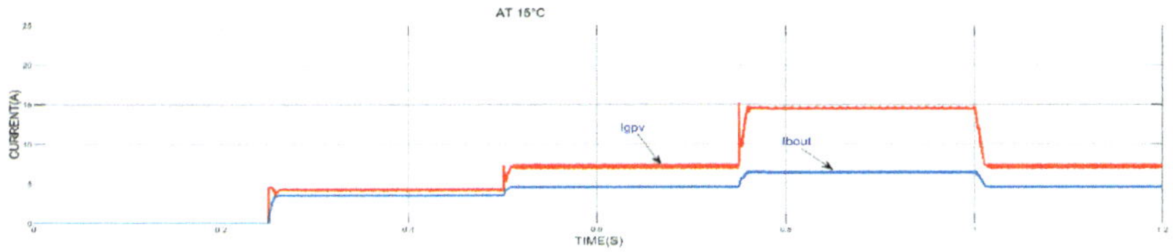

Fig. 12 Current on PV array side and current on boost converter side

Table 2 Compares input/output values using the INC and P&O MPPT approaches, at. 15 °C

S.no	Parameters	INC	P&O
1	Irradiance (W/m^2)	1000	1000
2	Temperature(°C)	15	15
3	PV voltage Vgpv(V)	62.84	64.25
4	PV current Igpv(A)	11.10	7.42
5	PV power Pgpv(W)	691.24	475.45
6	Boost side output voltage Vbout(V)	100.5	92.45
7	Boost side output current Ibout(A)	5.45	4.41
8	Boost side output power Pbout(W)	547.73	406.78

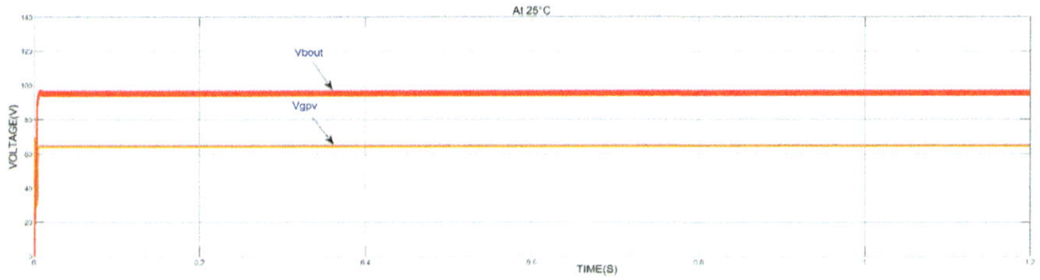

Fig. 13 Voltage across PV array and voltage across boost converter output

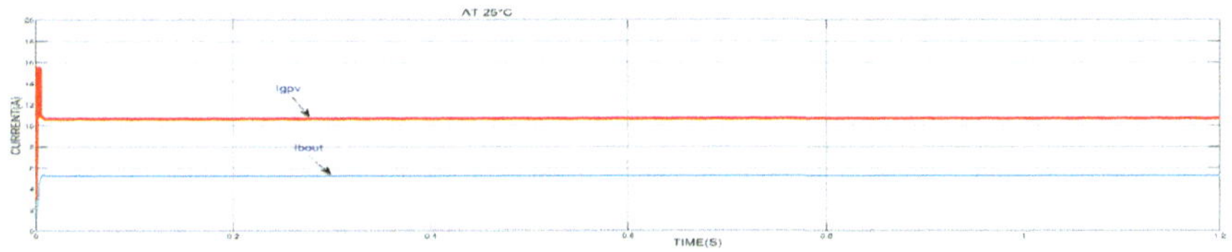

Fig. 14 Current on PV array side and current on boost converter side

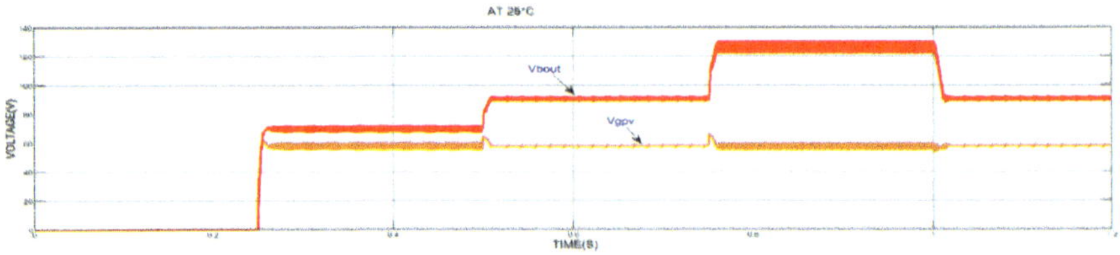

Fig. 15 Voltage across PV array and voltage across boost converter output

Table 3 compares input/output values using the INC and P&O MPPT approaches, at. 25°C

S.no	Parameters	INC	P&O
1	Irradiance(W/m^2)	1000	1000
2	Temperature(°C)	25	25
3	PV Voltage Vgpv (V)	63.56	61.52
4	PV Current Igpv(A)	10.34	7.62
5	PV Power Pgpv(W)	657.21	468.78
6	Boost output side voltage Vbout(V)	97.03	93.78
7	Boost output side current Ibout(I)	5.39	4.21
8	Boost output side power Pbout(W)	522.49	394.81

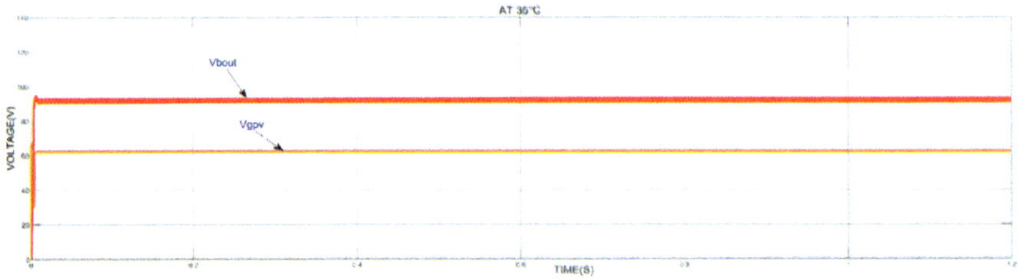

Fig. 16 Voltage across PV array and voltage across boost converter output

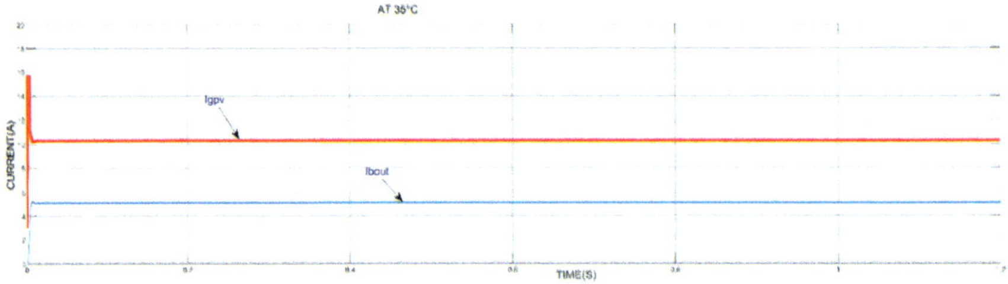

Fig. 17 current on PV array side and current on boost converter side

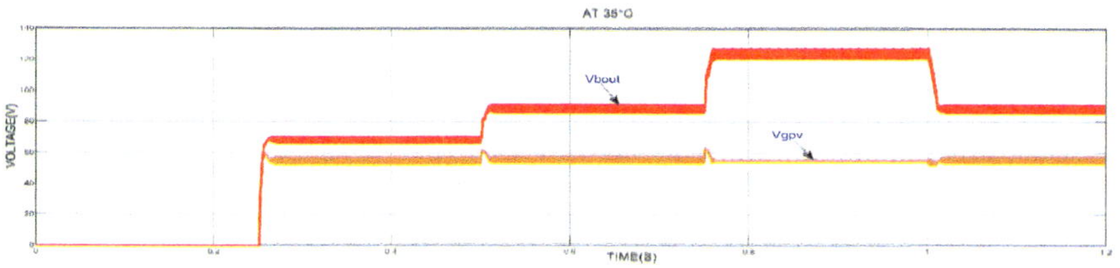

Fig. 18 Voltage across PV array and voltage across boost converter output

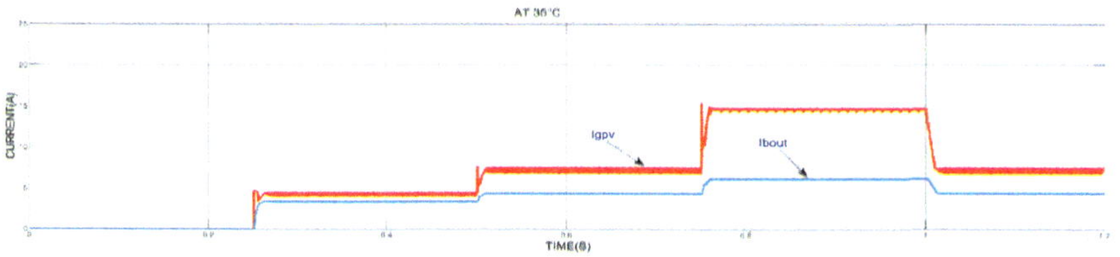

Fig. 19 Current on PV array side and current on boost converter side

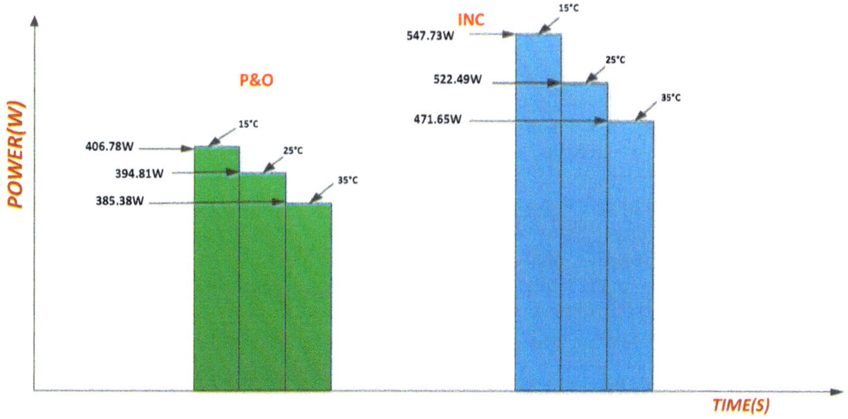

Fig. 20 For constant irradiation, an analysis of converter output power using the P&O and INC approaches is made

Table 4 Compares input/output values using the INC and P&O MPPT approaches, at. At 35°C

S.no	Parameters	INC	P&O
1	Irradiance(W/m^2)	1000	1000
2	Temperature(°C)	35	35
3	PV voltage Vgpv(V)	61.12	58.78
4	PV current Igpv(A)	10.12	7.56
5	PV power Pgpv(W)	618.53	444.376
6	Boost output side voltage Vbout(V)	92.12	91.54
7	Boost output side current Ibout(A)	5.12	4.21
8	Boost output side power Pbout(W)	471.65	385.38

4 Conclusion

This paper compares the INC technique to the P&O method with a boost converter for maximum power tracking. The MATLAB/Simulink simulation results indicate that the PV and load side results with time and temperature variation INC algorithm has become more accurate than the P&O method. Simulink result confirm that voltage and current oscillation if the INC is lower than the P&O with increased productivity. The INC method reaches the MPP quicker than the P&O method. The simulation result show that temperature increasing corresponding power will decrease in both technique.

During fluctuations in temperature, the INC MPPT algorithm outperforms P&O in both steady-state and intermittent conditions.

References

1. Mohamed SA, Abd El Sattar M (2019) A comparative study of P&O and INC maximum power point tracking techniques for grid-connected PV systems. SN Appl Sci 1:174
2. Wai RJ, Wang WH, Lin CY (2008) High-performance stand-alone Photovoltaic generation system. IEEE Trans Ind Electron 55(I):240–250
3. Saber H, Bendaouad A, Rahmani L, Radjeaib H (2022) A comparative study of the FLC, INC and P&O methods of the MPPT algorithm for a PV system. I978–1–6654–7108–4/22/$31.00 ©2022 IEEE
4. Radianto D, Dousouky GM, Shoyama M (2015) MPPT based on incremental conductance-fuzzy logic algorithm for photovoltaic system under variable climate conditions. In: 2015 IEEE international telecommunications energy conference (INTELEC). Osaka, Japan, pp 1–5
5. Kjaer SB Evaluation of the 'Hill climbing' and the 'Incremental conductance' maximum power point trackers for photovoltaic power systems. IEEE Trans Energy Convers 27(4):922–929
6. Tanvir A, Sharmin S, Faysal N (2016) Comparative analysis between single diode and double diode model of PV cell: concentrate different parameters effect on its efficiency. J Power Energy Eng. 4:31–46
7. Vidhya K, Arfan G, Erping Z (2016) Modelling and simulation of maximum power point tracking algorithms and review of MPPT techniques for PV applications. In: 5th international conference on electronic devices, systems and applications (ICEDSA), 6–8 Dec 2016, pp 1–4
8. Mitchell DM (1988) DC-DC switching regulator analysis. McGraw-Hill, New York
9. Soni A (1999) DC-DC switching boost converter. ILLINOIS College of Engineering
10. Seguier G (1993) Power electronic converters: DC-DC conversion. SpringerVerlag Inc., New York
11. Bakkar M, Abdel-Geliel M, Zied MA (2015) Photovoltaic maximum power point grid connected based on power conditioning technique employing fuzzy controller. Q Renew Energy Power Qual J (RE&PQJ) 1(13):339–344
12. Hairul N, Saad M (2010) Comparison study of maximum power point tracker techniques for PV systems. In: Proceedings of the 14th international Middle East power systems conference (MEPCON'10), Cairo University, Egypt, December 19–21, 2010
13. El-Sayed M, Leeb S (2014) Evaluation of maximum power point tracking algorithms for photovoltaic electricity generation in Kuwait. Renew Energy Power Qual J (RE&PQJ) 1(12):44–50
14. Lamnad M et al (2016) Comparative study of IC, P&O and FLC method of MPPT algorithm for grid connected PV module. J Theor Appl Inf Technol 89(1):242–253

15. Elgendy M, Zahawi B, Atkinson D (2012) Assessment of perturb and observe MPPT algorithm implementation techniques for PV pumping applications. IEEE Trans Sustain Energy 3(1):21–33
16. Sahnoun MA, et al (2013) Maximum power point tracking using P&O control optimized by a neural network approach: a good compromise between accuracy and complexity. Energy Procedia 42:650–659
17. Patel G, Patel DB, Paghdal KM (2016) Analysis of P&O MPPT algorithm for PV system. Int J Electr Electron Eng (IJEEE) 5(6):1–10
18. Manna S, Singh DK, Akella AK (2023) Design and implementation of a new adaptive MPPT controller for PV system. ELESEVIER Energy rep 9:1818–1829

SPRINGER NATURE

GPSR Compliance

The European Union's (EU) General Product Safety Regulation (GPSR) is a set of rules that requires consumer products to be safe and our obligations to ensure this.

If you have any concerns about our products, you can contact us on ProductSafety@springernature.com

In case Publisher is established outside the EU, the EU authorized representative is:

Springer Nature Customer Service Center GmbH
Europaplatz 3
69115 Heidelberg, Germany

The manufacturer's authorised representative in the EU is Springer Nature Customer Service Centre GmbH, Europaplatz 3, 69115 Heidelberg, Germany. If you have any concerns regarding our products, please contact ProductSafety@springernature.com

Printed and bound by CPI Group (UK) Ltd, Croydon, CR0 4YY
01/10/2026
02225860-0005